Lecture Notes in Mathematics

Edited by A. Dold and B. Eckmann

737

Volterra Equations

Proceedings of the Helsinki Symposium on Integral Equations, Otaniemi, Finland, August 11–14, 1978

Edited by
Stig-Olof Londen and Olof J. Staffans

Springer-Verlag
Berlin Heidelberg New York 1979

Editors

Stig-Olof Londen
Olof J. Staffans
Institute of Mathematics
Helsinki University of
Technology
SF-02150 Espoo 15

AMS Subject Classifications (1970): 34G05 45D05, 45J05, 45K05,
45M05, 45N05,

ISBN 3-540-09534-9 Springer-Verlag Berlin Heidelberg New York
ISBN 0-387-09534-9 Springer-Verlag New York Heidelberg Berlin

Library of Congress Cataloging in Publication Data
Helsinki Symposium on Integral Equations,
Otaniemi, Finland, 1978.
Volterra equations.
(Lecture notes in mathematics; 737)
Bibliography: p.
Includes index.
1. Volterra equations--Congresses.
I. Londen, Stig-Olof. II. Staffans, Olof, J., 1947- III. Title. IV. Series: Lecture notes in mathematics (Berlin); 737.
QA3.L28 no. 737 [QA431] 510'.8s [515'.45] 79-18836
ISBN 0-387-09534-9

Printing and binding: Beltz Offsetdruck, Hemsbach/Bergstr.
2141/3140-543210

P R E F A C E

This volume comprises the papers presented at the Helsinki Symposium on Integral Equations held at the Helsinki University of Technology in Otaniemi, Finland, during the four days 11 to 14 August 1978. The Symposium was devoted to current research on Volterra equations, the main emphasis being on the qualitative theory.

The papers have been arranged in alphabetical order according to the person by whom it was presented. With respect to the subject the papers might be divided into two classes:

i) those whose setting is a finite-dimensional space and where much of the interest is directed toward asymptotics,

ii) those who analyze abstract equations (including partial integrodifferential equations) where the setting is an infinite-dimensional space.

Adopting this classification one finds that the papers by Grossman, Hannsgen, Herdman, Jordan, Levin, Londen, Seifert, Sell and Staffans belong to the former category while the remaining articles are more or less clearcut examples of the second. Thus existence problems in infinite-dimensional spaces are considered by Aizicovici, Barbu, Deimling, Lakshmikanthan, Mizel, Pavel, Travis and Webb while Clément and Wheeler deal with asymptotics in Banach or Hilbert spaces. Questions of well-posedness of linear Volterra equations in Banach spaces are taken up by Grimmer and Miller and a semigroup approach is to be found in Brewer's article. The work of MacCamy deals with numerical procedures for Volterra equations in infinite dimensional spaces. Different partial integrodifferential problems are treated in the articles by Leitman, Nohel and Raynal. The existence of periodic solutions of some Volterra equations is analyzed in the paper of Cushing. Some probabilistic problems are considered by Bellomo.

The Symposium was organized by a committee consisting of Gustaf Gripenberg, Stig-Olof Londen and Olof J. Staffans. In addition the secretarial staff of the Institute of Mathematics at the Helsinki University of Technology provided a much appreciated help both before, during and after the Symposium.

The Symposium was supported by the Finnish Department of Education and by the foundation Magnus Ehrnrooths stiftelse. We gratefully acknowledge their financial help which enabled us to carry through the Symposium. We are also indebted to the Helsinki University of Technology for generously providing excellent facilities.

Finally our thanks go to all our colleagues who took part in the Symposium and contributed to its work.

Stig-Olof Londen

Olof J. Staffans

C O N T E N T S

LIST OF PARTICIPANTS

*Sergiu Aizicovici	Universitatea Al. I. Cuza, Iasi, Romania
*Viorel Barbu	Universitatea Al. I. Cuza, Iasi, Romania
*Nicola Bellomo	Politecnico di Torino, Torino, Italy
*Dennis W. Brewer	University of Arkansas, Fayetteville, Arkansas, USA
*Philippe P. Clément	Technische Hogeschool Delft, Delft, The Netherlands
*Jim M. Cushing	University of Arizona, Tucson, Arizona, USA
*Klaus K. Deimling	Gesamthochschule Paderborn, Paderborn, Germany
Matts R. Essén	Kungliga Tekniska Högskolan, Stockholm, Sweden
Evelyn Frank	Evanston, Illinois, USA
*Ronald C. Grimmer	Southern Illinois University, Carbondale, Illinois, USA
Gustaf Gripenberg	Tekniska Högskolan i Helsingfors, Otnäs, Finland
*Stanley I. Grossman	University of Montana, Missoula, Montana, USA
*Kenneth B. Hannsgen	Virginia Polytechnic Institute and State University, Blacksburg, Virginia, USA
Seppo V. Heikkilä	Oulun Yliopisto, Oulu, Finland
*Terry L. Herdman	Virginia Polytechnic Institute and State University, Blacksburg, Virginia, USA
*G. Samuel Jordan	University of Tennessee, Knoxville, Tennessee, USA
*V. Lakshmikantham	University of Texas at Arlington, Arlington, Texas, USA
S. Leela	SUNY, Geneseo, New York, USA
*Marshall J. Leitman	Case Western Reserve University, Cleveland, Ohio, USA
*Jacob J. Levin	University of Wisconsin, Madison, Wisconsin, USA
*Stig-Olof Londen	Tekniska Högskolan i Helsingfors, Otnäs, Finland
Eric R. Love	University of Melbourne, Parkville, Victoria, Australia
*Richard C. MacCamy	Carnegie-Mellon University, Pittsburgh, Pennsylvania, USA

* = lecture included in this volume

M.A. Malik	Concordia University, Montreal, Québec, Canada
*Richard K. Miller	Iowa State University, Ames, Iowa, USA
*Victor J. Mizel	Carnegie-Mellon University, Pittsburgh, Pennsylvania, USA
Olavi Nevanlinna	Oulun Yliopisto, Oulu, Finland
*John A. Nohel	University of Wisconsin, Madison, Wisconsin, USA
Shin-ichi Ohwaki	Kumamoto University, Kumamoto, Japan
*Nicolae H. Pavel	Universitatea Al. I. Cuza, Iasi, Romania
*Marie Lise Raynal	Université de Bordeaux I, Talence, France
*George Seifert	Iowa State University, Ames, Iowa, USA
Seppo Seikkala	Oulun Yliopisto, Oulu, Finland
*George R. Sell	University of Minnesota, Minneapolis, Minnesota, USA
*Olof J. Staffans	Tekniska Högskolan i Helsingfors, Otnäs, Finland
*Curtis C. Travis	Oak Ridge National Laboratory, Oak Ridge, Tennessee, USA
Jacek J. Urbanowicz	Politechnika Warszawska, Warszaw, Poland
*Glenn F. Webb	Vanderbilt University, Nashville, Tennessee, USA
*Robert L. Wheeler	University of Missouri, Columbia, Missouri, USA

* = lecture included in this volume

ON AN ABSTRACT VOLTERRA EQUATION

SERGIU AIZICOVICI
Institute of Mathematics
University of Iaşi
6600 Iaşi, Romania

1. INTRODUCTION

In this note we study the existence of solutions to a class of Volterra integrodifferential equations of the form

$$(1.1) \qquad u'(t) + \int_0^t a(t-s)g(u(s))ds \ni f(t), \quad 0 < t < T.$$

Here $T \in (0,\infty)$ is arbitrary, u and f take values in a real infinite dimensional Hilbert space H, a stands for a scalar convolution kernel, while g denotes a nonlinear monotone (possibly multivalued) operator acting in H. (See [3] and [5] for background material on monotone operators).

To realize the difficulty of this problem, let us remark that in the case when a=1, (1.1) formally reduces (by differentiation) to a nonlinear hyperbolic equation.

When approaching the existence of solutions to Eq.(1.1), one has to choose between two opposite ways. The first way (used by Londen [7],[8]) rests upon hard conditions on the convolution kernel, excluding the case a=1 and therefore an application to hyperbolic equations. The second alternative (cf. [1] ,[2]) does allow a broad class of kernels (including a=1), at the expense of strong restrictions on the admissible nonlinearities g. We are going to further illustrate this second way.

2. MAIN RESULT

Consider a real reflexive, separable Banach space V, such that $V \subset H$, with dense and continuous inclusion. We have

$$V \subset H \subset V' ,$$

where V' is the dual of V. The pairing between $v_1 \in V'$ and $v_2 \in V$ will be denoted by (v_1, v_2); it coincides with their inner product in H, whenever $v_1 \in H$. We use the notations $|\cdot|$ and $\|\cdot\|$ to indicate the norms in H and V, respectively. Assume that

(2.1) The injection $V \subset H$ is compact.

Let A be a cyclically maximal monotone operator in $V \times V'$. Hence, there exists a convex, lower semicontinuous (l.s.c.) and proper function $\varphi : V \to (-\infty, +\infty]$, such that

(2.2) $\quad A = \partial\varphi$, ($\partial$ = subdifferential).

We suppose that

(2.3) A is everywhere defined ($D(A) = V$), single-valued and maps bounded subsets of V into bounded subsets of V',

(2.4) A is weakly continuous, i.e., for any sequence $\{u_n\} \subset V$, such that $u_n \to u$, weakly in V, we have $Au_n \to Au$, weakly-star in V',

(2.5) $\quad \lim_{\|u\| \to \infty} \varphi(u) = +\infty$.

Remark 2.1. (i) Conditions (2.2)-(2.5) are clearly satisfied by each linear positive, symmetric and coercive operator $A : V \to V'$.

(ii) Let Ω be a bounded subset of $R^n (n \geq 3)$, with smooth boundary. If

$$H = L^2(\Omega) \ , \ V = H_0^1(\Omega) \ ,$$

then it is immediate that (2.1)-(2.5) hold, provided that the (nonlinear) operator A be given by

$$Au = -\Delta u + Mu \ , \ u \in V,$$

where $M : L^p(\Omega) \to L^q(\Omega)$, $2 \leq p \leq 2n/(n-2)$, $1/p + 1/q = 1$, is of the form

$$(Mu)(x) = \beta(u(x)), \ x \in \Omega, \ u \in L^p(\Omega) \ ,$$

with $\beta : R \to R$ satisfying

$$\beta \in C(-\infty, \infty), \beta \text{ monotone}, \beta(0) = 0,$$

$$|\beta(r)| \leq c(|r|^{p-1} + 1), \ c > 0, \ r \in R.$$

Consider next a convex, l.s.c., proper function $\psi : H \to (-\infty, \infty]$ and define the maximal monotone operator B in H by

(2.6) $\quad B = \partial\psi$

Denote by $D(\psi)$ the effective domain of ψ and suppose that

(2.7) $\quad V \cap \text{int.}D(\psi) \neq \phi$, (int. = interior).

Remark 2.2. It is obvious that (2.7) is fulfilled in the case in which ψ is the indicator function of a closed convex subset $K \subset H$, with

$$V \cap \text{int.}K \neq \phi \ .$$

Let $\underline{a} : [0,T] \to R$ satisfy (cf. [6, Cond. (a)])

(2.8) $\underline{a}$ is absolutely continuous on $[0,T]$,

(2.9)
$$\begin{cases} \text{There is } k>0 \text{ such that} \\ v\in L^2(0,T;H),\ d_1\ ,\ d_2\in[0,\infty) \text{ and} \\ \int_0^t (a*v(s),\ v(s))ds \le d_1 + d_2 \max_{0\le s\le t} \left|\int_0^s v(\tau)d\tau\right| ,\ 0\le t\le T\ , \\ (\text{where } a*v(t) = \int_0^t a(t-s)v(s)ds) \\ \text{imply} \\ \left|\int_0^t v(s)ds\right| \le k(d_1^{1/2} + d_2)\ ,\ 0\le t\le T\ , \\ \text{and} \\ \left|\int_0^t (a*v(s),\ v(s))ds\right| \le k(d_1 + d_2^2)\ ,\ 0\le t\le T\ . \end{cases}$$

Remark 2.3. According to [6, Prop.(a)], Conditions (2.8) and (2.9) hold if $\underline{a}$ satisfies either

(a_1) $a\ ,\ a'\in L^1(0,T)\ ;\ a(0)>0\ ,$
a' is of bounded variation over $[0,T]$,

or

(a_2) $a\in C^2(0,T]\cap C[0,T]\ ;\ a(0)>0\ ,$
a is nonnegative, decreasing and convex on $[0,T]$.

Finally, we require that $f : [0,T]\to H$ be subject to

(2.10) $f\ ,\ f'\in L^1(0,T;H)$.

Our existence result is contained in the following

THEOREM. Let conditions (2.1)-(2.10) hold. Then, for each $u_0\in V\cap D(\psi)$ the initial-value problem

$$u'(t) + a*(A+B)u(t) \ni f(t)\ ,\ 0<t<T\ ,$$
$$u(0) = u_0\ ,$$

has a generalized solution u in the sense

(2.11) $u\in C([0,T];H)\cap L^\infty(0,T;V)$,

(2.12) $u'\in L^\infty(0,T;H)$,

(2.13) $u(t)\in V\cap D(\psi)\ ,\ \forall t\in[0,T]$,

(2.14) $\int_0^t Au(s)ds\in L^\infty(0,T;H)$,

(2.15) <u>there exists a bounded measure</u> μ <u>on</u> $[0,T]$ <u>with values in</u> H ($\mu \in \mathcal{M}(0,T;H)$) , <u>such that</u>

(i) $\int_0^T (\psi(v(t)) - \psi(u(t)))dt \geq \mu(v-u)$,

<u>for any</u> $v \in C([0,T];H)$ <u>with</u> $\psi(v) \in L^1(0,T)$,

(ii) $u'(t) + \int_0^t a(t-s)Au(s)ds + \int_0^t a(t-s)d\mu(s) = f(t), \ 0 < t < T$,

(2.16) $u(0) = u_o$.

Remark 2.4. The idea of using the interiority condition (2.7) as well as the above notion of generalized solution was suggested to us by the work of M.Schatzman [9], where a nonlinear hyperbolic equation is studied in a finite dimensional space.

3. PROOF OF THE THEOREM

First introduce the operator A_H in H by

$$A_H x = Ax \ , \ x \in D(A_H) = \{ x \in V \ ; \ Ax \in H \} \ .$$

By (2.2) and (2.5) it is easily seen that $A_H = \partial \varphi_H$, with

$$\varphi_H(x) = \varphi(x) \ , \ x \in V$$
$$= +\infty \ , \ x \in H \setminus V$$

Consider then the approximating equation

(3.1) $u'_\lambda(s) + \lambda(A_H + B_\lambda)u_\lambda(s) + a * (A_H + B_\lambda)u_\lambda(s) = f(s), \lambda > 0, \ 0 < s < T$,

with the initial condition

(3.2) $u_\lambda(0) = u_o$.

Recall that $B_\lambda = \partial \psi_\lambda$, where

(3.3) $\psi_\lambda(x) = \psi(J_\lambda x) + |x - J_\lambda x|^2/(2\lambda)$, $x \in H$, $(J_\lambda = (I + \lambda B)^{-1})$.

Inasmuch as B_λ is continuous and everywhere defined on H, we have

(3.4) $A_H + B_\lambda = \partial(\varphi_H + \psi_\lambda)$.

It then follows (cf. e.g. [6,Thm.2]) that the problem (3.1),(3.2) has a solution u_λ satisfying

(3.5) $u_\lambda \in C([0,T];H)$, $u'_\lambda \in L^2(0,T;H)$,

(3.6) $g_\lambda \in L^2(0,T;H)$; $g_\lambda(t) = (A_H + B_\lambda)u_\lambda(t)$, $t \in (0,T)$.

Since B_λ is lipschitzian, (3.5) and (3.6) imply

(3.7) $A_H u_\lambda$, $B_\lambda u_\lambda \in L^2(0,T;H)$.

Multiply (3.1) by $g_\lambda(s)$ and integrate over $(0,t)$, $0 \le t \le T$. Using (2.10), (3.2), (3.4)-(3.6) and [5, Lemma 3.3], we get

$$(3.8)\quad \begin{aligned}&(\varphi + \psi_\lambda)(u_\lambda(t)) + \lambda \int_0^t |g_\lambda(s)|^2 ds + \int_0^t (a * g_\lambda(s), g_\lambda(s))ds = \\ &= \varphi(u_0) + \psi_\lambda(u_0) + (f(t), \int_0^t g_\lambda(s)ds) - \int_0^t (f'(s), \int_0^s g_\lambda(\tau)d\tau), \ 0 \le t \le T .\end{aligned}$$

Without loss of generality, we may assume that $\psi \ge 0$. Then (3.8), (2.5), (2.10) and the fact that $\psi_\lambda(u_0) \le \psi(u_0)$ yield

$$(3.9)\quad \int_0^t (a * g_\lambda(s), g_\lambda(s))ds \le C(1 + \max_{0\le s\le t} |\int_0^s g_\lambda(\tau)d\tau|) \ , \ t \in [0,T] .$$

Here and in the sequel, C denotes various finite positive constants independendent of λ .

By (2.9), we infer from (3.9) that

$$(3.10)\quad |\int_0^t g_\lambda(s)ds| \le C \ , \ 0 \le t \le T ,$$

$$(3.11)\quad |\int_0^t (a * g_\lambda(s), g_\lambda(s))ds| \le C \ , \ 0 \le t \le T .$$

Consequently, (3.8) leads to

$$(3.12)\quad \sup_{0\le t\le T} (\varphi(u_\lambda(t)) + \psi_\lambda(u_\lambda(t))) \le C ,$$

$$(3.13)\quad \lambda \int_0^T |g_\lambda(s)|^2 ds \le C .$$

From (3.12) it follows (see (2.5))

(3.14) $\{u_\lambda ; \lambda > 0\}$ is bounded in $L^\infty(0,T;V)$.

In view of (2.3), this implies

(3.15) $\{A_H u_\lambda ; \lambda > 0\}$ is bounded in $L^\infty(0,T;V')$.

Employing (2.8), (2.10), (3.10) and (3.13) in (3.1) yields

(3.16) $\{u'_\lambda ; \lambda > 0\}$ is bounded in $L^2(0,T;H)$.

Calling on (2.1), (3.14) and (3.16), we can assume that $(\lambda \to 0^+)$

(3.17) $u_\lambda \to u$, strongly in $C([0,T];H)$,

(3.18) $u_\lambda \to u$, weakly-star in $L^\infty(0,T;V)$,

(3.19) $u'_\lambda \to u'$, weakly in $L^2(0,T;H)$.

Notice that (3.2), (3.17) and (3.18) yield (2.11) and (2.16). Combining then (2.4), (3.15) and (3.18) gives

(3.20) $A_H u_\lambda \to Au$, wekly-star in $L^\infty(0,T;V')$.

Taking into account (3.3) and (3.12), we have

$$|u_\lambda(t) - J_\lambda u_\lambda(t)| \le (2\lambda C)^{1/2} \ , \quad \psi(J_\lambda u_\lambda(t)) \le C \ , \quad 0 \le t \le T \ .$$

By (3.17), this implies (since ψ is l.s.c.)

(3.21) $\psi(u(t)) \le C$, $t \in [0,T]$.

Analogously, from (3.12) and (3.18) it follows

(3.22) $\varphi(u(t)) \le C$, $0 \le t \le T$.

Obviously, (3.21) and (3.22) are equivalent to (2.13) .

Condition (2.7) comes now into play supplying a bound on $B_\lambda u_\lambda$. Our argument is close to that of [9,p.358].
Let $x_0 \in V \cap \mathrm{int}.D(\psi)$ and denote by $S(x_0,\rho)$ a closed ball of center x_0 and radius ρ, contained in the interior of $D(\psi)$. We can suppose that ψ is bounded on $S(x_0,\rho)$. For any $z \in C([0,T];H)$, such that $|z(t)| \le 1$, $t \in [0,T]$, one has

(3.23) $$(B_\lambda u_\lambda(t),\ x_0 + \rho z(t) - u_\lambda(t)) \le \psi_\lambda(x_0 + \rho z(t)) - \psi_\lambda(u\ (t)),\quad 0 \le t \le T.$$

Integrating (3.23) from 0 to T gives (see (3.6),(3.7))

(3.24) $$\rho \int_0^T (B_\lambda u_\lambda(t), z(t))dt \le C + \int_0^T (g_\lambda(t), u_\lambda(t) - x_0)dt - \int_0^T (A_H u_\lambda(t), u_\lambda(t) - x_0)dt \ .$$

Remark that

(3.25) $$\int_0^T (g_\lambda(t),\ u_\lambda(t))dt = (u_\lambda(T),\ \int_0^T g_\lambda(s)ds) - \int_0^T (u'_\lambda(t),\ \int_0^t g_\lambda(s)ds) \ .$$

Combining (3.10), (3.14)-(3.16) and (3.24),(3.25) we are led to

$$\rho \int_0^T (B_\lambda u_\lambda(t),\ z(t))dt \le C \ ,$$

and hence (cf. e.g. [5,Lemma A.3, p.149])

$$\int_0^T |B_\lambda u_\lambda(t)|dt \le C \ .$$

We may conclude that

(3.26) $B_\lambda u_\lambda \to \mu$, weakly-star in $\mathcal{M}(0,T;H)$.

This, (3.17) and the inequality

$$\int_0^T (\psi_\lambda(v(t)) - \psi_\lambda(u_\lambda(t)))dt \ge \int_0^T (B_\lambda u_\lambda(t),\ v(t) - u_\lambda(t))dt \ ,$$

$$(v \in C([0,T];H) \ , \ \psi(v) \in L^1(0,T))$$

yield (2.15)(i) (see [9, p.359]).
From (3.10), (3.20) and (3.26) it follows

$$(3.27)\quad \int_0^t A_H u_\lambda(s)ds \to \int_0^t Au(s)ds\ ,\ \text{weakly-star in } L^\infty(0,T;H)\ ,$$

$$(3.28)\quad \int_0^t B_\lambda u_\lambda(s)ds \to q(t)\ ,\ \text{weakly-star in } L^\infty(0,T;H)\ ,$$

where

$$(3.29)\quad q' = \mu,$$

in the sense of H-valued distributions over (0,T).

From (3.13), (3.19), (3.26)-(3.29) one finally obtains (2.12), (2.14) and (2.15)(ii) upon passage to the limit as $\lambda\to 0^+$. The proof is complete.

Remark 3.1.(Interpretation of μ) In view of (3.28), (3.29) the measure $\mu\in\mathcal{M}(0,T;H)$ can be expressed as the derivative of a function $q\in L^\infty(0,T;H)$. Therefore (see [4, p.47]) q actually is of bounded variation. Thus, we may write

$$(3.30)\quad \mu = \dot q + dq_0\ ,\ q_0(t) = q(t) - \int_0^t \dot q(s)ds\ ,\ 0\le t\le T\ ,$$

where $\dot q$ is the weak derivative of q and dq_0 denotes the distributional derivative of q_0. Then, by slightly modifying the argument used in [4, p.240] for the case when ψ is an indicator function, we conclude from (2.15)(i) and (3.30) that

$$\dot q(t)\in Bu(t)\ ,\ \text{a.e. on } (0,T)\ ,$$

and

$$dq_0(u-x)\ge 0\ ,$$

for every $x\in C([0,T];H)$ with $x(t)\in \overline{D(\psi)}$, $\forall\, t\in[0,T]$.

REFERENCES

1. S. AIZICOVICI , On a nonlinear integrodifferential equation, J.Math. Anal.Appl. 63(1978), 385-395.
2. S. AIZICOVICI , Existence theorems for a class of integro-differential equations, An.Şti.Univ."Al.I.Cuza" Iaşi, in press.
3. V. BARBU, "Nonlinear semigroups and differential equations in Banach spaces", Noordhoff, Leyden, 1976.
4. V.BARBU & TH.PRECUPANU, "Convexity and optimization in Banach spaces", Sijthoff & Noordhoff, 1978.
5. H. BREZIS , "Opérateurs maximaux monotones et semigroupes de contractions dans les espaces de Hilbert", Math.Studies 5, North-Holland, 1973.
6. M.G. CRANDALL, S.O. LONDEN & J.A. NOHEL , An abstract nonlinear Volterra integrodifferential equation, J.Math.Anal.Appl., to appear.
7. S.O. LONDEN , An existence result on a Volterra equation in a Banach space, Trans.Amer.Math.Soc. 235(1978), 285-305.
8. S.O. LONDEN , On an integrodifferential equation with a maximal monotone mapping, Report HTKK-MAT-A89, Helsinki, 1976.

9. M. SCHATZMAN , A class of nonlinear differential equations of second order in time, Nonlinear Anal. 2(1978), 355-373.

DEGENERATE NONLINEAR VOLTERRA INTEGRAL EQUATIONS IN HILBERT SPACE

V. Barbu

Faculty of Mathematics

University of Iaşi, Romania

1. INTRODUCTION

The principal object of this lecture, which is primarily based on the papers {3} and {4}, is to study the Volterra equation

$$(1.1)\qquad Bx(t) + \int_0^t a(t-s)Ax(s)ds \ni f(t) ; \qquad 0 < t < \infty$$

in a Hilbert space H, where A, B are nonlinear maximal monotone operators in H and a(.) is a scalar kernel.

In the special case $B \equiv I$ (the identity operator) equation (1.1) has been extensively studied, e.g. {1}, {2}, {12}, {15}, {17} (additional references may be found in the bibliographies of the cited papers).

The results we present in section 2 give existence for equation (1.1) under fairly mild assumptions on A and B. The term degenerate refers to the fact that the coercivity of B is not required (the special case $B \equiv 0$ is compatible with our assumptions). For other existence results regarding this equation we refer to the author's paper {3} (see also {4}).

The relevance of equation (E) in quasi-static viscoelasticity and in theories involving hereditary effects and fading memory is well-known. An example of this type is presented in section 4.

2. MAIN EXISTENCE RESULTS

We first list the notations and hypothese which will be in effect troughout the paper.

(i) H is a real Hilbert space and W is a real reflexive Banach space which is dense in H and

(2.1) $$W \subset H \subset W'$$

algebraically and topologically (W' is the dual space of W) Further, it will be assumed that

(2.2) The injection of W into H is compact.

We shall denote by $|\,.\,|$ and $\|.\|$ the norms in H and W. By (.,.) we shall denote the inner product in H and the pairing between W and W'.

(ii) $A = \partial\psi : W \longrightarrow W'$ and $B = \partial\varphi : H \longrightarrow H$ where ψ and φ are lower semicontinuous, proper convex functions from W and H, respectively to $\bar{R}^+ =]-\infty\ ,+\infty]$. Further, assume that

(2.3) $$\lim_{\|u\| \to +\infty} \psi(u)/\|u\| = +\infty$$

and

(2.4) A is bounded on every bounded subset of W.

By φ and ψ we have denoted the subdifferential of φ and ψ , respectively. It should be recalled that (2.3) implies that the operator A_H defined in H by

$$A_H u = Au \cap H \qquad u \in H$$

is maximal monotone in H. We denote by A_λ the Yosida approximation of A_H and recall that $A_\lambda = \operatorname{grad}\psi_\lambda$ where (see e.g. { 1 } p.57)

(2.5) $$\psi_\lambda(x) = \psi((I + \lambda A_H)^{-1}x) + \frac{\lambda}{2}|A_\lambda x|^2$$

for all $x \in H$ and $\lambda > 0$.

Suppose further,

(iii) There is a constant $M \geq 0$ such that

$$(A_\lambda x, y) \geq M \quad \text{for all } \lambda > 0 \text{ and } x \in D(B),\ y \in Bx. \tag{2.6}$$

As regards the kernel $a(.)$ we shall require that

(iv) $a \in C^2(R^+)$, a'' is a function of bounded variation locally on $R^+ = [0,+\infty[$, and

$$a(o) > o, \qquad a'(o) \leq o. \tag{2.7}$$

THEOREM 1. Let assumptions (i) up to (iv) be satisfied. Let $T > 0$ and let $f \in W^{1,2}(0,T;H)$ be given such that $f(0) \in B(W)$ and $f'' \in L^2(0,T;W')$. Then equation (1.1) has at least one solution $x \in L^\infty(0,T;W)$ in the following sense: there exist the functions v and w such that

$$w \in L^\infty(0,T;W'), \quad v \in C(0,T;H_w) \tag{2.8}$$

$$w(t) \in Ax(t), \quad v(t) \in Bx(t) \qquad \text{a.e. } t \in]0,T[. \tag{2.9}$$

$$v(t) + \int_0^t a(t-s)w(s)ds = f(t), \text{ for all } t \in [0,T] . \tag{2.10}$$

Furthermore, one has

$$v \in W^{1,\infty}(0,T;W'), \int_0^t w(s)ds \in L^\infty(0,T;H). \tag{2.11}$$

We have denoted by $W^{1,2}(0,T;X)$ (X is a Banach space) the space of all functions $x(t)$: $[0,T] \longrightarrow X$ with $x' \in L^p(0,T;X)$. By $C(0,T;H_w)$ we have denoted the space of all weakly continuous functions from $[O,T]$ in H.

Next we shall present a variant of Theorem 1 in which the requirement $a'(0) \leq 0$ in Assumptions (iv) was droped. Namely, the kernel $a(t)$ is only assumed to satisfy

(iv)' $a \in C^2(R^+)$; a'' is locally absolutely continuous on R^+ and $a(0) > 0$.

THEOREM 2. *Let A, B, a and f satisfy all the conditions of Theorem 1 except that assumption (iv) is replaced by (iv)'. Suppose further that there exist a positive constant μ such that*

$$(Ax,x) \leq \mu\, \phi(x) \quad \text{for all } x \in W. \tag{2.11}$$

Then the conclusions of Theorem 1 hold.

We remark that condition (2.11) is in particular satisfied if the function ϕ is Gâteaux differentiable on W and homogeneous in the following sense: There exist $d \geq 1$ and $C \geq o$ such that

$$\phi(\lambda x) \leq C\lambda^d \phi(x) \quad \text{for all } x \in W, \quad \lambda \geq 1.$$

Of course in this case the operator A must be single valued.

Proof of Theorem 1. Here the proof will be only outlined. (A complete proof may be found in {3}). Without no loss of generality we may assume that $Ao = o$ and $o \in Bo$. We start with approximating equations

$$\lambda(x_\lambda(t)+A_\lambda x_\lambda(t))+Bx_\lambda(t)+a * A_\lambda x_\lambda(t) \ni f(t), \quad t \in [o,T] \tag{2.12}$$

which admits for each $\lambda > o$ a unique solution x_λ. It should be observed that $x_\lambda \in W^{1,\infty}(o,T;H)$. Here $a * A_\lambda x_\lambda$ denotes the convolution product $\int_o^t a(t-s)A_\lambda x_\lambda(s)ds$. By Proposition (a) in {8} we have

$$\int_o^t (A_\lambda x_\lambda, a * A_\lambda x_\lambda)ds \geq \frac{1}{2}\, a(o)|F_\lambda(t)|^2 - C\int_o^t (|F_\lambda(s)|^2 + |F_\lambda(s)| \sup\{|F_\lambda(\tau)|\})ds \tag{2.13}$$

where $F_\lambda(t) = \int_o^t A_\lambda x_\lambda(s)ds$. By assumption (iii) and (2.13) it follows after some standard calculations involving equation (2.12) that there exists $C \geq o$ such that

$$\lambda\int_o^t |A_\lambda x_\lambda(t)|^2 dt + |F_\lambda(t)|^2 \leq C \quad \text{for } \lambda > o,\ t \in [o,T] \tag{2.14}$$

where T is suitable chosen (It suffices to prove the existence on an arbitrarily small interval $[o,T]$, see e.g. { 1 } p.243).

Next we differentiate (2.12), multiply by x_λ' and integrate over $[o,t]$. We get

$$(2.14)\qquad \lambda \int_0^t |x_\lambda'|^2 dt + a(o)\varphi_\lambda(x_\lambda(t)) \leq a(o)\,\varphi_\lambda(x_\lambda(o)) + $$
$$+ a'(o)\int_0^t (x_\lambda, A_\lambda x_\lambda)ds - (x_\lambda, a'*A_\lambda x_\lambda) +$$
$$+ \int_0^t (x_\lambda, a''*A_\lambda x_\lambda)ds + \int_0^t (f, x_\lambda')ds.$$

Let $w_0 \in W$ be such that $f(o) \in Bw_0$. Using the monotonicity of B it follows by equation (2.12) that

$$\frac{1}{2}|x_\lambda(o)|^2 + \varphi_\lambda(x_\lambda(o)) \leq \frac{1}{2}|w_0|^2 + \varphi(w_0), \qquad \lambda > o.$$

Hence by (2.14) and (2.15)

$$(2.15)\qquad \varphi_\lambda(x_\lambda(t)) \leq C_1(|x_\lambda(t)| + (\int_0^t |x_\lambda(s)|^2 ds)^{1/2}) + C_2 .$$

This and (2.5) imply that

$$(2.16)\qquad \|(I+\lambda A_H)^{-1}x_\lambda\| + |x_\lambda(t)|^2 \leq C, \text{ for } \lambda > o,\ t \in [o,T]$$

while by (2.12) we see that $v_\lambda(t) = f - a*A_\lambda x_\lambda - \lambda(x_\lambda + A_\lambda x_\lambda)$ remain in a bounded subset of $L^\infty(o,T;H)$. On the other hand it follows by conditions (2.2) and (2.4) and Ascoli's theorem that the family $\{F_\lambda\}$ is compact in $C(o,T;W')$. Thus extracting a subsequence if necessary we may assume that for $\lambda \longrightarrow o$, we have

$$(2.17)\qquad \begin{aligned} x_\lambda &\longrightarrow x && \text{weak-star in } L^\infty(o,T;H) \\ F_\lambda &\longrightarrow F && \text{strongly in } C(o,T;W') \\ (I+\lambda A_H)^{-1}x_\lambda &\longrightarrow x && \text{weak-star in } L^\infty(o,T;W) \\ A_\lambda x_\lambda &\longrightarrow g && \text{weak-star in } L^\infty(o,T;W') \end{aligned}$$

and

$$(2.18)\qquad v_\lambda \longrightarrow v \qquad \text{weak-star in } L^\infty(o,T;H)$$
$$\text{and strongly in } C(o,T;W').$$

In particular it follows by (2.18) that $v \in C(o,T;H_w)$ and

$$(2.19) \qquad v_\lambda(t) \longrightarrow v(t) \qquad \text{weakly in } H \text{ for each } t \geq o$$

and

$$(2.20) \qquad v(t) + (a * g)(t) = f(t) \quad \text{for all } t \in [o,T] .$$

Moreover, it follows by the maximality of B and (2.17) that $v(t) \in Bx(t)$ for each $t \in [o,T]$. It remains to prove that $g(t) \in Ax(t)$ a.e. $t \in]o,T[$. For this it suffices to show that

$$(2.21) \qquad \limsup_{\lambda \to o} \int_0^T (A_\lambda x_\lambda, x_\lambda)dt \leq \int_0^T (g,x)dt.$$

To this end we multiply equation (2.12) by x'_λ and integrate over $[o,T]$ to obtain after some calculations

$$(2.22) \qquad a(o) \limsup_{\lambda \to o} \int_0^T (A_\lambda x_\lambda, x_\lambda)dt \leq \varphi^*(f(o)) + \int_0^T (f',x)dt - \int_0^T (a' * g,x)dt - \liminf_{\lambda \to o} \varphi^*(v_\lambda(T))$$

where, φ^* is the conjugate function of φ. Since φ^* is weakly lower semicontinuous on H we deduce from (2.21) that

$$(2.23) \qquad a(o) \limsup_{\lambda \to o} \int_0^T (A_\lambda x_\lambda, x_\lambda)dt \leq \varphi^*(f(o)) - \varphi^*(v(T)) + \int_0^T (f' - a' * g, x)dt.$$

By (2.20) we see that the right hand side of (2.23) is just $a(o) \int_0^T (g,x)dt$ thereby completing the proof.

<u>Proof of Theorem 2</u>. A glance to the proof of Theorem 1 reveals that condition $a'(o) \leq o$ was used only to obtain estimate (2.15) from inequality (2.14). If $a'(o) > o$ we proceed as follows. By condition (2.11) we have

$$(A_\lambda x_\lambda, x_\lambda) \leq \mu\psi((I+\lambda A)^{-1}x_\lambda) + \lambda\, |A_\lambda x_\lambda|^2$$

and therefore

$$(A_\lambda x_\lambda, x_\lambda) \leq \eta\, \psi_\lambda(x_\lambda) \qquad \text{for all } \lambda > o$$

with a suitable chosen η. The later combined with inequality (2.14) and Gronwall's inequality yields (2.15) as desired. For the rest the proof coincides with that of Theorem 1.

We conclude this section with some remarks concerning the particular case in which W is a Hilbert space and (j) A is a linear continuous simmetric operator from W to W' satisfying

$$(2.24) \qquad (Ax,x) \geqslant \omega \|x\|^2 \qquad \text{for all} \quad x \in W$$

where $\omega > 0$.

This case was considered in some details in {4} where existence and uniqueness are studied. It is clear from the proof of Theorem 1 that assumption (2.2) is unnecessary in this case. Furthermore, the solution is unique. Here is the argument. Assume the contrary and let there exist two solutions x and y to equation (1.1). We set z = x-y and use Proposition a in {8} on the space W endowed with the inner product $((u,v)) = (A^{1/2}u, A^{1/2}v)$ to get

$$0 \geq \int_0^t (z(s),(a * Az)(s))ds \geq \frac{1}{2} a(0) \|G(t)\|^2 -$$

$$- C \int_0^t (\|G(s)\|^2 + \|G(s)\| \sup_{0 \leq \tau \leq 1} \|G(\tau)\|)ds$$

where $G(t) = \int_0^t z(s)ds$. The latter implies $z \equiv 0$ as claimed.

Thus we can state

THEOREM 3. *Let the operator* A *satisfy assumption* (j) *and* $B = \partial\varphi$ *satisfy assumption* (iii). *Suppose further that the kernel a satisfy assumption* (iv)'. *Then for each* $f \in W^{1,2}(0,T;H)$ *such that* $f(0) \in B(W)$, *equation* (1.1) *has a unique solution* x *satisfying* (2.8), (2.9), (2.10) *and* (2.11).

It is worth noting that the unilateral problem

$$(2.15) \qquad x(t) = P_K(f(t) - \int_0^t a(t-s)Ax(s)ds) \quad \text{for} \quad t \geq 0$$

where K is a closed convex subset of H and P_K is the projection operator on K, can be written in the form (1.1) where $B = \partial\varphi$ and

$$\varphi(x) = \frac{1}{2} |x|^2 \quad \text{for} \quad x \in K, \quad \varphi(x) = +\infty \quad \text{for } x \notin K.$$

Finally, we notice that equation (1.1) can be brought into the form

$$(1.1') \qquad y(t) + \int_0^t a(t-s)Qy(s)ds \ni f(t), \quad t \in R^+$$

where the nonlinear (multivalued) operator Q is given by $Q = AB^{-1}$ However, since under our assumptions on A and B the operator Q is neither monotone nor continuous on H the standard existence theory for equation (1.1') is not applicable (even if the space H is finite dimensional).

3. APPLICATION TO DEGENERATE NONLINEAR PARABOLIC EQUATIONS

The following illustrates the kind of problem to which the results of the previous section can be applied. The examples we have chosen can be generalized in several directions, but we do not here attempt maximum generality nor claim to be comprehensive in any sense.

Perhaps the simplest of integral equations of the form (1.1) are those described by differential equation

$$(3.1) \qquad (Bx(t))' + Ax(t) \ni g(t), \quad 0 < t < \infty ,$$

with the initial value condition

$$(3.2) \qquad x(0) = x_0$$

There exists an extensive literature on equations of the form (3.1), mainly concerned with the case in which B is linear. We mention in this connection the papers of Bardos and Brézis{ 5}, Showalter{ 21}, {22}, Lagnese{ 13}, Brill{ 6}.

As an immediate application of Theorem 1 (respectively Theorem 2) we get

THEOREM 4. Let A and B satisfy assumptions of Theorem 1 or Theorem 2. Let $T > o$, $g \in L^2(o,T;H)$, $x_0 \in W$, $v_0 \in H$ be given such that $v_0 \in Bx_0$ and $g' \in L^2(o,T;W')$. Then the initial value problems (3.1), (3.2) has at least one solution $x \in L^\infty(o,T;W)$ in the following sense: there exist functions v and w which satisfy

(3.3) $$v \in W^{1,\infty}(o,T;W') \cap C(o,T;H_w)$$

(3.4) $$w \in L^\infty(o,T;W'),\ \int_o^t w(s)ds \in L^\infty(o,T;H)$$

(3.5) $$v(t) \in Bx(t), w(t) \in Ax(t), \text{ a.e. } t \in]o,T[$$

(3.6) $$v'(t)+w(t) = g(t), \quad \text{a.e. } t \in]o,T[$$

(3.7) $$v(o) = v_0.$$

Theorem 4 extends in several directions the main existence result obtained by Grange and Mignot {11} by a different approach. If A is linear we may apply Theorem 3 to derive the uniqueness.

One example to which Theorem 3 applies neatly is the parabolic equation

(3.8) $$\frac{\partial}{\partial t}\ \gamma(y(t,x)) + Ay(t,x) \ni g(t,x);\ t \geqslant o, x \in \Omega$$

with boundary value conditions

(3.9) $$y(t,x) = o \quad \text{for } t > o,\ x \in \Gamma$$

and initial conditions

(3.10) $$v_0(x) \in \gamma(y(o,x)) \quad \text{for} \quad x \in \Omega$$

where Ω is a bounded and open domain in R^n with a sufficiently smooth boundary Γ.

Here γ is a (possible multivalued) maximal monotone graph in $R \times R$ such that $o \in \gamma(o)$ and A is a nonlinear operator on Sobolev space $W_0^{1,p}(\Omega)$ $(p \geq 2)$ such that

(3.11) $$A = \partial \phi, \quad \phi : W_0^{1,p}(\Omega) \longrightarrow R.$$

We shall assume further that A satisfies conditions (2.3) and (2.4) where $H = L^2(\Omega)$ and $W = W_0^{1,p}(\Omega)$.

We have in mind to apply Theorem 4 where $(By)(x) = \gamma(y(x))$ a.e. $x \in \Omega$ and A is defined as above. To this end we must further assume that

(3.12) $$\int_\Omega Ay(x) \quad \gamma_\lambda(y(x))dx \geq o \text{ for all } y \in W_0^{1,p}(\Omega) \text{ and } \quad \lambda > o.$$

Under these assumptions, Theorem 4 is applicable. Let $T > o$, $g \in L^2(o,T;L^2(\Omega))$ and $v_0 \in L^2(\Omega)$ be such that $\partial g/\partial t \in L^2(o,T;L^2(\Omega))$ and $v_0(x) \in \gamma(y_0(x))$ a.e. $x \in \Omega$ where $y_0 \in W_0^{1,p}(\Omega)$. Then problem (3.8), (3.9) and (3.9) has a solution $y \in L^\infty(o,T;W_0^{1,p}(\Omega))$ in the following sense: There exists $v \in W^{1,}(o,T;W^{-1,p'}(\Omega)) \cap C(o,T;(L^2(\Omega))_w)$ such that

(3.13) $$v(t,x) \in \gamma(y(t,x)) \quad \text{a.e. } t \in]o,T[,\ x \in \Omega$$

(3.14) $\partial v/\partial t + Ay = g$ on $[o,T] \times \Omega$

(3.15) $v(o,x) = v_0(x)$ a.e. $x \in \Omega$.

Equation (3.8) with γ and A satisfying the above assumptions serves as a general model for parabolic nonlinear boundary value problems of degenerate type (see {5}, {9}, {14}). In this case A is a partial differential operator of the form

$$Ay = - \sum_{i=1}^{n} \frac{\partial}{\partial x_i} A_i(x,y,\nabla y) \tag{3.16}$$

where $A_i(x,y,\xi)$ are continuous in y, ξ, measurable in x and satisfy

$$\sum_{i=1}^{n} (A_i(x,y,\xi)-A_i(x,y,\eta))(\xi_i - \eta_i) \geq o \tag{3.17}$$

for all $x,y \in \Omega \times R$ and $\xi, \eta \in R^N$ and $A_i(x,y,o) \equiv o$ for all $i=1,\dots n$ and $x,y \in \Omega \times R$. Of course some growth conditions of the form

$$|A_i(x,y,\xi)| \leq C(|y|^{p-1} + |\xi|^{p-1} + g(x)) \tag{3.18}$$

where $g \in L^{p'}(\Omega)$ $(1/p + 1/p' = 1)$, as well as a coercitivity condition must be added. It is easy to see that condition (3.12) as well as conditions (2.3) and (2.4) hold in the present situation.

4. EXISTENCE FOR HEAT FLOW IN RIGID CONDUCTORS WITH MEMORY

In recent years, several non Fourier models for heat conduction in solids have been proposed. Of particular interest is the theories proposed by Gurtin and Pipkin {10} and Coleman and Gurtin {7} (see also {20}) which are based on a memory effect in the conductor. It is our purpose here to derive an existence result for a nonlinear heat conductor Ω with memory whose constitutive equations for the flux q and internal energy e are given by

$$(4.1) \qquad e(t,x) = \beta(\vartheta(t,x)) + \int_{-\infty}^{t} \alpha(t-s)\ \beta(\vartheta(s,x))ds,$$

$$(4.2) \qquad q(t,x) = -k(o)\ \sigma(\nabla_x \vartheta(t,x)) - \int_{-\infty}^{t} k'(t-s)\ \sigma(\nabla_x(s,x))ds$$

for $t \in R$ and $x \in \Omega$. Here ϑ is the temperature. Of course, (4.1) and (4.2) represent a special case of Gurtin and Pipkin's constitutive equations. Along with the energy balance equation, (4.1) and (4.2) yield

$$\beta(\vartheta(t,x)) + \int_{-\infty}^{t} \alpha(t-s)\ \beta(\vartheta(s,x))ds$$

$$\int_{-\infty}^{t} k(t-s)\ \nabla_x \sigma(\nabla_x \vartheta(s,x))ds = f(t,x).$$

This equation is equivalent with

$$(4.3) \qquad \beta(\vartheta(t,x)) - \int_{0}^{t} a(t-s)\ \nabla_x \sigma(\nabla_x \vartheta(s,x))ds = g(t,x)$$

for $t \geq o$ and $x \in \Omega$ where

$$a(t) = k(t) + \int_{0}^{t} \rho\ (t-s)k(s)ds$$

and

$$g(t,x) = f_o(t,x) + \int_{0}^{t} \rho(t-s)f_o(1,x)ds.$$

Here ρ is the resolvant for α

$$f_o(t,x) = f(t,x) - \int_{-\infty}^{o} \alpha(t-s)\ \beta(\vartheta_o(s,x))ds + \int_{-\infty}^{o} k(t-s)\ \nabla_x \sigma(\nabla_x \vartheta_o(s,x))ds$$

and ϑ_o represents the temperature initial history.

We must impose also the boundary value condition

$$(4.4) \qquad \vartheta(t,x) = o \quad \text{on boundary } \Gamma \text{ of } \Omega \quad .$$

We intend to apply Theorem 2 where $H = L^2(\Omega)$, $W = W_0^{1,p}(\Omega)$, $(B\vartheta)(x) = \beta(\vartheta(x))$ a.e. $x \in \Omega$ and $A: W_0^{1,p}(\Omega) \longrightarrow W^{-1,p'}(\Omega)$ defined by

$$A\vartheta = -\nabla_x \sigma(\nabla_x \vartheta) \quad \text{on } \Omega \quad .$$

To this purpose we shall assume that σ and β are continuous and monotone (nondecreasing) on R. We suppose further that

(4.5) $$\beta(o) = \sigma(o) = o$$

(4.6) $$|\sigma(r)| \leq C_1(|r|^{p-1} + 1) \qquad r \in R$$

(4.7) $$j(r) \geq \omega |r|^p + C_2$$

where $j(r) = \int_o^r \sigma(s)ds$ and $\omega > o$,

(4.8) $k(o) > o$, k'' absolutely continuous on $[o,T]$.

Of course, a condition on b implying that the resolvent kernel is in $L^1_{loc}(R^+)$ must be added (see {17}, {18} for results of such type).

Under these conditions Theorem 2 can be applied to obtain existence for equation (4.3) but the details are left to the reader.

R E F E R E N C E S

1. BARBU, V. - Nonlinear Semigroups and Differential Equations in Banach Spaces, Noordhoff International Publishing & Ed.Academiei, Leyden - Bucharest 1976.

2. - Nonlinear Volterra Equations in a Hilbert Space, SIAM J. Math. Anal. Vol.6 (1975), p.728-741.

3. - Existence for Nonlinear Volterra Equations in Hilbert Spaces, SIAM J.Math.Anal. Vol.9 (1978).

4. BARBU, V. and MALIK, M.A. - Semilinear Integro-differential equations in Hilbert Space, J.Math.Anal.Appl. (to appear)

5. BARDOS, C. BREZIS, H. - Sur une classe de problemes d'evolution non linéaire, J. Differential Equations 6(1969), 343-345.

6. BRILL, S. - Sobolev Equations in Banach Space, J. Differential Equations 24(1977), 412-425.

7. COLEMAN, B.D. GURTIN, M.E. - Equipresence and constitutive equations for rigid heat conductors, ZAMP 18(1967), 199-208.

8. CRANDAL, M.G. LONDEN, S.O. NOHEL, J.A. - An abstract nonlinear Volterra integro-differential equation, J.Math.Anal.Appl. (to appear).

9. DUBINSKI, J. - Weak convergence in nonlinear elliptic and parabolic equations, Amer. Math.Soc.Transl. 67(1968), 226-258.

10. GURTIN, M.E. PIPKIN, A.C. - A general theory of heat condition with finite wave speeds, Arch.Rat. Mech.Anal. 31(1968), 113-126.

11. GRANGE, O. MIGNOT, F. - Sur la résolution d'une équation et d'une inéquation paraboliques non linéaires, J. Functional Analysis 11(1972), 77-92.

12. GRIPENBERG, G. - An existence result for a nonlinear Volterra integral equation in Hilbert space, Report HTKK-MAT-A 86, 1976.

13. LAGNESE, J. - General boundary value problems for differential equations of Sobolev type, SIAM J. Math.Anal. 3(1972), 105-119.

14. LIONS, J.L. - Quelques méthodes de résolution des problèmes aux limites non linéaires, Dunod, Paris,1969.

15. LONDEN, S.O. - On an integral equation in a Hilbert space, SIAM J.Math Anal.

16. LONDEN N, S.O. STAFFANS, O.J. - A note on Volterra

integral equations in a Hilbert space,Report HTKK MAT-A 90(1976).

17. MACCAMY, R.C. - Stability theorems for a class of functional differential equations, SIAM J.Appl.Math.

18. - An integro-differential equation with application in heat flow, Quart.Appl. Math. 35(1977),1-19.

19. MIGNOT, F. - Un théorème d'existence et d'unicité pour une équation parabolique non linéaire, Seminaire Brézis-Lions 1973/1974.

20. NUNZIATO, J. - On heat conduction in materials with memory, Quart.Appl.Math. 29(1971), 187-203.

21. SHOWALTER, R.E.-Existence and representation theorems for a semilinear Sobolev equation in Banach space, SIAM J. Math.Anal. 3(1972), 527-543.

22. - A nonlinear parabolic Sobolev equation, J. Math. Anal. Appl. 50(1975),183-190.

DIRECT SOLUTION METHODS FOR A CLASS OF INTEGRAL EQUATIONS WITH RANDOM DISTRIBUTION OF THE INHOMOGENEOUS PART

N. Bellomo
Istituto di Meccanica Razionale
Politecnico - Torino - Italy

G. Pistone
Istituto di Analisi Matematica
Università - Torino - Italy

1. Introduction

This paper deals with systems of integral Volterra equations with random distribution of the inhomogeneous part and indicates some direct methods of achieving an analytical modelling of the probabilistic solutions and their optimum control. More in particular, two coupled, non-linear equations with the inhomogeneous part distributed according to a given initial probability density are considered, and a basic theorem is proposed in order to construct the operator, which transforms the initial probability density into the density at fixed control value of the undependent variable. See for the general bibliography refs. [1,2,3,4,5].
As is known, integral equations of the Volterra type can also be an equivalent formulation of the ordinary differential equations. This simpler case is considered in secs.2-3, and the main results are given in secs.5-6, where the probabilistic solution is modelled by a short-range expansion, the coefficients of the expansion being optimized by a suitable minimization technique. The model solution is then utilized for the optimum control problem. Such a problem, in this paper, consists in finding the optimum initial condition, in terms of the density of the inhomogeneous part, to obtain an optimum output in terms of the probability density on the dependent variables at a given control step of the undependent variable.

2. Description of the problem

In this section integral equations of Volterra type, corresponding to ordinary differential equations, are considered. Some of the proposed results are afterwords, in sec.6, extended to the general case.
In particular, let us take into account the following system of coupled integral

$$x = x_0 + \int_0^t K_1(s,\ x(s),\ y(s);\ \underline{w})\ ds \tag{1a}$$

$$y = y_0 + \int_0^t K_2(s,\ x(s),\ y(s);\ \underline{w})\ ds \tag{1b}$$

where $\underline{w}$ is a parameter, $\underline{w} \in W \subset R^p$, and the initial condition (x_0, y_0) is distributed according to an initial probability density $P_0 = P_0(x_0, y_0)$.

Eq.(1) can also be written in the following form:

$$\underline{z} = \underline{z}_0 + \underline{A}\{\underline{z};\ \underline{w}\}\ ;\quad \underline{z} = (x,\ y)\ ;\quad \underline{A} = (A_1,\ A_2)\ ;\quad A_{1,2} = \int_0^t K_{1,2}\,ds \tag{2}$$

and let us suppose that the deterministic problem is well posed and bounded according to the following hypotheses:

Hypothesis I: For each $\underline{w} \in W$ the kernel $\underline{K} = (K_1, K_2)$ is of class C^1 in a bounded domain $D = D_t \times D_x \times D_y$, where D_t, D_x and D_y are open real intervals with $0 \in D_t$. In such a domain $\underline{K}$ satisfies a uniform Lipschits condition:

$$\forall t,\ \underline{z}_1,\ \underline{z}_2 :\qquad |\underline{K}(t,\ \underline{z}_1) - \underline{K}(t,\ \underline{z}_2)| < a|\underline{z}_1 - \underline{z}_2|$$

Hypothesis II: For every $w \in W$ the global solutions of eq.(2) exist in D.

Remark I: Let us denote by $\phi_t : D_z \longrightarrow D_z$ the solution map defined by:

$$\phi_t(\underline{z}_0) = \underline{z}(t)$$

according to the theory of differential systems [6] it follows that ϕ_t is a diffeomorphism in D_z.

With regard to the afore-mentioned mathematical description of the problem, the following problems will be dealt with in this paper:

a) *Let P_0 be a probability density in D_z and let μ_0 be the probability measure with density P_0, the problem of finding the density P_t of the immage probability measure $\phi_t(\mu_0)$ will be dealt with in sec.3.*

b) *Let $t \in D_t$ a given time and let Ψ be a functional given on the probability measure on D_z and W, the problem of finding a probability measure μ_0 and a parameter $\underline{w}_0$ such that $\Psi\{\phi_{t,\underline{w}_0}(\mu_0,\ \underline{w}_0)\}$ is minimal will be dealt with in sec.4.*

Problem a will be called evolution problem, problem b will be called control problem.

3. Some remarks on the evolution problem

The evolution problem of the system (2), under hypotheses I and II has been studied by T. Soong, see ref.|4|. In particular, in the already quoted ref.|4|, the following formula is proposed:

$$P_t(x, y) = P_0 \circ \Psi_t(x, y)\ J_t(x, y) \tag{3}$$

where $\Psi_t = \phi_t^{-1}$ and J_t is the jacobian of the transformation $\underline{z} \to \underline{z}_0$:

$$J_t(\underline{z}) = \det|d\Psi_t/d\underline{z}| \tag{4}$$

In practical problems Ψ_t is computed by backword numerical integration of eq.(2), on the other hand eq.(4) does not supply a tool suitable to compute J_t. Therefore the following result can be proposed:

Lemma I: If Hypotheses I and II hold, the time-evolution of J_t can be computed by solution of the following augmented differential equation:

$$\dot{\zeta} = \underline{F}(\zeta(s), s), \qquad 0 < s < t \tag{5}$$

with final condition:

$$\zeta(t) = (x, y, 1) \tag{6}$$

where:

$$\underline{F}(x, y, J, s) = \{K_1(s, x, y), K_2(s, x, y), -J\ \underline{\nabla}\cdot\underline{K}(s, x, y)\} \tag{7}$$

Ψ_t and J_t (see eq.(3)) are given as follows:

$$\zeta(0) = (\Psi_t(x, y), J_t(x, y)) \tag{8}$$

Proof:

Let $G_t = \det|d\phi_t/d\underline{z}|$. Applying the equation of the vaiations of eq.(2) and considering that det(exp[A]) = exp(tr.[A]), it follows:

$$G_t(x_0, y_0) = \exp\int_0^t \underline{\nabla}\cdot\underline{K}(s, \phi_s(x_0, y_0))ds \tag{9}$$

Moreover:

$$J_t(x, y) = 1/G_t \circ \Psi_t(x, y) \tag{10}$$

setting $z_{1,2}(s, x, y) = \phi_s \circ \Psi_t(x, y)$ it follows:

$$J_t(x, y) = \exp\{-\int_0^t (\underline{\nabla}\cdot\underline{K})(s, z_{1,2}(s, x, y))ds\} \tag{11}$$

and

$$\dot{z}_{1,2}(s, x, y) = K_{1,2}(s, z_{1,2}(s, x, y)) \tag{12}$$

moreover:

$$z_{1,2}(t,\ x,\ y) = (x,\ y) \tag{13}$$

therefore:

$$z_3(s,\ x,\ y) = \exp\left[-\int_0^t \underline{\nabla}\cdot\underline{K}(u,\ z_{1,2}(u,\ x,\ y))du\right] \tag{14}$$

The Lemma is then prooved by eqs.(11-14). See also ref.[7,8].

4. Some remarks on the optimum control problem

Let us note that eqs.(3,4) of the preceding section can also be written in the following equivalent form:

$$P_0(x_0,\ y_0) = P_t\circ\phi_t(x_0,\ y_0)\ G_t(x_0,\ y_0) \tag{15}$$

$$G_t(x_0,\ y_0) = \det|d\phi_t/d\underline{z}_0| \tag{16}$$

Remark II: ϕ_t is the forward evolution of eq.(2). If the initial value problem is considered, the result of Lemma I holds and the evolution of G is obtained by eq.(2) augmented with the equation: $\dot{G}_t = G_t(\underline{\nabla}\cdot\underline{K})$.

Let us now consider the problem b, and let $\bar{D}_z$ be the closure of the bounded set $D_z \in \mathbb{R}^2$. Moreover, let us denote, by Prob.($\bar{D}_z$), the set of probability measures on $\bar{D}_z$. A topology associated with the weak convergence can be joined to Prob.($\bar{D}_z$). A sequence (μ_n) is weakly conver-gent to μ if, for all bounded continuous f: $\bar{D}_z \to R$, we have: $\int f\ d\mu_n \longrightarrow \int f\ d\mu$.

Accordingly the following Lemma can be proposed:

Lemma II: Let W be compact and let Ψ_0: Prob.($\bar{D}_z$)$\times W \to \mathbb{R}$ be continuous. then Ψ_0 has a minimum.

Proof:

If $\bar{D}_z$ is compact, then Prob.($\bar{D}_z$) is compact for the weak convergence [9]. Therefore a continuous functional on a compact set has at least a minimum.

Proposition 1: Let Ψ : Prob.($\overline{\phi_t(D_z)}$)$\times W$ and let us assume that W is compact and that ϕ_t can be defined as a continuous function on $\bar{D}$. For every $\varepsilon > 0$ a probability density P_ε on D_z and a parameter $\underline{w}_\varepsilon$ exist such that:

$$\Psi\{\phi_t(P_\varepsilon\cdot m_{D_z}),\ \underline{w}_\varepsilon\} < \min_{\mu\in \mathrm{Prob.}(\bar{D}_z),\ \underline{w}\in W} \Psi\{\phi_t(\mu),\ \underline{w}\} + \varepsilon \tag{17}$$

where m_D is a Lebesgue measure on D_z.

Proof:

By Lemma I a minimum $(\bar{\mu}, \bar{\underline{w}})$ for the functional defined by:

$$\Psi_0(\mu, \underline{w}) = \Psi(\phi_t(\mu), \underline{w}) \tag{18}$$

exists. Moreover the set of absolutely continuous probability measures on D_z is dense in Prob.$(\bar{D}_z)$, see ref.[9].

Remark III: Assuming that the partial derivatives of $K_{1,2}$ in eq.(1) are bounded, then ϕ_t can be defined as a continuous function on $\bar{D}_z$. Moreover from the equation of the variations it follows that G_t in eq.(15) is bounded.

This Remark, together with Remark II and eqs.(15,16) can help, in some particular cases, to show that the optimum control problem has a density.

5. Direct solution methods

The term direct solution method denotes here any method for computing a solution of the afore-mentioned class of problems a-b, related to eq.(2), without both the direct use of $\phi_{t,w}$ as a result of the solution of a differential equation and the use of the equations derived in sec.3.

Let us firstly assume that eq.(2) has not a dependence on parameters and let us consider a very particular control problem, which is here called inverse problem, where at given probability density P_t, the initial density P_0, such that the immage with respect to ϕ_t of the probability measure $P_0(\underline{z}_0)d\underline{z}_0$ is $P_t(\underline{z})d\underline{z}$, is sought for, see eq.(15). This objective will be pursued by means of the search for a suitable sequence ϕ^n of approximating functions converging to ϕ. The obtained results will be afterwards extended to the more general case of flow ϕ depending on parameters.

Remark IV: Let Φ_t be a diffeomorphism on the open set D_z and let be given a sequence (ϕ^n) of diffeomorphism such that $\phi^n \to \phi$ in the sense of the Lebesgue-measure. The sequence of probability measures $\mu_n(d\underline{z}_0) = P_t \circ \phi^n(\underline{z}_0) G_n(\underline{z}_0) d\underline{z}_0$, $G_n(\underline{z}_0) = |\det(d\phi^n/d\underline{z}_0)|$ is weakly convergent to a probability measure $P_0(\underline{z}_0)d\underline{z}_0$ which solve the inverse problem for given P_t as a final density.

In general, not all the approximation procedures assure that the approximating sequence (ϕ^n) is such that each ϕ^n is a diffeomorphism and that is invertible. In this case, the problem consists in proving that the natural candidates:

$$\mu_n(d\underline{z}_0) = P_t \circ \phi^n(\underline{z}_0) G_n(\underline{z}_0) d\underline{z}_0 \tag{19}$$

for approximating the solution of the initial value problem, converge. In practice, the following procedure has been tested, ref.[10], and has given realiable results:

$$x = x_0 + \sum_{j=1}^{n} a_j t^j \qquad\qquad y = y_0 + \sum_{j=1}^{n} b_j t^j \tag{20}$$

where the coefficients a_j and b_j are given by:

$$a_j = \sum_{k=1}^{m} \sum_{h=1}^{m} a_{khj}\, x_0^k\, y_0^h \qquad\qquad b_j = \sum_{k=1}^{m} \sum_{h=1}^{m} b_{khj}\, x_0^k\, y_0^h \tag{21}$$

the operator G_t, according to its definition, is then given by:

$$G_t = (1 + \sum_{j=1}^{n} \alpha_j t^j)(1 + \sum_{j=1}^{n} \beta_j t^j) - \sum_{j=1}^{n} \gamma_j t^j \sum_{j=1}^{n} \sigma_j t^j \tag{22}$$

where:

$$\alpha_j = \sum_{k=1}^{m} \sum_{h=1}^{m} k\, a_{khj}\, x_0^{(k-1)} y_0^h \qquad\qquad \beta_j = \sum_{k=1}^{m} \sum_{h=1}^{m} h\, b_{khj}\, x_0^k\, y_0^{(h-1)}$$

$$\gamma_j = \sum_{k=1}^{m} \sum_{h=1}^{m} h\, a_{khj}\, x_0^k\, y_0^{(h-1)} \qquad\qquad \gamma_j = \sum_{k=1}^{m} \sum_{h=1}^{m} k\, b_{khj} x_0^{(k-1)} y_0$$

and after some algebrical manipulations, and limiting the power expansion to n:

$$G_t = 1 + (\alpha_1 + \beta_1)t + \sum_{j=2}^{n} (\alpha_j + \beta_j + \omega_j)\, t^j \tag{23}$$

where:

$$\omega_j = \sum_{p=0}^{j-2} (\alpha_{j-p-1}\, \beta_{p+1} - \gamma_{j-p-1}\, \sigma_{p+1})$$

which is an anlytical form suitable to apply the operator G_t upon P_0 in order to evaluate P_t.

In particular let us formulate the following remark:

Remark V: $P_t = G_t P_0$ is a function, at fixed t, of x and y, which for every initial condition (x_0, y_0) are given by the power expansion defined by eqs.(20-21).

A natural way for obtaining the expansion defined by eqs.(20,21) is to transform eq.(2) into an equivalent formulation in terms of unconstrained minimization of the following functional:

$$\Gamma = \Gamma(\underline{z}) = \sup\{\Theta_1, \Theta_2\} \ ; \quad \Theta_1 = \int_D (A \ - x_0 - x)^2 ds dx_0 dy_0$$

$$\Theta_2 = \int_D (A_2 - y_0 - y)^2 ds dx_0 dy_0 \tag{24}$$

where, if the expansion of eqs.(20,21) is applied, the equivalent problem becomes the minimization of a known function of a set of variables:

$$\text{Min.}\{\Gamma(\underline{\Omega})\} \ ; \quad \underline{\Omega} = (\underline{a}, \underline{b}) \quad ; \quad \underline{a} = \{a_{khj}\} \ ; \quad \underline{b} = \{b_{khj}\} \tag{25}$$

This procedure is equivalent to the Bubnov-Galerkin's methods, see ref.[3], sec.5, and can be dealt with on the basis of a suitable minimization technique [11].

Remark VI: G_t can be expanded in the same fashion as eqs.(20,21) as follows:

$$G_t = 1 + \sum_{j=1}^{n} c_j t^j \ ; \qquad c_j = \sum_{k=1}^{m} \sum_{h=1}^{m} c_{khj} x_0^k y_0^h \tag{26}$$

and the transformed problem is:

$$\text{Min.}\{\Gamma\} \ ; \ \Gamma = \Gamma(\underline{z}, G) = \sup.\{\Theta_1, \Theta_2, \Theta_3\} \tag{27}$$

where Θ_1 and Θ_2 have been defined in eq.(24) and Θ_3 is given by:

$$\Theta_1 = \int_D (A_3 - G_s - 1)^2 ds dx_0 dy_0 \ ; \qquad A_3 = \int_0^s (-G \ \underline{\nabla} \cdot \underline{K}) du \tag{28}$$

which utilizes the result of the Lemma proposed in the third section.

The discussion of the convergence of the numerical procedures proposed in this section must get through the following points:

a) Qualitative analysis of the evolution equation. (An example is given in ref. [15]).

b) Analysis of the error bounds in the considered procedures.

For this last aim let us supply some preliminary definitions:

Def.1: A mapping g: $\bar{D}_z \in \mathbb{R}^2$ belongs to the class Lip.$(\bar{D}_z)$ if a constant L exists such that for every $\underline{z}'_0$, $\underline{z}''_0$ $\bar{D}_z$ the quantities $|g(\underline{z}_0)|$ and $|g(\underline{z}'_0) - g(\underline{z}''_0)| / |\underline{z}'_0 - \underline{z}''_0|$ are bounded by L. The smallest constant with such a value will be denoted by $||g||_{Lip.}$.

Def.2: A mapping $\phi: \overline{D}_z \to R^2$ belongs to the class $C^1(\overline{D}_z)$ if a continuously weakly differentiable mapping $\hat{\phi}: R^2 \to R^2$ exists such that $\hat{\phi}(\underline{z}_0) = \phi(\underline{z}_0)$ for $\underline{z}_0 \in \overline{D}_z$. The restriction to $\overline{D}_z$ of $d\hat{\phi}/d\underline{z}_0$ will be denoted by $d\phi/d\underline{z}_0$ and the number $\inf_{\underline{z}_0 \in \overline{D}_z} |d\phi/d\underline{z}_0|$ by k_ϕ.

Accordingly the following theorem can be proposed:

Theorem: Let $\phi, \Psi \in C^1(\overline{D}_z)$ and G_ϕ, G_Ψ the jacobian determinants of ϕ and Ψ. If P is a given probability density with support in $\phi(\overline{D}_z)$ and ϕ is a diffeomorphism of positive type, then:

$$\forall\, \underline{z}_0 \in \overline{D}_z: \quad |\Psi(\underline{z}_0) - \phi(\underline{z}_0)| < \varepsilon \Rightarrow \left| \int_{D_z} g(\underline{z}_0) G_\Psi P \circ \Psi(\underline{z}_0) d\underline{z}_0 - \int_{D_z} g(\underline{z}_0) G_\phi P \circ \phi(\underline{z}_0) d\underline{z}_0 \right|$$

$$< \varepsilon \|g\|_{Lip.} k_\phi^{-1} \int_{D_z} |G_\Psi(\underline{z}_0) P \circ \Psi(\underline{z}_0)| d\underline{z}_0 + \|g\|_{Lip.} \int_{\{\underline{z}_0 : \, dist.(\underline{z}_0, \partial D_z) < 2\varepsilon k_\phi^{-1}\}} |G_\Psi(\underline{z}_0) P \circ \Psi(\underline{z}_0)| d\underline{z}_0$$

$$+ \|g\|_{Lip.} \int_{\{\underline{z} \in \phi(D_z): \, dist.(\underline{z}, \phi(\partial D_z)) < \varepsilon\}} P(\underline{z}) d\underline{z} \tag{29}$$

where ∂D_z is the boundary of D_z.

Proof:

Some preliminaries, see ref. |12|, are necessary for the proof, see also ref.[13,14]. For all $\Psi \in C^1(\overline{D}_z)$ let us denote by L_Ψ the set:

$$L_\Psi = R^2 - \Psi\{G_\Psi = 0\} - \Psi(\partial D_z) \tag{30}$$

The degree of Ψ is a function defined on L_Ψ by:

$$\deg._\Psi(\underline{z}) = \sum_{\underline{z}_0 \in \Psi^{-1}(\underline{z})} \mathrm{sign}.G_\Psi(\underline{z}_0) \tag{31}$$

$\deg._\Psi$ can be defined, for all $\underline{z} \notin \Psi(\partial D_z)$, by continuity in a way such that the following change of variable holds:

$$\int_{D_z} f \circ \Psi(\underline{z}_0) G_\Psi(\underline{z}_0) P \circ \Psi(\underline{z}_0)\, d\underline{z}_0 = \int_{\mathbb{R}^2 - \Psi(\partial D_z)} f(\underline{z}) \deg._\Psi(\underline{z})\, d\underline{z} \tag{32}$$

$\deg._\Psi$ as a function of the two variables Ψ and $\underline{z}$ is locally constant.

Let us now define the sets A_ε and B_ε as follows:

$$A_\varepsilon = \{\underline{z}: dist.(\underline{z}, \phi(\partial D_z)) > \varepsilon\} \quad ; \quad B_\varepsilon = \{\underline{z}: dist.(\underline{z}, \phi(\partial D_z)) \le \varepsilon\} \tag{33}$$

and consider the family of functions $\Psi_\theta = (1-\theta)\phi + \theta\Psi$, $0 < \theta < 1$; the hypothesis of the theorem implies that:

$$\Psi_\theta(\partial D_z) \subset B_\varepsilon \tag{34}$$

Eq.(34) and the locally constant value of $\deg._\Psi$ give:

$$\deg._\Psi = \deg._\phi = \mathbf{1}_{\phi(D_z)} \qquad \text{on } A_\varepsilon \tag{35}$$

where $\mathbf{1}_{\phi(D_z)}$ is the indicator function of the set $\phi(D_z)$, see ref.|12|.
Let us now consider the set $\Psi^{-1}(B_\varepsilon)$. If $\underline{z}_0 \in \Psi^{-1}(B_\varepsilon)$ and $\overline{z} \in \partial D_z$ then, from eq.(33),

$$|\underline{z}_0 - \overline{z}| \le K_\phi^{-1}|\phi(\underline{z}_0) - \phi(\overline{z})| \le K_\phi^{-1}(|\phi(\underline{z}_0) - \Psi(\underline{z}_0)| + |\Psi(\underline{z}_0) - \phi(\overline{z})|) \le 2K_\phi^{-1}\varepsilon \tag{36}$$

Consequently:

$$\Psi^{-1}(B_\varepsilon) \subset \{\underline{z}_0 \in D_z : \text{dist.}(\underline{z}_0, \partial D_z) \le 2k_\phi^{-1}\varepsilon\} \tag{37}$$

Setting $g = f\circ\phi$ and for all $g \in \text{Lip.}(\overline{D}_z)$,

$$L = |\int_{D_z} g(\underline{z}_0)G_\Psi(\underline{z}_0)P\circ\Psi(\underline{z}_0)\, d\underline{z}_0 - \int_{D_z} g(\underline{z}_0)G_\phi(\underline{z}_0)P\circ\phi(\underline{z}_0)\, d\underline{z}_0|$$

$$\le |\int_{D_z} (g(\underline{z}_0) - f\circ\Psi(\underline{z}_0))G_\Psi(\underline{z}_0)P\circ\Psi(\underline{z}_0)\, d\underline{z}_0| + |\int_{D_z} f\circ\Psi(\underline{z}_0)G_\Psi(\underline{z}_0)P\circ\Psi(\underline{z}_0)\, d\underline{z}_0$$
$$- \int_{D_z} g(\underline{z}_0)G_\phi(\underline{z}_0)P\circ\phi(\underline{z}_0)\, d\underline{z}_0| \tag{38}$$

moreover, according to Def.1 and 2,

$$|g(\underline{z}_0) - f\circ\Psi(\underline{z}_0)| \le \varepsilon\|f\|_{\text{Lip.}} \le \varepsilon\|g\|_{\text{Lip.}} k_\phi^{-1} \tag{39}$$

then the first term of the inequality of eq.(38) is bounded by:

$$\varepsilon\|g\|_{\text{Lip.}} k_\phi^{-1} \int_{D_z} |G_\Psi(\underline{z}_0)| P\circ\Psi(\underline{z}_0)\, d\underline{z}_0 \tag{40}$$

On the other hand the second term of eq.(38), after the change of variable, according to eq.(32) and the decomposition of the integral can be written as follows:

$$\left|\int_{A_\varepsilon} f(\underline{z})P(\underline{z})\deg._\Psi(\underline{z})\, d\underline{z} + \int_{B_\varepsilon} f(\underline{z})P(\underline{z})\deg._\Psi(\underline{z})\, d\underline{z} - \int_{\phi(D)} f(\underline{z})P(\underline{z})\, d\underline{z}\right| \tag{41}$$

and from eq.(35),

$$\left|\int_{A_\varepsilon} f(\underline{z})P(\underline{z})\mathbf{1}_{\phi(D)}(\underline{z})\, d\underline{z} + \int_{\Psi^{-1}(B_\varepsilon)} g(\underline{z}_0)P\circ\Psi(\underline{z}_0)d\underline{z}_0 - \int_{\phi(D)} f(\underline{z})P(\underline{z})d\underline{z}\right| =$$

$$= \left| \int_{\Psi^{-1}(B_\varepsilon)} g(\underline{z}_0) G(\underline{z}_0) P_\circ(\underline{z}_0)\, d\underline{z}_0 - \int_{\phi(D) - A_\varepsilon} f(\underline{z}) P(\underline{z})\, d\underline{z} \right| \tag{42}$$

This quantity, because of eq.(37) is bounded by:

$$\|g\|_{\mathrm{Lip.}} \int_{\{\underline{z}_0 :\ \mathrm{dist.}(\underline{z}_0, \partial D_z) \le 2k_\phi^{-1}\varepsilon\}} |G_\psi(\underline{z}_0)| P_\circ(\underline{z}_0)\, d\underline{z}_0 + \int_{\{\underline{z} \in \phi(D) :\ \mathrm{dist.}(\underline{z}, \phi(\partial D_z)) \le \varepsilon\}} P(\underline{z})\, d\underline{z} \tag{43}$$

Then the error L is bounded by the sum of the expression given by eq.(40) and eq.(43). Which proves the theorem.

If on the other hand, the control problem is the general one defined in sec.4, namely with the flow depending on $\underline{w}$, the problem itself can be dealt with by expanding P_0 in the form:

$$P_0 = \sum_{i=1}^{q} \alpha_i\, g_i(\underline{z}) \ ; \qquad \int g_i = 1 \quad ; \quad \sum_{i=1}^{q} \alpha_i = 1 \tag{44}$$

and by finding the optimum set $\underline{\Sigma} = \{\underline{\alpha}, \underline{w}\}$ which minimize a given functional. The procedure is justified by Proposition 1. and the optimization is transformed in a problem of minimization of a known function of a set of variables. In this class of problems the question of the definition of the error bounds is not dealt with in this paper.

6. Conclusion and discussion

Some solution methods of a system of two non-linear Volterra-type integral equations, with their inhomogeneous parts distributed according to an initial probability density, have been studied in this paper for the class of integral equations arising from ordinary differential equations.

It is worthwhile mentioning that all the results presented in sec.5 can be applied, besides the one defined by Remark VI, to the Volterra integral equation of the more general type, namely with kernels of the type: $K = K(s, x, y, t)$. Therefore the analysis carried out in this work acquires a wider validity.

As a matter of interest, let us finally remark that the proposed methods have recently been applied [10] to an optimum control problem where the integral equation defines the evolution of a system of propellent-droplets and the control problem consists in finding the optimum initial distribution and parameters towards a desired combustion process.

Acknowledgements

This research has been realized within the activities of the Italian Council for the Research, C.N.R., G.N.F.M. (Gruppo Nazionale Fisica Matematica) and G.N.A.F.A. (Gruppo Nazionale Analisi Funzionale e Applicazioni).

References

1. J. Cochran, The Analysis of Linear Integral Equations, Mc Graw Hill, New York, 1972.
2. L. Delves & J. Walsh, Numerical Solutions of Integral Equations, Clarendon Press, Oxford, 1974.
3. R. Bellman, Methods of Non-linear Analysis, Academic Press, New York, 1970.
4. T. Soong, Random Differential Equations, Academic Press, New York, 1973.
5. A. Barucha-Reid, Random Integral Equations, Academic Press, New York, 1972.
6. R. Abram & J. Marden, Foundations of Mechanics, Benjamin, New York, 1967.
7. N. Bellomo, R. Loiodice & G. Pistone, Time-evolution of a Multidroplet System in the Kinetic Theory of Vaporization and Condensation, in Rarefied Gas Dynamics, Ed J. Potter, AIAA, New York, 2, (1977), 651.
8. N. Bellomo & G. Pistone, Dynamical Systems with a Large Number of Degrees of Freedom: A Stochastic Mathematical Analysis for a Class of Deterministic Problems, in Atti IV Congresso AIMETA, Pitagora, Bologna, Italy, 1, (1978), 1.
9. P. Billingsley, Convergence of Probability Measures, Wiley, New York, 1968.
10. A. Pellegrino & G. Ripa, Sistemi Dinamici con Condizione Iniziale Random, Thesis of the Politecnico of Torino, Ed. N. Bellomo & R. Monaco, 1978.
11. C. Nelson Dorny, A Vector Space Approach to Models & Optimization, Wiley, New York, 1975.
12. J. Schwartz, Nonlinear Functional Analysis, Gordon & Breach, New York, 1969.
13. K. Yosida, Functional Analysis, Springer, Berlin, New York, 1974.
14. J. Dieudonné, Eléments d'Analyse, Gauthier-Villars, Paris, 1970.
15. N. Bellomo, R. Loiodice & G. Pistone, Qualitative and Numerical Analysis of a nonlinear differential equation with random initial conditions in statistical fluid mechanics, l'Aerotecnica J. of AIDAA, 1, (1977), 12.

A NONLINEAR CONTRACTION SEMIGROUP FOR A FUNCTIONAL DIFFERENTIAL EQUATION

Dennis W. Brewer

Department of Mathematics
University of Arkansas
Fayetteville, Arkansas 72701/USA

Our objective is to represent solutions of the initial value problem for the functional differential equation of infinite delay

$$(1)\qquad \begin{cases} u'(t) + \alpha u(t) + Bu(t) = F(u_t), & t \geq 0, \\ u(t) = \phi(t), & t \leq 0, \end{cases}$$

as a semigroup of nonlinear contraction operators on the space of initial data. Here $u_t(x) = u(t+x)$ for $x \leq 0$, $t \geq 0$, α is a positive number, and B is a single-valued, m-accretive operator on a Banach space E with norm $||\cdot||$. The initial data are taken from a Banach space X constructed as follows:

Choose $r \varepsilon (0,\infty)$ and a positive, nondecreasing function p on $(-\infty, 0]$. Let X be the set of strongly measurable functions ϕ from $(-\infty, 0]$ to E such that ϕ is continuous on $[-r,0]$ and $p\phi$ is integrable on $(-\infty, -r)$. The space X becomes a Banach space under the norm

$$(2)\qquad ||\phi||_X = \sup_{-r\leq x\leq 0} p(x)||\phi(x)|| + \int_{-\infty}^{-r} p(x)||\phi(x)||dx.$$

The mapping F is a possibly nonlinear functional from X into E.

By a strong solution $u = u(t;\phi)$ of (1) we will mean that $u: R \to E$, $u = \phi$ on $(-\infty,0]$, and there is a continuous function q on $[0,\infty)$ such that

$$u(t) = u(0) + \int_0^t q(s)ds, \quad t \geq 0,$$

where $q(t) = F(u_t) - \alpha u(t) - Bu(t)$ for $t \geq 0$.

Following the construction in [1] for a linear Volterra equation with infinite delay, we define an operator A in X by

$$(3)\qquad A\phi = -\phi', \quad D(A) = \{\phi \varepsilon X: \ \phi' \varepsilon X, \ \phi'(0) = F(\phi) - \alpha\phi(0) - B\phi(0)\}.$$

More precisely, we mean that if $\phi \varepsilon D(A)$ then there exists $\tilde{\phi} \varepsilon X$ such that

$$(4) \qquad \begin{cases} \phi(x) = \phi(0) + \int_0^x \tilde{\phi}(s)ds, & x \leq 0, \\ \tilde{\phi}(0) = F(\phi) - \alpha\phi(0) - B\phi(0). \end{cases}$$

Then we define $A\phi = -\tilde{\phi}$. Note that this implies that ϕ has a strong derivative ϕ' a.e. and $\phi' = \tilde{\phi}$ a.e. All integrals are in the sense of Bochner. This definition is motivated by the fact that the mapping $w(t) = u_t$ becomes a solution of the abstract Cauchy problem

$$(CP) \qquad \begin{cases} w'(t) + Aw(t) = 0, & t \geq 0, \\ w(0) = \phi, \end{cases}$$

in the sense of a strong solution of (CP) as defined in [4]. The problem (CP) is a natural setting for applications of nonlinear semigroup theory.

The main objective of this note is to prove the following representation theorem for solutions of the initial value problem (1).

THEOREM 1. *Let* A *and* X *be as defined above. Suppose* $p(x)e^{-\alpha x}$ *is non-decreasing on* $(-\infty,0]$, $p(0) = 1$, F *is uniformly Lipschitz continuous from* X *into* E *with Lipschitz constant* M *and*

$$(5) \qquad \alpha^2 \geq \alpha(M + 1) + M$$

Then $\overline{D(A)} = \{\phi \in X: \ \phi(0) \in \overline{D(B)}\}$, *and* $-A$ *generates a nonlinear contraction semigroup* S *on* $\overline{D(A)}$. *Furthermore* S *represents solutions of* (1) *in the sense that if* $u = u(t;\phi)$ *is a strong solution of* (1) *with* $\phi \in D(A)$, *then* $S(t) = u_t$ *for* $t \geq 0$.

The proof of Theorem 1 will be carried out by applying the following special case of the basic result in [6].

THEOREM 2. (*M. Crandall and T. Liggett*) *Let* A *be an accretive operator on a Banach space, and let* $\lambda_0 > 0$. *If* $R(I+\lambda A)$ *contains* $\overline{D(A)}$ *for* $0 < \lambda < \lambda_0$, *then* $-A$ *generates a nonlinear contraction semigroup on* $\overline{D(A)}$.

An operator A is accretive if $(I+\lambda A)^{-1}$ is a contraction on its domain for every $\lambda > 0$, and is m-accretive if $R(I+\lambda A)$ is the entire space for every $\lambda > 0$. The reader may consult [5] for definitions of other terms of semigroup theory used in this note. A link between strong solutions of (CP) and the semigroup guaranteed by Theorem 2 is the following theorem. The statement and proof of a more general case may be found in [5].

THEOREM 3. (H. Brezis and A. Pazy) *Let A be a closed, accretive operator and suppose −A generates the semigroup S. If (CP) has a strong solution w for some $\phi \in D(A)$, then $w(t) = S(t)\phi$ for $t \geq 0$.*

REMARK 4. An important special case of (1) is the Volterra integro-differential equation with infinite delay

$$(6) \quad \begin{cases} u'(t) + \alpha u(t) + Bu(t) = -\int_{-\infty}^{t} a(t-s)g(u(s))ds, & t \geq 0, \\ u(t) = \phi(t), & t \leq 0, \end{cases}$$

where $\phi \in X$ is prescribed. By defining

$$(7) \qquad F(\phi) = -\int_{-\infty}^{0} a(-s)g(\phi(s))ds,$$

one may transform (1) to (6). It is not difficult to verify that if $a\colon (0,\infty) \to R$, $a \in L^1(0,\infty)$, $|a(-x)| \leq Kp(x)$ for a.e. $x \in (-\infty,-r)$ and some constant $K > 0$, and $g\colon E \to E$ is uniformly Lipschitz continuous on E, then F as defined by (7) satisfies the hypotheses of Theorem 1.

REMARK 5. The initial data space X defined above is almost identical to the space employed by J. Hale in [8] to obtain exponential estimates for solutions of linear functional differential equations with infinite delay. This space is of "fading memory type" and was studied by B. D. Coleman and V. J. Mizel [3]. In [2] the author developed a quasi-contraction semigroup representation for the equation $u'(t) = F(u_t)$ in this space.

REMARK 6. The contraction semigroup obtained here yields the estimate

$$||u_t(\phi_1) - u_t(\phi_2)||_X \leq ||\phi_1 - \phi_2||_X$$

for every $t \geq 0$, $\phi_1, \phi_2 \in D(A)$, under the hypotheses of Theorem 1. This is a stability result for (1) and (6) in X. Nonlinear semigroup theory has been employed by several authors to obtain stability and asymptotic stability for various types of functional differential equations of finite delay. The papers of C. C. Travis and G. F. Webb [9], G. F. Webb [10], and J. R. Haddock [7] employ hypotheses and techniques similar to those used here.

THE PROOF OF THEOREM 1. The proof of Theorem 1 relies on the following lemmas:

LEMMA 7. *The operator* $(I+\lambda A)^{-1}$ *is a contraction on* X *for every* $\lambda > 0$.

LEMMA 8. $\overline{D(A)} = \{\phi \in X: \ \phi(0) \in \overline{D(B)}\}$.

LEMMA 9. *The mapping* $w: [0,\infty) \to X$ *defined by* $w(t) = u_t$ *is a strong solution of* (CP) *for* $\phi \in D(A)$ *whenever* u *is a strong solution of* (1) *as defined above*.

Lemmas 7 and 8 together imply by Theorem 2 that $-A$ generates a contraction semigroup S on $\overline{D(A)} = \{\phi \in X: \ \phi(0) \in \overline{D(B)}\}$. The operator A is closed since A is m-accretive by Lemma 7. Therefore Lemma 9 and Theorem 3 together imply that if u is a strong solution of (1) for $\phi \in D(A)$ then $S(t)\phi = u_t$ for every $t \geq 0$. This completes the proof of Theorem 1, assuming the truth of the lemmas.

THE PROOF OF LEMMA 7. Choose $h \in X$, $\lambda > 0$. We wish to find $\phi \in D(A)$ such that $(I+\lambda A)\phi = h$. By definition this means that

$$(8) \qquad \begin{cases} \phi(x) - \lambda\phi'(x) = h(x), \ x \leq 0, \\ \phi(0) - \lambda F(\phi) + \lambda\alpha\phi(0) + \lambda B\phi(0) = h(0). \end{cases}$$

Therefore ϕ must be of the form

$$(9) \qquad \phi(x) = \phi(0)e^{x/\lambda} + \frac{1}{\lambda}\int_x^0 \exp[\tfrac{1}{\lambda}(x-\eta)]h(\eta)d\eta, \quad x \leq 0.$$

For brevity let $\rho(x) = p(x)e^{-\alpha x}$ for $x \leq 0$ and let $\gamma = 1/\lambda$. From (9) we obtain

$$p(x)||\phi(x)|| \leq \rho(x)||\phi(0)||e^{(\alpha+\gamma)x} + \gamma\rho(x)e^{\alpha x}\int_x^0 e^{\gamma(x-\eta)}||h(\eta)||d\eta.$$

Since ρ is nondecreasing this yields

$$\begin{aligned} p(x)||\phi(x)|| &\leq \rho(0)||\phi(0)||e^{(\alpha+\gamma)x} + \gamma e^{\alpha x}\int_x^0 e^{\gamma(x-\eta)}\rho(\eta)||h(\eta)||d\eta \\ &\leq ||\phi(0)||e^{(\alpha+\gamma)x} + \frac{1 - e^{(\alpha+\gamma)x}}{1 + \alpha\lambda}\sup_{-r\leq x\leq 0} p(x)||h(x)||. \end{aligned}$$

Therefore

$$(10)\qquad \sup_{-r\le x\le 0} p(x)||\phi(x)|| \le \max\{||\phi(0)||,\ \frac{1}{1+\alpha\lambda}\sup_{-r\le x\le 0} p(x)||h(x)||\}.$$

From (9) and Fubini's Theorem we obtain

$$\int_{-\infty}^{-r} p(x)||\phi(x)||dx \le ||\phi(0)||\int_{-\infty}^{-r}\rho(x)e^{(\alpha+\gamma)x}dx$$

$$+\ \gamma\int_{-r}^{0}||h(\eta)||e^{\alpha\eta}\int_{-\infty}^{-r}\rho(x)e^{(\alpha+\gamma)(x-\eta)}dxd\eta$$

$$+\ \gamma\int_{-\infty}^{-r}||h(\eta)||e^{\alpha\eta}\int_{-\infty}^{\eta}\rho(x)e^{(\alpha+\gamma)(x-\eta)}dxd\eta.$$

Since ρ is nondecreasing we have

$$\int_{-\infty}^{-r} p(x)||\phi(x)||dx \le ||\phi(0)||\rho(-r)\frac{1}{\alpha+\gamma}e^{-(\alpha+\gamma)r}$$

$$+\ \gamma\sup_{-r\le x\le 0}||h(x)||\rho(-r)\int_{-r}^{0}e^{\alpha\eta}\int_{-\infty}^{-r}e^{(\alpha+\gamma)(x-\eta)}dxd\eta$$

$$+\ \gamma\int_{-\infty}^{-r}||h(\eta)||e^{\alpha\eta}\rho(\eta)\int_{-\infty}^{\eta}e^{(\alpha+\gamma)(x-\eta)}dxd\eta.$$

This estimate simplifies to

$$(11)\qquad \int_{-\infty}^{-r} p(x)||\phi(x)||dx \le \frac{\lambda p(-r)e^{-r/\lambda}}{1+\alpha\lambda}||\phi(0)||$$

$$+\ \frac{\lambda(1-e^{-r/\lambda})}{1+\alpha\lambda}\sup_{-r\le x\le 0} p(x)||h(x)|| + \frac{1}{1+\alpha\lambda}\int_{-\infty}^{-r} p(x)||h(x)||dx.$$

Estimates (10) and (11) show that $\phi \in X$ whenever $h \in X$. Straightforward calculations also show as expected that

$$\phi(x) = \phi(0) + \int_0^x \dot{\phi}(s)ds$$

where $\dot{\phi} = \frac{1}{\lambda}\phi - \frac{1}{\lambda}h \in X$. It remains to find $\phi(0)$ satisfying (8). This is equivalent to

$$\phi(0) = (I + \frac{\lambda}{1+\alpha\lambda}B)^{-1}(\frac{h(0)}{1+\alpha\lambda} + \frac{\lambda}{1+\alpha\lambda}F(\phi)).$$

In other words $\phi(0)$ must be a fixed point of the mapping $T\colon\ E \to E$

given by

$$T\theta = Q\left(\frac{h(0)}{1+\alpha\lambda} + \frac{\lambda}{1+\alpha\lambda}F\left(\theta e^{x/\lambda} + \frac{1}{\lambda}\int_x^0 \exp\left[\frac{1}{\lambda}(x-\eta)\right]h(\eta)d\eta\right)\right)$$

where $Q = (I + \frac{\lambda}{1+\alpha\lambda}B)^{-1}$ is a contraction on E since B is m-accretive. The mapping T satisfies

$$||T\theta_1 - T\theta_2|| \leq \frac{\lambda M}{1+\alpha\lambda}||(\theta_1 - \theta_2)e^{x/\lambda}||_X$$

since F is Lipschitz continuous. By calculating $||e^{x/\lambda}||_X$ one obtains

$$||T\theta_1 - T\theta_2|| \leq \frac{M}{\alpha+\gamma}||\theta_1 - \theta_2||(1 + \frac{p(-r)}{\alpha + \gamma}).$$

So T is a strict contraction on E if

(12) $$\alpha^2 \geq \alpha M + Mp(-r).$$

Therefore if (12) holds, T has a unique fixed point θ in E. Setting $\phi(0) = \theta$ one has $(I + \lambda A)\phi = h$ and $R(I + \lambda A) = X$ for every $\lambda > 0$. Choose h_1, $h_2 \in X$ and let $\phi_i = (I + \lambda A)^{-1}h_i$ for $i = 1, 2$. We wish to show that $||\overline{\phi}||_X \leq ||\overline{h}||_X$ where $\overline{\phi} = \phi_1 - \phi_2$ and $\overline{h} = h_1 - h_2$. Since

$$\phi_i(0) = Q\left(\frac{h_i(0)}{1+\alpha\lambda} + \frac{\lambda}{1+\alpha\lambda}F(\phi_i)\right), \quad i = 1,2,$$

(13) $$||\overline{\phi}(0)|| \leq \frac{C}{1+\alpha\lambda} + \frac{\lambda M}{1+\alpha\lambda}||\overline{\phi}||_X$$

where $C = \sup_{-r \leq x \leq 0} p(x)||\overline{h}(x)||$. It is easy to see that (10) and (11) also hold with ϕ replaced by $\overline{\phi}$ and h replaced by $\overline{h}$ so combining (10), (11), and (13) we obtain

$$||\overline{\phi}||_X \leq \left(\frac{C}{1+\alpha\lambda} + \frac{\lambda M}{1+\alpha\lambda}||\overline{\phi}||_X\right)\left(1 + \frac{\lambda p(-r)e^{-r/\lambda}}{1 + \alpha\lambda} + \frac{\lambda}{1+\alpha\lambda}(1 - e^{-r/\lambda})\right)$$
$$+ \frac{1}{1+\alpha\lambda}\int_{-\infty}^{-r} p(x)||\overline{h}(x)||dx.$$

This yields the estimate

$$||\bar{\phi}||_X(1 - \frac{\lambda M}{1+\alpha\lambda}(1 + \frac{\lambda}{1+\alpha\lambda})) \leq \frac{C}{1+\alpha\lambda}(1 + \frac{\lambda}{1+\alpha\lambda})$$

$$+ \frac{1}{1+\alpha\lambda}\int_{-\infty}^{-r} p(x)||\bar{h}(x)||dx.$$

The desired result will be true if

$$1 - \frac{\lambda M}{1+\alpha\lambda}(1 + \frac{\lambda}{1+\alpha\lambda}) \geq \frac{1}{1+\alpha\lambda}(1 + \frac{\lambda}{1+\alpha\lambda})$$

which holds if

(14) $$\alpha \geq M + 1 + \frac{M}{\alpha}.$$

In summary Lemma 7 is true when α is large enough to satisfy (12) and (14). These inequalities hold by hypothesis (5).

THE PROOF OF LEMMA 8. Choose $h \in X$ with $h(0) \in \overline{D(B)}$. We may assume without loss of generality that h is continuous with compact support on $(-\infty, 0]$. By Lemma 7 we may define

$$\phi_\lambda = (I + \lambda A)^{-1}h \quad \text{for } \lambda > 0,$$

so that $\phi_\lambda \in D(A)$ for every $\lambda > 0$. Recall that ϕ_λ satisfies (8). Let $\phi_\lambda(0) = \theta_\lambda$. Then by (8)

$$\theta_\lambda - \lambda F(\phi_\lambda) + \alpha\lambda\theta_\lambda + \lambda B\theta_\lambda = h(0)$$

or

$$\theta_\lambda = (I + \frac{\lambda}{1+\alpha\lambda}B)^{-1}(\frac{h(0)}{1+\alpha\lambda} + \frac{\lambda}{1+\alpha\lambda}F(\phi_\lambda)).$$

From this it is not difficult to prove, using the m-accretiveness of B, that $||\theta_\lambda - h(0)|| \to 0$ as $\lambda \to 0^+$. One can then show that

$$||\phi_\lambda - h||_X \to 0 \quad \text{as} \quad \lambda \to 0^+$$

which proves that $h \in \overline{D(A)}$. See the proof Lemma 2.2 of [2] for the details of a similar argument.

THE PROOF OF LEMMA 9. Suppose $u(t;\phi)$ is a strong solution of (1) with $\phi \in D(A)$. Then

$$u_t(x) = \begin{cases} \phi(0) + \int_0^{t+x} q(s)ds, & -t \leq x \leq 0, \\ \phi(t+x), & x \leq -t, \end{cases}$$

where $q(t) = F(u_t) - \alpha u(t) - Bu(t)$ is continuous on $[0,\infty)$. We wish to show that $w(t) = u_t$ is a strong solution of (CP) as defined in [7]. Since p is nondecreasing on $(-\infty,0]$, calculation yields the estimate

$$||u_t||_X \leq (1 + tp(-r)) \sup_{-r\leq x\leq t} ||u(x)|| + \int_{-\infty}^{-r} p(x)||\phi(x)||dx$$

so $u_t \in X$ for all $t \geq 0$. Choose $t \geq s \geq 0$, and let $h = t - s$, then similar computation shows that

$$||u_t - u_s||_X \leq (1 + tp(-r)) \sup_{-r\leq x\leq t} ||u(x) - u(x-h)||$$

$$+ \int_{-\infty}^{-r} p(x)||\phi(x) - \phi(x-h)||dx.$$

Therefore $w(t) = u_t \in C([0,\infty); X)$. We must prove that $u_t \in D(A)$ for $t \geq 0$. Since $\phi \in D(A)$ there exists $\tilde{\phi} \in X$ such that

$$\begin{cases} \phi(x) = \phi(0) + \int_0^x \tilde{\phi}(s)ds, & x \leq 0, \\ \tilde{\phi}(0) = F(\phi) - \alpha\phi(0) - B\phi(0) = q(0). \end{cases}$$

Hence

$$(15) \qquad u_t(x) = \begin{cases} \phi(0) + \int_0^{t+x} q(s)ds, & -t \leq x \leq 0, \\ \phi(0) + \int_0^{t+x} \tilde{\phi}(s)ds, & x \leq -t. \end{cases}$$

Define a function γ on $(-\infty,+\infty)$ by

$$\gamma(s) = \begin{cases} q(s), & s \geq 0, \\ \tilde{\phi}(s), & s < 0. \end{cases}$$

Since $q(0) = \tilde{\phi}(0)$, γ is continuous on $[-r,\infty)$. This is where the single-valuedness of B is used. One can show by arguments similar to those used earlier that $\gamma_t \in X$ for $t \geq 0$ and by (15)

$$(16) \qquad u_t(x) = u(t) + \int_t^{t+x} \gamma(s)ds = u_t(0) + \int_0^x \gamma_t(s)ds, \quad -\infty < x \leq 0.$$

Furthermore

$$(17) \qquad \gamma_t(0) = \gamma(t) = q(t) = F(u_t) - \alpha u_t(0) - Bu_t(0), \quad t \geq 0.$$

Therefore $u_t \in D(A)$ for $t \geq 0$. From (16) it follows that

(18) $$u_t - u_s = \int_s^t \gamma_\tau d\tau, \quad t, s \geq 0.$$

It is not difficult to show that $||\gamma_t||_X$ is bounded uniformly on bounded t intervals. Therefore $\gamma_t \in L^1_{loc}(0,\infty;X)$ and $w(t) = u_t \in W^{1,1}_{loc}(0,\infty;X)$ for $t \geq 0$.

It remains to show that $w(t) = u_t$ satisfies (CP) a.e. By (16) and (17) $Au_t = -\gamma_t$ for $t \geq 0$, and by (18)

$$\frac{dw}{dt} = \frac{du_t}{dt} = \gamma_t \quad \text{for a.e. } t \in (0,\infty).$$

Clearly $w(0) = u_0 = \phi$ and by the preceeding

$$\frac{dw}{dt} + Aw(t) = 0 \quad \text{for a.e. } t \in (0,\infty).$$

This completes the proof of Lemma 9.

REFERENCES

[1] V. Barbu and S. I. Grossman, Asymptotic behaviour of linear integrodifferential systems, Trans. Amer. Math. Soc. 173(1972), 277-288. MR 46 #7826.

[2] D. W. Brewer, A nonlinear semigroup for a functional differential equation, Trans. Amer. Math. Soc. 236(1978), 173-191.

[3] B. D. Coleman and V. J. Mizel, On the stability of solutions of functional differential equations, Arch. Rational Mech. Anal. 30(1968), 173-196. MR 37 #5499.

[4] M. G. Crandall, An introduction to evolution governed by accretive operators, Proc. Internat. Sympos. Dynamical Systems (Brown Univ., Providence, R. I., 1974).

[5] M. G. Crandall, Semigroups of nonlinear transformations in Banach spaces, Contributions to Nonlinear Functional Analysis (E. Zarantonello, editor), Academic Press, New York, 1971, 157-179.

[6] M. G. Crandall and T. M. Liggett, Generation of semigroups of nonlinear transformations on general Banach spaces, Amer. J. Math. 93(1971), 265-298. MR 44 #4563.

[7] J. R. Haddock, Asymptotic behaviour of solutions of nonlinear functional differential equations in Banach space, Trans. Amer. Math. Soc. 231(1977), 83-92.

[8] J. K. Hale, Functional differential equations with infinite delays, J. Math. Anal. Appl. 48(1974), 276-283. MR 51 #1067.

[9] C. C. Travis and G. F. Webb, Existence and stability for partial functional differential equations, Trans. Amer. Math. Soc. 200(1974), 395-418.

[10] G. F. Webb, Asymptotic stability for abstract nonlinear functional differential equations, Proc. Amer. Math. Soc. 54(1976), 225-230.

On Abstract Volterra Equations with Kernel of Positive Resolvent.

Ph.Clément
Technische Hogeschool Delft

1. Introduction.

Let X be a real Banach space with norm $||.||$. Let $A \subset X \times X$ be a m-accretive operator in X, [3], i.e. for every $\lambda > 0$, $J_\lambda := (I + \lambda A)^{-1}$ is a contraction everywhere defined on X. We consider the following Volterra equation of convolution type:

$$(1.1) \qquad u(t) + a \star Au(t) \ni f(t) \qquad t \geq 0$$

where a is a given real kernel and f is a given function with values in X. Since for every $\lambda > 0$, the Yosida approximation of A, $A_\lambda := \lambda^{-1}(I - J_\lambda)$ is Lipschitz continuous, the equation

$$(1.1)_\lambda \qquad u_\lambda(t) + a \star A_\lambda u_\lambda(t) = f(t) \qquad t \geq 0$$

possesses a unique solution $u_\lambda \in C([0,T];X)$ if $a \in L^1(0,T)$ and $f \in C([0,T];X)$. In [4], Crandall and Nohel have proved that if the assumption

$$(\text{H1}) \qquad \begin{array}{l} a \in W^{1,1}(0,T) \ , \ a(0) > 0, \ \dot{a} \in BV[0,T] \\ f \in W^{1,1}(0,T;X) \ , \ f(0) \in \overline{D(A)} \end{array}$$

is satisfied, then there exists $u \in C([0,T];X)$ such that $\lim_{\lambda \downarrow 0} u_\lambda = u$ in $C([0,T];X)$. u is called the <u>generalized solution</u> of (1.1). Note that if (H1) is satisfied, then there exist a unique $u_0 \in \overline{D(A)}$ and a unique $g \in L^1(0,T;X)$ such that

$$(1.2) \qquad f(t) = u_0 + a \star g(t) \qquad 0 \leq t \leq T.$$

Indeed $u_0 = f(0)$ and g is the unique solution of the equation

$$a(0)\ g(t) + \dot{a} \star g(t) = \dot{f}(t) \qquad 0 \leq t \leq T.$$

The proof of the existence of a generalized solution of (1.1) shows that (1.1) is closely related to the equation

$$(1.3) \qquad \begin{cases} \dot{u}(t) + A\ u(t) \ni g(t) & 0 < t \leq T \\ u(0) = u_0 \end{cases}$$

which is (1.1) with $a = 1$. It is known [3], that if u_1 and u_2 are the generalized solutions of (1.3) corresponding to the data $u_{0,1}$, $u_{0,2}$ and g_1, g_2, then the following estimate holds:

$$(1.4) \qquad ||u_1(t)-u_2(t)|| \le ||u_{0,1}-u_{0,2}|| + a*||g_1-g_2||(t)$$

on $[0,T]$, with $a = 1$. In this paper we consider a class of kernels satisfying (H1), containing the kernel $a = 1$, for which the estimate (1.4) still holds. Such class of kernels were introduced in [2, assumptions H4, H5]. Moreover we prove that if the kernel belongs to this class and is in $L^1(0,\infty)$, then the generalized solution of (1.1) converges strongly to a limit u_∞ provided that g itself converges to some limit g_∞. If $a \notin L^1(0,\infty)$, it is well known that u may not converge to a limit (take $X = \mathbb{R}^2$, $A = \begin{pmatrix} 0 & -1 \\ 1 & 0 \end{pmatrix}$, $a = 1$, $g = 0$, $u_0 \neq 0$).

In order to state our main assumption on the kernel a we need the following definitions. For $a \in L^1(0,T)$, let us denote by $r(a)$ the resolvent of a, i.e. the unique solution in $L^1(0,T)$ of the equation

$$(1.5) \qquad r(t) + a*r(t) = a(t) \qquad 0 \le t \le T$$

and by $s(a)$, the unique solution in $AC[0,T]$ of the equation

$$(1.6) \qquad s(t) + a*s(t) = 1 \qquad 0 \le t \le T.$$

Then our basic assumption on the kernel a is

(H2) For every $\lambda > 0$, $r(\lambda a) \ge 0$, a.e. on $[0,T]$ and $s(\lambda a) \ge 0$ on $[0,T]$.

It is known [6], [5], [2] that if a is positive, nonincreasing and $\log a$ is convex on $(0,T)$ then a satisfies (H2). Observe that (H2) implies $a \ge 0$.

2. Statement of results.

<u>Theorem 1</u>. <u>Let</u> a, f_1, f_2 <u>satisfy</u> (H1) <u>and</u> (H2) <u>on</u> $[0,T]$, <u>with</u> $f_i = u_{0,i}+a*g_i$, $i = 1, 2$. <u>Let</u> u_1, u_2 <u>be the corresponding generalized solutions of</u> (1.1) <u>on</u> $[0,T]$. <u>Then</u>

$$(2.1) \qquad ||u_1(t) - u_2(t)|| \le ||u_{0,1} - u_{0,2}|| + a*||g_1 - g_2||(t) \qquad 0 \le t \le T$$

<u>holds</u>.

Theorem 2. Let a, f satisfy (H1) and (H2) on [0,T] for every T > 0, with $f = u_0 + a*g$. If $a \in L^1(0,\infty)$, $\lim_{t\to\infty} g(t) = g_\infty$ exists and $g \in L^\infty(R^+,X)$, then

(2.2) $$||u(t) - u_\infty|| \le \int_t^\infty a(s)ds.\ (\int_0^\infty a(s)ds)^{-1}||u_0 - u_\infty|| + a*\|g-g_\infty\|(t)$$

for $t \ge 0$, where u is the generalized solution of (1.1) and $u_\infty = (I+\bar{a}A)^{-1}(u_0+\bar{a}g_\infty)$ with $\bar{a} = \int_0^\infty a(s)ds$.

For sake of completeness we state without proof a result of [2], which motivated the introduction of the assumption (H2).

Theorem 3 [2] let a, f satisfy (H1) and (H2) on [0,T] with $f = u_0 + a*g$. Let P be a closed convex cone in X, i.e. $P + P \subset P$, $\lambda P \subset P$ for $\lambda > 0$ and $P = \bar{P}$. If $J_\lambda(P) \subset P$ for $\lambda > 0$, $u_0 \in P$ and $g(t) \in P$ a.e. on [0,T], then u the generalized solution of (1.1) satisfies $u(t) \in P$, $t \in [0,T]$.

---

3. Proofs.

In proofs of theorems 1 and 2, we use the following

Lemma 4 let $a \in L^1(0,T)$, $a \ge 0$, u, $f \in C[0,T]$ be such that

(3.1) $$u(t) \le r*u(t) + f(t) \quad \text{holds on } [0,T]$$

where r is the resolvent of a. (see 1.5) Then

(3.2) $$u(t) \le f(t) + a*f(t) \quad t \in [0,T] \text{ holds.}$$

Remark. If $a(t) = e^t$, $r(t) = 1$ and $f(t) = K$, this is the simplest form of Gronwall Inequality. Lemma 4 is easily generalized to the nonconvolution case where $a(t,s)$, $r(t,s)$ satisfy $r(t,s) + \int_s^t a(t,u)r(u,s)du = a(t,s) \ge 0$. Then if $u, f \in C[0,T]$ satisfy $u(t) \le \int_0^t r(t,s)u(s)ds + f(t)$, $t \in [0,T]$, we have $u(t) \le f(t) + \int_0^t a(t,s)f(s)ds$, $t \in [0,T]$ (provided that the integrals make sense). In particular if $a(t,s) = g(s)\exp(\int_s^t g(u)du)$, $r(t,s) = g(s)$ with $g \in L^1(0,T)$, $g \ge 0$ and $f(t) = K$, we get Gronwall Inequality [1, page 31].

Proof of Lemma 4.

Define $g(t) = r*u(t) + f(t) - u(t)$ and $v(t) = f(t) + a*f(t)$. Then g, $v \in C[0,T]$, $g \ge 0$, $v(t) = r*v(t) + f(t)$, and $(v-u)(t) - r*(v-u)(t) = g(t)$. Hence $(v-u)(t) = g(t) + a*g(t) \ge 0$ and $u(t) \le v(t) = f(t) + a*f(t)$ on [0,T].

Proof of Theorem 1.

For $\lambda > 0$, let $u_{i,\lambda}$ be the solutions of $u + a*A_\lambda u = f_i$, $i = 1, 2$ on $[0,T]$. We have $u_{i,\lambda} + \lambda^{-1}a*u_{i,\lambda} = \lambda^{-1}a*J_\lambda u_{i,\lambda} + f_i$, hence $u_{i,\lambda} = r(\lambda^{-1}a)*J_\lambda u_{i,\lambda} + s(\lambda^{-1}a)u_{0,i} + \lambda r(\lambda^{-1}a)*g_i$, $i = 1, 2$, on $[0,T]$. Define $v_\lambda = ||u_{1,\lambda} - u_{2,\lambda}||$. Since J_λ is a contraction and since $r(\lambda^{-1}a)$, $s(\lambda^{-1}a)$ are nonnegative, we have $v_\lambda \leq r(\lambda^{-1}a)*v_\lambda + s(\lambda^{-1}a)||u_{0,1} - u_{0,2}|| + \lambda r(\lambda^{-1}a)*||g_1 - g_2||$. By using Lemma 4, we have $v_\lambda \leq (s(\lambda^{-1}a) + (\lambda^{-1}a)*s(\lambda^{-1}a))||u_{0,1} - u_{0,2}|| + \lambda(r(\lambda^{-1}a) + (\lambda^{-1}a)*r(\lambda^{-1}a))*||g_1 - g_2||$. By definition of r and s, we get $||u_{1,\lambda} - u_{2,\lambda}|| \leq ||u_{0,1} - u_{0,2}|| + a*||g_1 - g_2||$ on $[0,T]$. The conclusion of theorem 1 follows by letting λ go to 0.

Proof of Theorem 2.

For $\lambda > 0$, let u_λ satisfy $u_\lambda + a*A_\lambda u_\lambda = f$, for $t \geq 0$. Let $u_{\lambda\infty} = (I+\bar{a}A_\lambda)^{-1}(u_0+\bar{a}g_\infty)$ with $\bar{a} = \int_0^\infty a(s)ds$. Then $u_{\lambda\infty}$ satisfies $u_{\lambda\infty} + a*A_\lambda u_{\lambda\infty} = f + a*(g_\infty - g) - \xi w_\lambda$, where $\xi(t) = \int_t^\infty a(s)ds$ and $w_\lambda = A_\lambda u_{\lambda\infty} - g_\infty$. Define $v_\lambda = u_\lambda - u_{\lambda\infty}$. Then $v_\lambda = v_{\lambda,1} + v_{\lambda,2}$ where $v_{\lambda,1}$ satisfies $v_{\lambda,1} + \lambda^{-1}a*v_{\lambda,1} = \lambda^{-1}a*(J_\lambda u_\lambda - J_\lambda u_{\lambda\infty}) + a*(g - g_\infty)$ and $v_{\lambda,2}$ satisfies $v_{\lambda,2} + \lambda^{-1}a*v_{\lambda,2} = \xi w_\lambda$.
Note that $v_{\lambda,2} = G(\lambda)w_\lambda$ where $G(\lambda)$ satisfies $G(\lambda) + \lambda^{-1}a*G(\lambda) = \xi$. It follows that $G(\lambda)(t) = \lambda\int_t^\infty r(\lambda^{-1}a)(s)ds$ which is nonnegative. Consequently $||v_{\lambda,2}|| = G(\lambda)||w_\lambda||$ and $||v_{\lambda,2}|| + \lambda^{-1}a*||v_{\lambda,2}|| = \xi||w_\lambda||$.
By definition of r and $v_{\lambda,1}$, we have $v_{\lambda,1} = r(\lambda^{-1}a)*(J_\lambda u_\lambda - J_\lambda u_{\lambda\infty}) + \lambda r(\lambda^{-1}a)*(g-g_\infty)$ hence $||v_{\lambda,1}|| \leq r(\lambda^{-1}a)*||v_{\lambda,1}|| + r(\lambda^{-1}a)*||v_{\lambda,2}|| + \lambda r(\lambda^{-1}a)*||g - g_\infty||$. From Lemma 4, we get $||v_{\lambda,1}|| \leq (\lambda^{-1}a)*||v_{\lambda,2}|| + a*||g - g_\infty||$. Hence $||v_\lambda|| \leq ||v_{\lambda,2}|| + (\lambda^{-1}a)*||v_{\lambda,2}|| + a*||g - g_\infty|| = \xi||w_\lambda|| + a*||g - g_\infty||$.
It is standard that $\lim_{\lambda\downarrow 0}(I + \bar{a}A_\lambda)^{-1}(u_0 + \bar{a}g_\infty) = (I + \bar{a}A)^{-1}(u_0 + \bar{a}g_\infty)$. Thus $\lim_{\lambda\downarrow 0}||v_\lambda||(t) = ||u(t) - u_\infty||$, for every $t \geq 0$. Observe that $A_\lambda u_{\lambda\infty} = A_\lambda(I + \bar{a}A_\lambda)^{-1}(u_0 + \bar{a}g_\infty) = A_{\bar{a}+\lambda}(u_0 + \bar{a}g_\infty)$. Thus $\lim_{\lambda\downarrow 0} A_\lambda u_{\lambda\infty} = A_{\bar{a}}(u_0 + \bar{a}g_\infty)$ and $\lim_{\lambda\downarrow 0}||w_\lambda|| = ||A_{\bar{a}}(u_0 + \bar{a}g_\infty) - g_\infty||$.
But $\bar{a}||A_{\bar{a}}(u_0 + \bar{a}g_\infty) - g_\infty|| = ||(u_0 + \bar{a}g_\infty) - J_{\bar{a}}(u_0 + \bar{a}g_\infty) - \bar{a}g_\infty|| = ||u_0 - u_\infty||$, hence $||u(t) - u_\infty|| \leq \int_t^\infty a(s)ds.(\int_0^\infty a(s)ds)^{-1}\ ||u_0 - u_\infty|| + a*\|g-g_\infty\|(t)$, $t \geq 0$.

References.

[1] F. Brauer and J.A. Nohel: "Qualitative Theory of Ordinary Differential Equations", W.A. Benjamin, New York, 1969.

[2] Ph. Clément and J.A. Nohel, Abstract linear and nonlinear Volterra equations preserving positivity, SIAM J. Math. Anal. (to appear).

[3] M.G. Crandall, An introduction to evolution governed by accretive operators. Dynamical Systems, vol. 1, Academic Press, New York, (1976), pp. 131-165.

[4] M.G. Crandall and J.A. Nohel, An abstract functional differential equation and a related nonlinear Volterra equation, Israel J. Math., 1978.

[5] J.J. Levin, Resolvents and bounds for linear and nonlinear Volterra equations. Trans. Amer. Math. Soc. 228 (1977), 207-222.

[6] R.K. Miller, On Volterra integral equations with nonnegative integrable resolvents. J. Math. Anal. Appl., 22 (1968).

NONTRIVIAL PERIODIC SOLUTIONS OF SOME VOLTERRA INTEGRAL EQUATIONS

J. M. Cushing §
Department of Mathematics
University of Arizona
Tucson, Arizona 85721/USA

1. Introductory Remarks. My main purpose in this paper is to prove a bifurcation theorem for nontrivial periodic solutions of a general system of Volterra integral equations. The motivation for considering this problem can be found in models which arise in population dynamics, epidemiology and economics [1,3,6], an example of which appears in §5. The approach taken is that which is usually referred to as the method of Liapunov-Schmidt (or often called the method of alternative problems). This method, which is applicable in a very general setting, is outlined in §2 in a way suitable for the type of problems I have in mind. The fundamental ingredient for this approach in its application to many problems is a Fredholm alternative. A Fredholm alternative for systems of Volterra integral equations is proved in §3. The main bifurcation result (Theorem 4) appears in §4 and an application is given in §5 to a scalar model which has arisen in the mathematical theory of population growth and of epidemics.

2. Some General Remarks. Let X and Y be real normed linear spaces and suppose $L:X\to Y$ is a bounded linear operator with range $R(L)$ and nullspace $N(L)$. The following assumption will in force throughout.

H1: $R(L)$ and $N(L)$ are both closed and admit bounded projections.

Let $P:Y\to R(L)$ be a bounded projection as guaranteed by H1. It follows from H1 that $X = N(L)\oplus M$ where M is closed and that L has a bounded right inverse $A:R(L)\to M$, $LA = I$. Let $B(X,r)$ denote the open ball of radius $r>0$ centered at zero in a normed space X. We consider the operator equation

---

§ This work was performed while the author was an Alexander von Humboldt Foundation Fellow on sabbatical leave at the Lehrstuhl für Biomathematik der Universität Tübingen, Auf der Morgenstelle 28, D-7400 Tübingen, West Germany.

(1) $$Lx = T(x,\lambda) \quad \text{for} \quad (x,\lambda) \in B(X \times \Lambda, r)$$

where Λ is a real Banach space and $T:B(X \times \Lambda, r) \to Y$ is a continuous operator about which more is assumed below. It is easy to see that (x,λ) solves (1) if and only if

(2) $$z - APT(y+z,\lambda) = 0$$
(3) $$(I-P)T(y+z,\lambda) = 0$$

where $x = y+z$, $y \in N(L)$, $z \in M$.

The problem of interest here is the existence of nontrivial solutions near $x = 0$ (i.e. small amplitude solutions) under the assumption that $(x,\lambda) = (0,\lambda)$ is a solution for all small $\lambda \in \Lambda$ and thus it is natural to set $x = \varepsilon(y+z)$ for a small real parameter ε in (1) and hence in (2)-(3). Assume

<u>H2</u>: $$N(L) \neq \{0\} .$$

Let E^n denote real Euclidean n-space and $|\cdot|_X$ denote the norm in X. The following will be assumed about the operator T.

<u>H3</u>: $T(\varepsilon x,\lambda) = \varepsilon \overline{T}(x,\lambda,\varepsilon)$ for $0 \leq |\varepsilon| < \varepsilon_1 \leq +\infty$ and $x \in B(X,r)$ where $\overline{T}:B(X,r) \times B(\Lambda \times E^1, r) \to Y$ is $q \geq 1$ times continuously Fréchet differentiable in (x,λ,ε). Furthermore, for some $0 \neq y \in N(L)$, $|y|_X < r$, suppose $\overline{T}(y,0,0) = \overline{T}_x(y,0,0) = 0$.

Note that <u>H3</u> implies that $(x,\lambda) = (0,\lambda)$ solves (1) for all small ε. Letting $x = \varepsilon(y+z)$ in (2)-(3) one gets, after cancellation of an ε, the following equations

(4) $$z - AP\overline{T}(y+z,\lambda,\varepsilon) = 0$$
(5) $$(I-P)\overline{T}(y+z,\lambda,\varepsilon) = 0$$

which are to be solved for $(z,\lambda) \in M \times \Lambda$ as functions of ε (where y is fixed as in <u>H3</u>.) An existence theorem for (4)-(5) could now easily be stated by means of a straightforward application of the implicit function theorem, provided the Fréchet derivative of the left hand side of (5) with respect to λ at $(z,\lambda,\varepsilon) = (0,0,0)$ is a homeomorphism. This can be seen by applying the implicit function theorem to (4) to obtain a solution $z = z(\lambda,\varepsilon)$, $z(0,0) = 0$, which when substituted into (5) yields an equation solvable for $\lambda = \lambda(\varepsilon)$ again by the implicit function theorem. (Or alternatively one can simply apply the implicit function theorem to the operator defined by the left hand sides of (4) and (5).)

If $\dim \Lambda$ and $\operatorname{codim} R(L)$ are finite, then (5) becomes a real algebraic problem of a finite number of real equations for a finite number of real unknowns. Specifically, if

<u>H4</u>: $$m = \operatorname{codim} R(L) < +\infty$$

<u>H5</u>: $$\Lambda = E^m$$

then (5) reduces to m real equations in m real unknowns and states that the components $c = c(z,\lambda,\varepsilon) \in E^m$ of the left hand side of (5) (with respect to any finite set of elements which span a complement of $R(L)$) vanish. The requirement that the Fréchet derivative of the left hand side of (5) with respect to λ at $(z,\lambda,\varepsilon) = (0,0,0)$ be a homeomorphism then becomes the nondegeneracy condition that (note that $c_z(0,0,0) = 0$ by <u>H3</u>)

<u>H6</u>: $$D := \det c_\lambda(0,0,0) \neq 0 .$$

The real matrix $c_\lambda(0,0,0)$ is $m \times m$. Under these conditions one gets the following result from the implicit function theorem.

THEOREM 1. <u>Assume that all of the hypotheses H1 through H6 hold. For some</u> $\varepsilon_o > 0$, $\varepsilon_o \leq \varepsilon_1$, <u>there exist operators</u> $z: B(E^1,\varepsilon_0) \to M$, $\lambda: B(E^1,\varepsilon_0) \to E^m$ <u>which are</u> $q \geq 1$ <u>times continuously differentiable and are such that</u> $(x,\lambda) = (\varepsilon y + \varepsilon z(\varepsilon), \lambda(\varepsilon))$ <u>solves</u> (1) <u>for all</u> $|\varepsilon| \leq \varepsilon_0$ <u>where</u> $(z(0),\lambda(0)) = (0,0)$.

Note that the solutions in Theorem 1 are nontrivial since $x \neq 0$ for $\varepsilon \neq 0$ (since $y \neq 0$ in <u>H3</u>). Theorem 1 is a bifurcation result since the two solution branches $(0,\lambda)$ and $(x(\varepsilon),\lambda(\varepsilon))$ intersect at $(0,0)$. (Moreover, it can be shown that all solutions near $(0,0)$ have the form in Theorem 1 so that the implicit function theorem implies that these nontrivial solutions are unique.)

Before proceding to some applications to Volterra integral equations it may be worthwhile to discuss briefly the hypotheses <u>H1</u>-<u>H6</u>. First of all, <u>H1</u> and <u>H4</u> are motivated by (and are really abstract properties of) so-called Fredholm alternatives for the linear equation $Lx = f$, $f \in Y$. Thus, in applications, the establishment of a Fredholm alternative on suitable spaces would be the preliminary step in making use of Theorem 1. This, of course, is not necessarily easy, but has been done for many types of operators L on certain Banach spaces. For example, for spaces of periodic functions the Fredholm alternative is well known for ordinary differential systems and has been proved for more general Stieltjes-integrodifferential systems [2]. Examples for partial differential equations can be found in [5]. A Fredholm alternative for Volterra integral systems is proved here in §3.

Secondly, H2 is the familiar requirement in bifurcation theory that the linearization at the critical value $\lambda = 0$ be singular.

If Λ is viewed as a finite space of real parameters in equation (1) then the condition H5 requires that the problem have "enough" parameters, namely $m = \operatorname{codim} R(L)$. In many applications (for example, those concerning periodic solutions of autonomous systems) $m = 2$ in which case H5 requires that two parameters be available in the equation. Either parameters which explicitly appear can be used (in most applications there are usually more than enough parameters appearing explicitly) or parameters can be introduced by means of rescalings of independent variables (such as the unknown period in Hopf bifurcation, e.g. see [4,5] and §5 below).

Finally, H6 is a technical, but necessary assumption involving T and a knowledge of $R(L)$. It is necessary in the sense that it is well-known in bifurcation theory that bifurcation does not necessarily occur at the ciritcal value of the linearization and that some further nondegeneracy condition (such as H6) is required to insure that bifurcation take place. In applications, the description of $R(L)$ (and hence c in H6) usually involves an adjoint operator and its nullspace. Note that higher order terms in x which appear in T do not contribute to D.

Hypothesis H3 on T is rather routine to check in applications and in fact is usually verified by simple inspection. Besides a minimal amount of smoothness it requires that T be, roughly speaking, higher order in x and λ.

Once a Fredholm alternative has been established for the linear operator L the key hypotheses (the ones requiring the most analysis in applications) in Theorem 1 are H2 and H6.

3. Linear Volterra Operators. Assume that

A1: $\begin{cases} k(t) \text{ is a real } n\times n \text{ matrix valued function defined} \\ \text{and piecewise continuous on an interval } 0\le t\le a,\ 0<a<+\infty. \end{cases}$

Then the operator

$$(6)\qquad Ly := y(t) - \int_{t-a}^{t} k(t-s)y(s)\,ds \quad , \quad 0<a<+\infty$$

is a bounded linear operator from $X(p)$ into itself where $X(p)$ is the Banach space of real continuous, n-vector valued p-periodic functions under the supremum norm $|y|_p = \sup_{0\le t\le p}|y(t)|$. The integral appearing in L is continuous in t for any p-periodic function $y\in L^1[0,a]$.

Any function $y\in X(p)$ is square integrable on $0\le t\le p$ so that it is associated with a unique Fourier series $\sum_{-\infty}^{+\infty} a_j \exp(ijwt)$, $w = 2\pi/p$ where $a_{-j} = a_j'$, $j\ge 0$ ($'$ denotes complex conjugation) and $\sum_{-\infty}^{+\infty} |a_j|^2 < +\infty$. The Fourier coefficients of Ly are

$$(7)\qquad (I_n - k_j(p))a_j \quad \text{where} \quad k_j(p) := \int_0^a k(s)\exp(-ijws)\,ds \quad .$$

Here I_n is the $n\times n$ identity matrix. If $f\in R(L)$ then $Ly = f$ for some $y\in X(p)$ which implies

$$(A_j)\qquad\qquad (I_n - k_j(p))a_j = f_j \quad , \quad j\ge 0$$

(where the f_j are the Fourier coefficients of f) are solvable for a_j for all $j\ge 0$. The following lemma establishes the converse.

LEMMA 1. $f\in R(L)$ <u>if and only if</u> (A_j) <u>is solvable for</u> a_j <u>for all</u> $j\ge 0$, <u>in which case the</u> a_j <u>(with</u> $a_{-j} = a_j'$ <u>for</u> $j>0$) <u>are the Fourier coefficients of a</u> $y\in X(p)$ <u>such that</u> $Ly = f$ <u>for all</u> t .

<u>Proof</u>. It is only necessary to prove the converse. Let a_j solve (A_j) and define $a_{-j} = a_j'$, $j>0$. A (Stieltjes) integration of $k_j(p)$ by parts shows that

$$|k_j(p)| \le C/j \ , \ j > 0$$

for some constant $C > 0$ independent of j (but depending on p). Here $|k_j|$ means any matrix norm, say the largest absolute value of the entries in k_j. This means

(8) $$(I_n - k_j(p))^{-1} \text{ exists for } j \ge j_0 \ge 0$$

for some $j_0 \ge 0$ and is bounded uniformly in j so that

$$|a_j| \le \tilde{C}|f_j| \quad , \quad j \ge j_0 \ .$$

From this and the fact that $f \in X(p)$ implies the square summability of the f_j follows the square summability of the a_j. Thus these a_j determine a p-periodic function $y \in L^2[0,p]$ (by the Riesz-Fischer theorem). The Fourier coefficients of Ly are easily seen to be given by (7) and consequently, by the definition of a_j as a solution of (A_j), the functions Ly and f have identical Fourier coefficients and must be equal almost everywhere. However, the integral in L is continuous (as is f) so we conclude that y is an L^2 function which is equal almost everywhere to a continuous function. The function y can then be redefined on a set of measure zero so as to be continuous and hence lie in $X(p)$. Since this redefinition does not change the integral in L one obtains a function $y \in X(p)$ such that $Ly = f$ everywhere. §§

The operator

$$L_A y := y(t) - \int_t^{t+a} k^T(s-t)y(s)\, ds$$

will be called the <u>adjoint</u> of L. Lemma 1 applies with L replaced by L_A and (A_j) replaced by

(A_j^T) $$(I_n - k_j^T(p))a_j = f_j \quad , \quad j \ge 0$$

where k_j^T is the conjugate transpose of k_j. Thus

$$k_j^T(p) := \int_0^a k^T(s)\exp(ijws)\, ds \quad , \quad w = 2\pi/p \ .$$

Since the coefficient matrix in (A_j^T) is the conjugate transpose of that in (A_j) one obtains the following result from (8) and Lemma 1 with $f_j = 0$.

THEOREM 2. _The nullspaces of both_ L _and_ L_A _in_ $X(p)$ _are finite dimensional and_ $m = \dim N(L) = \dim N(L_A) < +\infty$.

The number m can be computed by finding the nullity ν_j of each singular matrix $I_n - k_j(p)$ (by (8) there are at most a finite set J of j's for which this matrix is singular). Each independent complex vector a_j in the nullspace of this matrix gives rise to two independent real solutions of $Ly = 0$ if $j > 0$ (namely the real and imaginary parts of $a_j\exp(ijwt)$) and one if $j = 0$ (namely a_0 itself). Thus, $m = \nu_0 + 2\sum_{0 \neq j \in J} \nu_j$.

Define

$$(x,y) := p^{-1} \int_0^p x(t)\cdot y(t)\, dt$$

for $x,y \in X(p)$. The next theorem describes the range of L and the complements of the range and nullspace of L.

THEOREM 3. $X(p) = N(L) \oplus M$ _and_ $X(p) = R(L) \oplus N(L_A)$ _where_

$$M = \{x \in X(p) : (x,y) = 0 \ \text{for all} \ y \in N(L)\}$$

$$R(L) = \{x \in X(p) : (x,y) = 0 \ \text{for all} \ y \in N(L_A)\}$$

are closed.

Proof. That M and $R(L)$ are closed is obvious, as is the decomposition $X(p) = N(L) \oplus M$. It remains only to establish that $R(L)$ is given as described in the theorem. By Lemma 1 and well-known algebraic facts, $f \in R(L)$ if and only if the Fourier coefficients f_j are all orthogonal to the nullspaces of $I_n - k_j^T(p)$ which is the same as to say if and only if f is orthogonal to solutions of $L_A y = 0$ (by Lemma 1 for L_A and by (A_j^T) with all $f_j = 0$). §§

It follows from Theorem 3 that $R(L)$ and $N(L)$ admit continuous projections. Thus, with regard to the hypotheses of the previous section §2, this theorem yields the following result.

COROLLARY 1. *The linear operator* $L:X(p)\rightarrow X(p)$ *defined by* (6) *satisfies* *H1* *and* *H4* *when* $k(t)$ *satisfies* *A1* .

4. *A Bifurcation Theorem for Volterra Integral Equations.* Consider the systems

$$x(t) - \int_{t-a}^{t} k(t-s)x(s)ds = T(x,\lambda) \quad , \lambda \in E^m \, , \; 0 < a < +\infty \tag{9}$$

$$y(t) - \int_{t-a}^{t} k(t-s)y(s)ds = 0 \tag{10}$$

where $k(t)$ satisfies *A1* and T satisfies *H3* with $X = Y = X(p)$ for some $p > 0$ and $\Lambda = E^m$. In order to apply Theorem 1 it follows from Corollary 1 that one needs only *H2*, *H5* and *H6*. Hypotheses *H2* and *H5* are equivalent to the following assumption.

A2: For some period $p > 0$ the linear homogeneous system (10) has exactly $m > 0$ independent, nontrivial p-periodic solutions $y_j \in X(p)$, $1 \leq j \leq m$.

By Theorem 2 the adjoint equation

$$y(t) - \int_{t}^{t+a} k^T(s-t)y(s)ds = 0$$

also has exactly m independent solutions $y_j^A \in X(p)$ which span $N(L_A)$. Assume without loss in generality that

$$(y_j^A, y_k^A) = \delta_{jk} \; .$$

(Such adjoint solutions can be found by choosing independent solution vectors of (A_j^T) with $f_j = 0$ to be orthonormal.) The components of the projection $I-P$ of any function $x \in X(p)$ onto $N(L_A)$ are then (x, y_j^A) so that c in *H6* is given by

$$c(z,\lambda,\varepsilon) = \mathrm{col}((\bar{T}(y+z,\lambda,\varepsilon),y_j^A)) .$$

Thus H6 becomes

$$\text{A3:} \begin{cases} D \neq 0 \quad \text{for some solution} \quad y \in X(p),\ y \neq 0,\ |y|_p < r \quad \text{of} \quad (10) \\ \text{where} \\ D := \det((\bar{T}_i(y,0,0),y_j^A)) ,\quad \bar{T}_i = (\partial/\partial\lambda_i)\bar{T} ,\quad \lambda = \mathrm{col}(\lambda_i) \\ \text{and} \quad 1 \le i,j \le m. \end{cases}$$

Theorem 1 of §1 now gives the following result.

THEOREM 4. Assume A1, A2 and A3 hold and that T satisfies H3 with $X = Y = X(p)$ and $\Lambda = E^m$. Then there is an $\varepsilon_0 > 0$ such that (9) has nontrivial p-periodic solutions of the form $x = \varepsilon(y+z)$, $\lambda = \lambda(\varepsilon)$ for $|\varepsilon| \le \varepsilon_0$ where $(z,y_j) = 0$, $1 \le j \le m$, $z(0) = 0$, $\lambda(0) = 0$ and $z(\varepsilon)$, $\lambda(\varepsilon)$ are $q \ge 1$ times Fréchet differentiable in ε.

The key hypotheses in any application of Theorem 4 (see §5) are A2 and A3. Note that in order for A2 to hold it is necessary (see Lemma 1 and (A_j)) that the "characteristic function" of (10)

$$\det(I_n - \int_0^a k(s)\exp(-zs)ds)$$

have purely imaginary roots. Also note that any higher order (i.e. $O(|x|_p^2)$) terms in T yield order ε terms in $\bar{T}$ and hence make no contribution to D in A3.

With the Fredhom alternative for Stieltjes-integrodifferential systems given in [2], a theorem similar in content to Theorem 4 can easily be stated for such systems by means of Theorem 1. This theorem would be a considerable generalization of that given in [2] (and would include as a special case the existence part of the well-known Hopf bifurcation theorem for ordinary differential systems).

Note that T need not be independent of t so that the period p can be prescribed by the equation. In autonomous problems p can be used as one of the parameters (as in Hopf type bifurcation) although the analysis is frequently greatly simplified if explicit parameters are used.

5. <u>An Application</u>. The scalar integral equation

$$(11)\qquad N(t) = r\int_{t-1}^{t-d} N(s)(1-N(s))ds \quad , \quad 0 < r \quad , \quad 0 \leq d < 1$$

has arisen in the mathematical theory of epidemics and population dynamics [1,3,6]. For $d = 0$ there are no nonconstant periodic solutions [1,6], but since d represents a delay due to a certain indubation period one might expect nonconstant periodic solutions for at least some $d > 0$. Numerical evidence of such periodic solutions was reported in [1]; we will confirm their existence here.

In order to apply Theorem 4 several changes of variable must be made. If

$$(12)\qquad r > 1/(1 - d)$$

then (11) has a positive equilibrium $e = 1 - r^{-1}(1-d)^{-1}$. Let $x = N - e$. First (11) will be studied with d held fixed and r treated as a bifurcation parameter. Since the null space of the linearization turns out to have dimension $m = 2$ ((11) is autonomous in the sense that time t translates of solutions are solutions) a second parameter is needed which will be taken (in the classical Hopf bifurcation manner) to be an unknown period p. After a change of independent variable from t to tp^{-1} is made, solutions are then sought in $X(1)$. If these changes of dependent and independent variables are made in (11) and if, for notational simplicity, a variable

$$\beta := \frac{2 - (1-d)r}{1-d}$$

is defined, then (11) reduces to

$$(13)\qquad x(t) = p\int_{t-1/p}^{t-d/p} (\beta x(s) - f(\beta)x^2(s))\, ds$$

$$f(\beta) := \frac{2 - (1-d)\beta}{1-d} \quad .$$

Bifurcation from the trivial solution $x = 0$ will occur at some critical values of β (i.e. r) and the period p. Let β_o and p_o

denote these (yet to be determined) critical values, set $\lambda_1 = p-p_0$, $\lambda_2 = \beta-\beta_0$ and rewrite (13) in the form of (9) with $n = 1$, $a = 1/p_0$ and

$$k(s) = \begin{cases} 0 , & 0 \leq s < d/p_0 \\ \beta_0 p_0 , & d/p_0 \leq s \leq 1/p_0 \end{cases}$$

$$T(x,\lambda) = (\lambda_1+p_0)\int_{t-1/(\lambda_1+p_0)}^{t-d/(\lambda_1+p_0)} ((\lambda_2+\beta_0)x(s) - f(\lambda_2+\beta_0)x^2(s))\, ds$$

$$- \int_{t-a}^{t} k(s)x(s)\, ds \quad .$$

This kernel $k(s)$ satisfies <u>A1</u>. It is also easy to see that this T satisfies <u>H3</u> with $X = Y = X(1)$, $\Lambda = E^2$ and $q = 1$. (The operator $\overline{T}$ is just T above with an ε placed in front of $x^2(s)$.) Thus, to apply Theorem 4 one need only justify <u>A2</u> and <u>A3</u> with $m = 2$.

<u>(1)</u> To find solutions $y \in X(1)$ of (10) one can turn to the scalar equations (A_j) with $f_j = 0$. There exist exactly two independent 1-periodic solutions of (10) if and only if $1-k_j(1) \neq 0$ for $j \neq 1$ and $1-k_1(1) = 0$. A straightforward calculation shows that $1-k_0(1) = 1-\beta_0(1-d)$ and for $j \geq 1$ that $1-k_j(1)$ equals

$$(1 - (\beta_0 p_0/\pi j)\cos\pi j(1+d)/p_0 \sin\pi j(1-d)/p_0)$$

$$+ i\ (\beta_0 p_0/\pi j)\sin\pi j(1+d)/p_0 \sin\pi j(1-d)/p_0 \ .$$

From this it is seen that $m = 2$ if and only if p_0 is chosen so that

$$\sin\pi(1+d)/p_0 = 0 \quad , \quad \sin\pi(1-d)/p_0 \neq 0$$

(14)

$$(-1)^{\alpha}\sin\pi j(1-d)/p_0 \neq j(-1)^{j\alpha}\sin\pi(1-d)/p_0 \quad \text{for all } j \geq 2,$$

β_0 is given by

$$\beta_0 = (\pi/p_0)(-1)^{\alpha}\csc\pi(1-d)/p_0 \quad , \quad \alpha = (1+d)/p_0$$

and $\beta_0 \neq 1/(1-d)$. Note that (14) implies that α is a positive integer. Moreover since it is required that (12) hold, β_o must satisfy $\beta_o < 1/(1-d)$. The problem then is: given d, choose p_o such that all of these conditions hold. Note that (14) fails to hold for any p_o if $d = 0$. This means no bifurcation occurs when no "incubation" or "delay" is present.

An in depth study of this problem for p_o will not be taken up here. One can however easily observe that (14) and all of the conditions on β_o hold if $p_o = 1+d$ and if d satisfies

(15) $$\sin\pi(1-d)/(1+d) > 1/2 \quad .$$

In this case

(16) $$\beta_o = -\pi(1+d)^{-1}\csc\pi(1-d)/(1+d) < 0 \quad .$$

Then $m = 2$ and $y_1(t) = \sin 2\pi t$, $y_2(t) = \cos 2\pi t$. Since (A_j) and (A_j^T) are scalar equations the adjoint solutions are the same, that is $y_j(t) = y_j^T(t)$.

Note that (15) holds if and only if $1/11 < d < 5/7$. Inequality (15) is sufficient, but not necessary for $m = 2$. I conjecture that $m = 2$ for suitably chosen p_o and β_o for in fact any d satisfying $0 < d < 1$. §§

<u>(2)</u> Equation (11) is autonomous in the sense that time translates of solutions are solutions and as a result one loses no generality in assuming that $y = \sin 2\pi t$ in <u>A3</u> (as opposed to any other nontrivial linear combination of y_1 and y_2). A straightforward, but rather lengthy calculation shows that D in <u>A3</u> is a nonzero constant multiple of $1 - \cos 2\pi(1-d)/(1+d)$ and consequently $D \neq 0$ for d satisfying (15). §§

In terms of the original variables Theorem 4 now yields the following result for equation (11).

THEOREM 5. *If $1/11 < d < 5/7$ then (11) has solutions of the following form for small $|\varepsilon|$:*

(17) $$N(t) = e + \varepsilon \sin 2\pi t/p + \varepsilon z(t/p,\varepsilon)$$

for $p = 1 + d + \lambda_1(\varepsilon)$, $r = 2(1-d)^{-1} - \beta_o - \lambda_2(\varepsilon)$ *and*

$e = 1-r^{-1}(1-d)^{-1} > 0$ *where* $\beta_o < 0$ *is given by* (16); *where* $z(\cdot,\varepsilon)$ *is 1-periodic, is Fréchet continuously differentiable in* ε *with* $z(t,0) \equiv 0$ *and is orthogonal to* $\sin 2\pi t$ *and* $\cos 2\pi t$; *and where each* $\lambda_j(\varepsilon)$ *is a continuously differentiable real valued function of* ε *satisfying* $\lambda_j(0) = 0$.

Note that the solution (17) of (11) is p-periodic in t and that (12) holds (i.e. $e > 0$) for small $|\varepsilon|$. If $d = 1/2$ Theorem 5 proves the existence of those nonconstant periodic solutions found numerically in [1] for p near 3/2 and r near $4 + 4\pi/3\sqrt{3}$.

In Theorem 5 the constant d is held fixed in (11) and the two parameters r and p are used in the application of Theorem 4. It is also possible to use the explicitly appearing parameters d and r as the bifurcation parameters in Theorem 4, in which case p is held fixed. If this is done the details are similar to (and in fact simpler than) those given above for the proof of Theorem 5 and are consequently omitted here. The following theorem results from this alternative approach.

THEOREM 6. *Given any* p *satisfying* $12/11 < p < 12/7$ *there exist p-period solutions of* (11) *for small* $|\varepsilon|$ *of the form*

$$N(t) = e + \varepsilon \sin 2\pi t/p + \varepsilon z(t,\varepsilon) \ , \ e = 1 - r^{-1}(1-d)^{-1}$$

for $d = p - 1 + \lambda_1(\varepsilon)$, $r = 2(2-p)^{-1} - \beta_o - \lambda_2(\varepsilon)$ *where*

$$\beta_o = -\pi p^{-1} \csc \pi(2-p)/p < 0 ;$$

where $z(\cdot,\varepsilon)$ *is p-periodic and orthogonal to both* $\sin 2\pi t/p$, $\cos 2\pi t/p$ *and is Fréchet continuously differentiable in* ε *with* $z(t,0) \equiv 0$; *and where each* $\lambda_j(\varepsilon)$ *is a continuously differentiable real valued function of* ε *satisfying* $\lambda_j(0) = 0$.

Theorem 5 and 6 are easily seen to yield the same bifrucation phenomenon and to be identical to lowest order in ε .

It is of interest to know the signs of λ_j as functions of ε in the above results, for they determine the "direction of bifurcation". I hope to deal with this problem as well as to apply Theorem 4 to other more general models (in particular, to systems of Volterra integral equations such as appear in [1,3]) in future work.

6. Some Further Applications to Scalar Equations. Consider the scalar (n = 1) equation

$$(18)\quad x(t) = \int_{t-a}^{t-b} (k(t-s,\beta)x(s) + g(t-s,\beta,x(s)))ds \,, \quad \beta \in E^1$$

$0 \le b < a < +\infty$ where g satisfies

A4: $g: [a,b] \times E^1 \times B(E^1,r) \to E^1$ is continuously differentiable and $|g(t,\beta,x)| = o(|x|)$ uniformly for $a \le t \le b$ and on compact β sets.

Suppose that the linearized equation

$$(19)\qquad y(t) = \int_{t-a}^{t-b} k(t-s,\beta_0)y(s)\, ds$$

has exactly two independent p_0-periodic solutions for some period $p_0 > 0$ and some $\beta_0 \in E^1$. If without loss in generality p_0 is the minimal period then these solutions are $\sin 2\pi t/p_0$, $\cos 2\pi t/p_0$, which because (19) is scalar are also solutions of the adjoint equation. If the changes of variables $\bar{t} = t/p$ and $\bar{x}(\bar{t}) = x(\bar{t}p)$ are made in (18), then (18) can be written in the form (9) with $n = 1$, with $k(t)$ replaced by $p_0 k(p_0 t,\beta_0)u(t;b/p_0)$ where $u(t;c)$ is the unit step function at c, with a replaced by a/p_0, with $\lambda_1 = \beta-\beta_0$ and $\lambda_2 = p-p_0$, and finally with

$$T(x,\lambda) := (\lambda_2+p_0)\int_{t-a/(\lambda_2+p_0)}^{t-b/(\lambda_2+p_0)} (k((\lambda_2+p_0)(t-s),\lambda_1+\beta_0)x(s) +$$

$$g((\lambda_2+p_0)(t-s),\lambda_1+\beta_0,x(s)))ds - p_0\int_{t-a/p_0}^{t-b/p_0} k(p_0(t-s),\beta_0)x(s)ds$$

(the bars on x and t having been dropped for convenience). By A4 this T satisfies H3 for $q = p = 1$ provided k is continuously differentiable in its arguments. Theorem 4 can now be applied with $p = q = 1$ and $m = 2$ provided $D \neq 0$ in A3 with $v(t) = \kappa_1 \sin 2\pi t + \kappa_2 \cos 2\pi t$, $\kappa_1^2 + \kappa_2^2 \neq 0$. In the usual Hopf-type bifurcation theorems this nondegeneracy condition is related to the transversal crossing of the imaginary axis by a conjugate pair of roots of the characteristic equation. This can also be done here as follows. In order that (19) have exactly $m = 2$ independent

p_0-periodic solutions it is sufficient (but not necessary) that the characteristic equation

$$(20) \qquad h(z,\beta) := 1 - \int_b^a k(s,\beta)\exp(-zs)ds = 0$$

have two and only two (conjugate) purely imaginary roots $z = \pm 2\pi i/p_0$ when $\beta = \beta_0$. If the root $z = 2\pi i/p_0$ is simple (i.e. $h_z(2\pi i/p_0,\beta_0) \neq 0$) then the implicit function theorem implies that (20) has a unique continuously differentiable branch of solutions $z = z(\beta)$, $z(\beta_0) = 2\pi i/p_0$ near $\beta = \beta_0$. By implicit differentiation one can compute $z'(\beta_0)$. A lengthy, but straightforward calculation shows that D, as calculated in A3, is a non-zero constant multiple of $\operatorname{Re} z'(\beta_0)$. The result is the following Hopf-type bifurcation theorem for the scalar equation (18).

THEOREM 7. Assume g satisfies A4 and that $k(t,\beta)$ is continuously differentiable in t and β on $0 \leq b \leq t \leq a < +\infty$ and for β near β_0. Assume that (20) has two and only two purely imaginary roots $\pm 2\pi i/p_0$, $p_0 > 0$ for $\beta = \beta_0$, that these roots are simple and that $\operatorname{Re} z'(\beta_0) \neq 0$ where $z(\beta)$ is the unique branch of roots of (20) satisfying $z(\beta_0) = 2\pi i/p_0$. Then for small real ε equation (18) has p-periodic solutions of the form

$$x(t) = \varepsilon y(t/p) + \varepsilon z(t/p,\varepsilon) \quad , \quad \beta = \beta_0 + \lambda_1(\varepsilon) \ , \quad p = p_0 + \lambda_2(\varepsilon)$$

where $y(t) = \kappa_1 \sin 2\pi t + \kappa_2 \cos 2\pi t$, $\kappa_1^2 + \kappa_2^2 \neq 0$, where $z(t,\varepsilon)$ is a 1-periodic function of t and where z and λ_i have the properties in Theorem 4 with $q = 1$.

As a final application consider the scalar equation with two parameters given by

$$(21) \qquad x(t) = \int_{t-a}^{t} (\beta_1 k_1(t-s) + \beta_2 k_2(t-s))x(s)\, ds \ + \ R(x,\beta)$$

$$\beta = \operatorname{col}(\beta_1,\beta_2) \in E^2 \quad , \quad \int_0^a k_i(s)ds = 1 \ \text{ for } \ i = 1,2 \ .$$

Suppose that the linear equation

$$(22) \qquad y(t) = \int_{t-a}^{t} (\beta_1 k_1(t-s) + \beta_2 k_2(t-s))y(s)\, ds$$

has, for an isolated pair $\beta_0 = \operatorname{col}(\beta_1^0,\beta_2^0) \in E^2$, exactly two independent p-periodic solutions (namely $\sin 2\pi t/p$ and $\cos 2\pi t/p$ which because (22) is scalar are also the adjoint solutions) for some period $p>0$. A simple Fourier analysis shows that this is true if and only if

$$W := C_1(1)S_2(1)-S_1(1)C_2(1) \neq 0 \ , \ (C_1(1)-1)S_2(1) \neq (C_2(1)-1)S_1(1)$$

$$(23) \qquad \beta_1^0 = S_2(1)/W \quad \text{and} \quad \beta_2^0 = -S_1(1)/W$$

and for every integer $n \geq 2$ either $S_2(1)S_1(n)-S_1(1)S_2(n) \neq 0$ or $S_2(1)C_1(n)-S_1(1)C_2(n) \neq W$. Here

$$S_i(n) := \int_0^a k_i(s)\sin 2\pi ns/p \, ds \ , \quad C_i(n) := \int_0^a k_i(s)\cos 2\pi ns/p \, ds.$$

The primary reason for mentioning these details here is to show that the assumption that (22) has, for an isolated $\beta_0 \in E^2$, exactly two independent p-periodic solutions implies $W \neq 0$. As will be seen below this in turn will imply the nondegeneracy condition $D \neq 0$.

Equation (21) has the form (9) with $k(t) = \beta_1^0 k_1(s)+\beta_2^0 k_2(s)$ and

$$T(x,\lambda) := \int_{t-a}^{t} (\lambda_1 k_1(t-s) + \lambda_2 k_2(t-s))x(s) \, ds \quad + \quad S(x,\lambda)$$

where $\lambda = \beta-\beta_0 \in E^2$ and $S(x,\lambda) := R(x, \lambda+\beta_0)$. Assume

<u>A5</u>: $R: B(X(p),r) \times E^2 \to X(p)$ is $q \geq 1$ times continuously Fréchet differentiable and $|R(x,\beta)|_p = o(|x|_p)$ near $x = 0$ uniformly on compact β sets.

Then T satisfies <u>H3</u>. Inasmuch as an easy calculation shows that D given in <u>A3</u> is equal to $D = -(\kappa_1^2 + \kappa_2^2)W/4 \neq 0$ for $y(t) = \kappa_1 \sin 2\pi t/p + \kappa_2 \cos 2\pi t/p$, $\kappa_1^2 + \kappa_2^2 \neq 0$, Theorem 4 now yields the following result for the scalar equation (21).

THEOREM 8. <u>Assume that the kernels</u> k_i <u>satisfy</u> <u>A1</u> <u>and that</u> R <u>satisfies</u> <u>A5</u>. <u>If the linear equation</u> (22) <u>has, for an isolated</u> $\beta = \mathrm{col}(\beta_1^0,\beta_2^0) \in E^2$, <u>exactly two independent p-periodic solutions for some period</u> $p>0$, <u>then</u> (21) <u>has a branch of nontrivial p-periodic</u>

solutions for $\beta_i = \beta_i^0 + \lambda_i(\varepsilon)$, β_i^0 *given by* (23), *as described in Theorem 4 with* $m = 2$.

All of the bifurcation theorems in this paper are purely existence results. It would also, of course, be of interest to study the stability of the nontrivial periodic solutions found in the theorems above, a problem not addressed here.

REFERENCES

1 K. L. Cooke and J. A. Yorke, Some equations modelling growth processes and gonorrhea epidemiology, Math. Biosci. 16(1973), 75-101.

2 Frank Hoppensteadt, Mathematical Theories of Populations: Demographics, Genetics and Epidemics, SIAM Regional Conference Series in Applied Mathematics (1975), SIAM, Philadelphia, Pa.

3 J. M. Cushing, Bifurcation of periodic solutions of integro-differential systems with applications to time delay models in population dynamics, SIAM J. Appl. Math. 33, no. 4(1977), 640-654.

4 David H. Sattinger, Topics in Stability and Bifurcation Theory, Lecture Notes in Mathematics 309(1973), Springer-Verlag, New York.

5 A. B. Poore, On the theory and application of the Hopf-Friedrichs bifurcation theory, Arch. Rat. Mech. Anal. 60 (1976), 371-393.

6 Paul Waltman, Deterministic Threshold Models in the Theory of Epidemics, Lecture Notes in Biomathematics 1(1974), Springer-Verlag, New York.

Fixed points of condensing maps

Klaus Deimling

Let X be a Banach space, $D \subset X$ closed bounded convex, $T: D \to X$ continuous. We prove that T has a fixed point if T is α-condensing and weakly inward. Then we consider perturbations of condensing maps by certain dissipative maps and we find fixed points of such maps under further conditions on D and/or X . This is followed by special results for the case $D = \{x \in K : |x| \leq r\}$, where K is a cone in X . Finally, we indicate possible applications in the theory of integral equations.

1. Notations and definitions

1.1 X will always be a real Banach space. X^* denotes the normed dual of X . We let $B_r(x_o) = \{x \in X : |x-x_o| < r\}$. For $D \subset X$, we denote by $\mathring{D}$, ∂D , $\bar{D}$ and $\overline{\text{conv}}\, D$ the interior, boundary, closure and convex hull of D , respectively. $\rho(x,D)$ denotes the distance from x to D .
If, for each $x \in X$, there is a unique $Px \in D$ such that $|x-Px| = \rho(x,D)$, then $P: X \to D$ will be called the metric projection of X onto D .
For bounded subsets $B \subset X$, the measures of noncompactness $\alpha(B)$ and $\beta(B)$ are defined by

$\alpha(B)=\inf\{d>0: B$ can be covered by finitely many sets of diameter $\leq d\}$.
$\beta(B)=\inf\{r>0: B$ can be covered by finitely many $B_r(x_i)\}$.

Properties of α and β may be found e.g. in § 2 of [6] , § 4.3 of Fenske [12] or Sadovskii [24] .

1.2 In connection with these measures of noncompactness one has defined certain classes of maps. $T: D \to X$ will be said to be α-Lipschitz with constant k if $\alpha(TB) \leq k\alpha(B)$ for all bounded $B \subset D$. T is called α-condensing if $\alpha(TB) < \alpha(B)$ whenever $\alpha(B) > 0$.
1.3 As a generalization of an inner product, one has defined semi-inner products $(\cdot,\cdot)_\pm: X \times X \to \mathbb{R}$, for an arbitrary real B-space X , by means

of the formulas

$$(x,y)_{\pm} = \max_{\min} \{x^*(x) : x^* \in X^* , x^*(y) = |y|^2 = |x^*|^2\} .$$

For example, if $X = C(J)$, the space of all continuous functions $J = [0,a] \to \mathbb{R}$ with the max-norm, then

$$(x,y)_{\pm} = \max_{\min} \{x(t)\operatorname{sgn} y(t) : t \text{ such that } |y(t)| = |y|\}|y| .$$

Properties and further examples may be found e.g. in §3 of [6] and in [9].

1.4 In connection with fixed point theorems for $T: D \to X$ we need certain conditions on T at the boundary ∂D . In case $\mathring{D} \neq \emptyset$, the weakest one is the Leray-Schauder-condition

(1) There exists an $x_0 \in \mathring{D}$ such that $Tx-x_0 = \lambda(x-x_0)$ and $x \in \partial D$ imply $\lambda \leq 1$.

This condition is the right one if topological degree theory can be used. Another condition which had its origin in the study of flow-invariant sets for differential equations is

(2) $\liminf_{\lambda \to 0+} \lambda^{-1} \rho(x+\lambda(Tx-x),D) = 0$ for $x \in \partial D$.

In case D is convex this is equivalent to

(3) $x \in \partial D$, $x^* \in X^*$ and $x^*(x) = \sup_D x^*(y) \Rightarrow x^*(Tx-x) \leq 0$;

see e.g. § 4 of [6] . For an bounded open convex set D condition (3) is stronger than (1) . In fact, suppose $Tx = x_0 + \lambda(x-x_0)$ for some $x \in \partial D$; then there exists $x^* \in X^*$ such that $x^*(x) = \sup_D x^*(y) > x^*(x_0)$ and therefore

$$(\lambda-1)x^*(x-x_0) = x^*(Tx-x) \leq 0 , \quad \text{i.e. } \lambda \leq 1 .$$

Therefore, (3) is most interesting for convex sets with $\mathring{D} = \emptyset$. People with more geometric feeling have defined another condition. For $x \in D$ they call $\{(1-\lambda)x+\lambda y : \lambda \geq 0 , y \in D\}$ the inward set of x w.r. to D . Then a map $T: D \to X$ is called inward if for all $x \in D$, Tx belongs to the inward set of x , and T is said to be weakly inward if Tx belongs to the closure of the inward set of x ; see e.g. Halpern/Bergman [13] . It is easy to see, c.p. Caristi [3] , that "weakly inward" is equivalent to (3) .

2. Fixed points of condensing maps

2.1 Let us start with a well known result which is an easy consequence of degree theory for condensing maps ; see e.g. Fenske [12] , Nußbaum

[18] , Sadovskii [24]

Proposition 1. Let $D \subset X$ open bounded, $T: \bar{D} \to X$ continuous, α-condensing and such that the LS-condition (1) is satisfied. Then T has a fixed point.

A simple consequence is

Proposition 2. Let $D \subset X$ compact convex, $T: D \to X$ continuous and such that condition (2) is satisfied. Then T has a fixed point.

Proof. Since the essential sets D and $T(D)$ are compact, we may assume that X is separable. Since (2) does not change if we replace the norm by an equivalent one, we may then assume that $|\cdot|$ is strictly convex, by Clarkson's result [4] . Therefore, the metric projection $P: X \to D$ is continuous. Hence $TP: X \to X$ is compact. Now it is easy to see that TP satisfies (2) with D replaced by $D_\delta = \{x \in X : \rho(x,D) \leq \delta\}$. Since $\check{D}_\delta \neq \emptyset$, Proposition 1 gives us an $x_\delta = TPx_\delta \in D_\delta$. Now, we may let $\delta \to 0$ to obtain a fixed point of T .

q.e.d.

Remark. Since "weakly inward" is the same as condition (2) , Proposition 2 is a special case of Theorem 4.1 in Halpern/Bergman [13] .

2.2 Now, we can prove the main result of this paragraph.

Theorem 1. Let $D \subset X$ be closed bounded convex, $T: D \to X$ continuous, α-condensing and such that (2) holds. Then T has a fixed point.

Proof. 1. Since all conditions are invariant under simultaneous translation, we may assume $0 \in D$. We may also assume that T is α-Lipschitz with constant $k < 1$, since kT satisfies (3) too and $k_n T x_n = x_n \in D$ for $k_n < 1$ and $k_n \to 1$ implies $\alpha(\{x_n : n \geq 1\}) = 0$ and therefore $x_{n_i} \to x_0 = Tx_0$ for some subsequence (x_{n_i}) and an $x_0 \in D$.

2. Consider the sequence (D_n) defined by

$$D_0 = D \ , \ D_n = \overline{\text{conv}}\,[(T(D_{n-1}) + B_{\delta_n}(0)) \cup \{0\}] \cap D_{n-1} \ \text{ for } n \geq 1 \ ,$$

where the $\delta_n > 0$ are such that $\sum_{n \geq 1} \delta_n < \infty$.

Evidently, D_n is closed bounded convex, and $D_n \neq \emptyset$ since $0 \in D_n$ for every $n \geq 0$. Furthermore, $D_n \subset D_{n-1}$ and

$$\begin{aligned} \alpha(D_n) &\leq \alpha(T(D_{n-1})) + 2\delta_n \leq \dots \\ &\leq k^n \alpha(D) + 2(k^{n-1}\delta_1 + \dots + k\delta_{n-1} + \delta_n) \to 0 \text{ as } n \to \infty \ . \end{aligned}$$

Therefore, $D_* = \bigcap_{n \geq 0} D_n$ is nonempty compact and convex. We know that T satisfies (2) for D_o. Let us show that T satisfies (2) for D_n if it satisfies (2) for D_{n-1}. We have for fixed $x \in D_n \subset D_{n-1}$

$$\lambda_p^{-1}|x+\lambda_p(Tx-x)-z_p| \leq \delta < \delta_n \quad \text{for some } 1 \geq \lambda_p \to 0+,\ z_p \in D_{n-1}.$$

Let $u_p = x+\lambda_p^{-1}(z_p-x)$. Then $|u_p-Tx| \leq \delta < \delta_n$, in other words $u_p \in T(D_{n-1})+B_{\delta_n}(0)$, and $(1-\lambda_p)x+\lambda_p u_p = z_p \in D_{n-1}$. Therefore, $z_p \in D_n$, and this implies $\lambda_p^{-1}\text{dist}(x+\lambda_p(Tx-x),D_n) \leq \delta$ for $p \geq p_o(\delta)$, i.e. T satisfies (2) for D_n.

3. Now, let us prove that T satisfies (2) for D_*. Let $x \in D_*$. Then $x \in D_n$ and since T satisfies (2) for D_n we know that the IVP $u' = Tu-u$, $u(0) = x$ has a solution u_n on $[0,\infty)$ with range in D_n; see Theorem 4.1 in [6]. The sequence (u_n) has a subsequence that converges uniformly on compact intervals to a solution u of the same IVP. Since $\alpha(D_n) \to 0$ we have $\sup\{\rho(x,D_*) : x \in D_n\} \to 0$ as $n \to \infty$, and therefore the range of u is in D_*. Thus

$$0 = \rho(u(\lambda),D_*) = \rho(x+\lambda(Tx-x)+o(\lambda),D_*) \text{ as } \lambda \to 0+$$

and this clearly implies (2) for D_*. Since D_* is compact convex, we can now apply Proposition 2.

q.e.d.

Remarks. The sequence D_n and the fact that T satisfies the boundary condition for D_n is taken from the proof to Proposition 5.8 in Caristi's thesis [2]. However, he was not able to show that T satisfies (2) for D_* and we also can not do this without differential equations. With Theorem 4.8 in [6] we have proved Theorem 1 under the strong additional hypothesis that the IVPs $u' = Tu-u$, $u(0) = x \in D$ have at most one solution. Reich's Theorem C in [22] is Theorem 1 under the stronger condition that T be inward.

3. Fixed points of perturbations of condensing maps

3.1 Let us start with a conjecture which we can not prove.

Conjecture. Let $D \subset X$ be closed bounded convex; $T_1,T_2: D \to X$ continuous and bounded; $\alpha(T_1B) \leq k_1\alpha(B)$ for $B \subset D$ and $(T_2x-T_2y,x-y)_- \leq k_2|x-y|^2$ for $x,y \in D$. Let $T = T_1+T_2$ satisfy the boundary condition (2). Then T has a fixed point provided that $k_1+k_2 < 1$.

Remark. If the conjecture is true then it is also true that T has a fixed point if T_1 is only condensing and $k_2 \leq 0$. In fact, we may assume $0 \in D$ and then we see that (2) holds for $k_n T$ instead of T, where we choose $k_n < 1$ such that $k_n \to 1$. Therefore there exists $x_n = Tx_n + y_n$ with $y_n = (k_n - 1)Tx_n \to 0$ as $n \to \infty$. This implies

$$|x_n - x_m| \leq |T_1 x_n - T_1 x_m| + |y_n - y_m|$$

and therefore $\alpha(\{x_n : n \geq 1\}) = 0$. Thus T has a fixed point.

We have seen that the Conjecture is true if $T_2 = 0$. It is also true in case $T_1 = 0$ since then the Poincaré-operator U_ω: $D \to D$ for $u' = T_2 u - u$ satisfies

$$|U_\omega x - U_\omega y| \leq e^{-(1-k_2)\omega} |x-y| \quad \text{for} \quad x, y \in D$$

and therefore $u' = T_2 u - u$ has a constant solution, i.e. T_2 has a fixed point.

3.2 Let us now indicate some special situations where the conjecture can be proved.

Proposition 3. The Conjecture is true if in addition one of the following hypotheses is satisfied.

(i) T uniformly continuous ; $\mathring{D} \neq \emptyset$.

(ii) T uniformly continuous ; for every $x \in D$, the IVP $u' = Tu - u$, $u(0) = x$ has at most one solution.

Proof. If (i) or (ii) holds, we can apply Theorem 2 and Theorem 1 of [9] , respectively, to obtain an ω-periodic solution u_ω of $u' = Tu - u$, for every $\omega > 0$. If $U(t)$: $D \to 2^D$ denotes the evolution operator which maps x onto the values at t of all solutions through x then $\alpha(U(t)B) \leq e^{-Lt}\alpha(B)$ for $B \subset D$ and $L = 1 - k_1 - k_2 > 0$. Therefore we obtain a constant solution of $u' = Tu - u$ like in the proof to Theorem 4.8 in [6] .

q.e.d.

Proposition 4. The Conjecture is true if T is uniformly continuous, X^* is uniformly convex and the metric projection P: $X \to D$ exists and is continuous.

Proof. In Theorem 3 of [9] we claimed that $u' = Tu - u$ has an ω-periodic solution, without the assumption that X^* be uniformly convex. The theorem is perhaps true, but the proof is only correct if X^* is also uniformly convex. The essential difficulty has been to show that Sx , the set of all solutions on $[0,\omega]$ of $u' = Tu-(1+\varepsilon)u$ with $u(0) = x$, is the limit w.r. to the Hausdorff-metric of sets $A_n \supset Sx$ which are homeomorphic to compact convex sets. The proof of $A_n \to Sx$ amounts to prove the following (see [9] for details): If (v_n) is such that

$$v_n' = TPv_n-(1+\varepsilon)v_n+z_n(t) \ , \ v_n(0)=x \in D \ , \ z_n(t) \to 0 \text{ uniformly}$$

and $\rho(v_n(t),D) \to 0$ uniformly on $[0,\omega]$ then (v_n) has a uniformly convergent subsequence the limit of which is in Sx . Now, the hypothesis that X^* be uniformly convex implies that $(\cdot,\cdot)_\pm$ is uniformly continuous on bounded subsets of $X \times X$, see e.g. Lemma 3.2 (v) in [6] ; thus

$$(TPv_n-TPv_m,v_n-v_m)_- \leq (TPv_n-TPv_m,Pv_n-Pv_m)_- + \varepsilon_{nm}$$

with $\varepsilon_{nm} \to 0$ as $n,m \to \infty$. Now, we can proceed as in the proof to Theorem 1 in [10] to get what we want. Since we then have ω-periodic solutions for every $\omega > 0$, we obtain a constant solution as before.

q.e.d.

Remarks. Condition (ii) of Proposition 3 is satisfied if T is uniformly continuous and e.g. $(T_1x-T_1y,x-y)_+ \leq k|x-y|^2$ for some $k \in \mathbb{R}$. The unnatural assumption "T uniformly continuous" is always due to the fact that we cannot prove existence for the IVP without this condition. Clearly the condition concerning the metric projection in Proposition 4 is always satisfied if X is uniformly convex.

Recently, Schöneberg [26] has developped a degree theory for so called semicondensing vector fields, a particular case of which is a vector field $T = T_1+T_2$ with T_1 and T_2 as in the Conjecture. By means of this degree it is possible to improve Proposition 3 (i) to the following special case of Corollary 1.3 in [26] .

Proposition 5. The Conjecture is true if $\check{D} \neq \emptyset$ and T satisfies the weaker LS-condition (1) .

Proof. Without loss of generality we have $0 \in \check{D}$ and $tTx \neq x$ for $x \in \partial D$ and $0 \leq t \leq 1$. Now, the set

$$M = \{(t,y) \in [0,1]\times\check{D} : \text{there exists } x \in \check{D} \text{ such that } x-tT_2x = tT_1y\}$$

is not empty since $\{(0,y) : y \in \check{D}\} \subset M$, and M is open in $[0,1]\times X$ since $I-tT_2$ is a strongly accretive map and therefore open ; see Theorem 3 in [8] . For the same reason, the solution map $H: M \to \mathring{D}$, defined by $(I-tT_2)H(t,y) = tT_1y$ is defined and continuous. Furthermore, one has

(4) $$|H(t,x)-H(s,y)| \leq \frac{1}{1-k_2}|T_1x-T_1y| + \frac{|t-s|}{1-k_2}|T_2H(s,y)+T_1y| ,$$

and this estimate clearly implies $\alpha(H(M \cap ([0,1]\times B))) \leq k\alpha(B)$ for $B \subset \mathring{D}$, where $k = k_1(1-k_2)^{-1} < 1$. Therefore, we have

$$\mathrm{Deg}(I-H(t,\cdot),\mathring{D},0) = \mathrm{Deg}(I,\mathring{D},0) = 1 \quad \text{for small } t > 0 ,$$

since $H(0,\cdot) = 0$ and H defines an admissible homotopy for strict α-contracting perturbations of the identity ; see e.g. Fenske [12] . Since $\mathrm{Deg}(I-H(t,\cdot),\mathring{D},0) = 1$, we find $x \in \mathring{D}$ such that $H(t,x) = x$, i.e. $x = tTx$, for t small. Now, (4) with $t = s$ implies that $N(t) = \{x \in \mathring{D} : H(t,x) = x\}$ is compact. Therefore $N(\tau) \neq \emptyset$ for $\tau = \sup\{t \in [0,1] : N(t) \neq \emptyset\}$. This implies $\tau = 1$ since $\tau < 1$ and "M open" would imply $\tau_1 = \sup\{t : (t,y) \in M \text{ for some } y \in \mathring{D}\} > \tau$ and therefore $\mathrm{Deg}(I-H(t,\cdot),\mathring{D},0) = 1$ for $t \in [0,\tau_1)$, a contradiction to the definition of τ .

q.e.d.

Since we did not use the convexity of D , $\mathring{D}$ may be an arbitrary open bounded set.

4. Fixed points in cones

4.1 Let $K \subset X$ be a cone, i.e. a closed convex set such that $\lambda K \subset K$ for $\lambda \geq 0$ and $K \cap (-K) = \{0\}$. Let $K^* = \{x^* \in X^* : x^*(x) \geq 0 \text{ for all } x \in K\}$. For a cone K the boundary condition

$$\lim_{\lambda \to 0+} \lambda^{-1}\mathrm{dist}(x+\lambda Fx,K) = 0 \quad \text{for } x \in \partial K$$

is equivalent to

(5) $$x \in \partial K , \; x^* \in K^* \text{ and } x^*(x) = 0 \;\Rightarrow\; x^*(Fx) \geq 0 .$$

See Example 4.1 in [6] .

Proposition 6. Let $K \subset X$ be a cone, $r > 0$ and $D = \{x \in K : |x| \leq r\}$. Let $T: D \to X$ be either α-condensing or $T = T_1+T_2$ uniformly continuous with $\alpha(T_1B) \leq k_1\alpha(B)$ for $B \subset D$ and $(T_2x-T_2y,x-y)_- \leq k_2|x-y|^2$ for $x,y \in D$. Suppose that

(i) $(Tx,x)_- \leq |x|^2$ for $|x| = r$

(ii) $x \in \partial D$, $|x| < r$, $x^* \in K^*$ and $x^*(x) = 0 \Rightarrow x^*(Tx) \geq 0$

holds. Then T satisfies (2) .

Proof. For $|x_o| < r$ the IVP $u' = Tu-u$, $u(0) = x_o$ has a local solution in K , by Theorem 4.1 in [6] and Theorem 2 in [10] . Therefore (2) is true for x_o . Now, suppose that $|x_o| = r$. Let $\varepsilon > 0$ and $T_\varepsilon = T-\varepsilon I$. Then T_ε satisfies the same conditions as T but $(T_\varepsilon x-x,x)_- \leq -\varepsilon|x|^2$ for $|x| = r$. Consider $x_n \in D$ with $|x_n| < r$ and $x_n \to x_o$. The IVPs $u' = T_\varepsilon u-u$, $u(0) = x_n$ have solutions u_n on some common interval $[0,a]$ and (u_n) has a uniformly convergent subsequence the limit of which is a solution of $u' = T_\varepsilon u-u$, $u(0) = x_o$. Now, we may let $\varepsilon \to 0$ to obtain a solution of $u' = Tu-u$, $u(0) = x_o$, and therefore (2) holds for x_o .

q.e.d.

By Proposition 6 it is clear what we get from Theorem 1 , Prop. 3 and Prop. 4 in the special case $D = \{x \in K : |x| \leq r\}$. Let us mention that a class of maps that satisfy (5) but need not map ∂K into K is the class of quasimonotone operators satisfying $F(0) \in K$, where $F: K \to X$ is said to be quasimonotone w.r. to K if it has the property

$$x \in K\ ,\ z \in K \text{ and } x^*(z) = 0 \text{ for some } x^* \in K^* \implies x^*(Fx) \leq x^*(F(x+z)) \quad .$$

More generally, F satisfies (5) if $F = F_1+F_2$ with F_1 quasimonotone, $F_1(0) \in K$ and $F_2: K \to K$.

4.2 Let us now assume that X has an admissible projectional scheme $\{X_n,P_n\}$, i.e. there exists a sequence of finite dimensional subspaces X_n of X and continuous linear projections $P_n: X \to X_n$ such that $P_n x \to x$ for each $x \in X$. Then it is natural to look for criteria such that $P_n T$ has a fixed point in $D \cap X_n$ and such that these fixed points converge to a fixed point of T as $n \to \infty$.

Theorem 2. Let X have an admissible projectional scheme $\{X_n,P_n\}$. Let $K \subset X$ be a cone, $D = \{x \in K : |x| \leq r\}$ for some $r > 0$ and $T: D \to X$ continuous. Suppose also that

(i) $|P_n| = 1$ for each n ; $(Tx,x)_+ \leq |x|^2$ for $|x| = r$.

(ii) $P_n K \subset K$ for each n and T satisfies condition (ii) of Prop. 6 .

(iii) $x_n \in D \cap X_n$ and $P_n T x_n = x_n$ for $n \geq n_o \implies (x_n)$ has a convergent subsequence.

Then T has a fixed point in D .

Proof. Since $P_n K \subset K$, we have $P_n K = K \cap X_n$. Therefore $P_n T$ satisfies (ii) of Prop. 6 with $D_n = D \cap X_n$ instead of D . Since $|P_n| = 1$, we have $(P_n Tx,x)_- \leq (Tx,x)_+$ for $x \in D_n$; see e.g. Prop. 7.1 in [6] . Thus,

Prop. 6 tells us that $P_nT|_{D_n}$ satisfies (2) for D_n . Then P_nT has a fixed point $x_n \in D_n$, by Prop. 2 , and by condition (iii) we obtain an $x \in D$ such that $x = Tx$.

q.e.d.

Remarks. (i) $|P_n| = 1$ and $P_nK \subset K$ is satisfied if $\{X_n, P_n\}$ is defined by

$$X_n = \text{span}\{e_1,\dots,e_n\} \text{ and } P_n(\sum_{i\geq 1} x_ie_i) = \sum_{i\leq n} x_ie_i ,$$

where (e_n) is a montone Schauder base for X and K is the standard cone $K = \{x \in X : x_i \geq 0 \text{ for every } i \geq 1\}$.
Another example is $X = C(J)$, the space of continuous functions on $J = [0,a]$, with the standard cone K of nonnegative functions and $\{X_n,P_n\}$ defined by first order splines: Let $0 = t_{no} < t_{n1} < \dots < t_{nn} = a$ such that $\delta_n = \max_i (t_{ni} - t_{n,i+1}) \to 0$ as $n \to \infty$; let $e_{nk} \in X$ be the polygon defined by $e_{nk}(t_{ni})$ = Kronecker's δ_{ki} . Then $X_n = \text{span}\{e_{no},\dots,e_{nn}\}$ and $P_nx = \sum_{k=o}^{n} x(t_{nk})e_{nk}$; the partitions are equidistant.
(ii) Condition (iii) is certainly satisfied if T-I is A-proper w.r. to $\{X_n,P_n\}$; see e.g. Petryshyn [20] or [7] for standard examples. Let us mention the following particular case

Corollary 1. Suppose that (i) and (ii) of Theorem 2 are satisfied. Let $T = T_1+T_2$, where $T_1 : D \to X$ is compact and $T_2 : X \to X$ satisfies $(T_2x-T_2y,x-y)_+ \leq k|x-y|^2$ for some $k < 1$ and all $x,y \in X$. Then T has a fixed point.

Proof. We have to show that (iii) of Theorem 2 is satisfied. Let $S = I-T_2$. Then $(Sx-Sy,x-y)_+ \geq (1-k_2)|x-y|^2$ and therefore S is a homeomorphism from X onto X ; see Theorem 3.6 in [6] which is a particular case of Corollary 3 in [8] . Since T_1 is compact we have w.l.o.g. $P_nT_1x_n \to y$ as $n \to \infty$. Since S is onto, we have $y = Sx_o$ for some $x_o \in X$. Therefore $P_n(Sx_n - SP_nx_o) \to 0$ as $n \to \infty$. This implies

$$|x_n-P_nx_o|^2 \leq k|x_n-P_nx_o|^2 + |z_n||x_n-P_nx_o| \text{ with } |z_n| \to 0 ,$$

and therefore $x_n \to x_o$.

q.e.d.

4.3 The following result is a generalization to Banach spaces with a Schauder base of Miranda's fixed point theorem which has been useful in the study of periodic solutions to finite dimensional systems of ODEs ; see e.g. p. 236 of Rouche/Mawhin [23] or p. 59 in [7] . If K is a cone then it is usual to write $x \leq y$ iff $y-x \in K$. If $v,w \in X$ are such that $v \leq w$ then the order interval $[v,w]$ is the set $\{x \in X : v \leq x \leq w\}$.

Theorem 3. Let X be a Banach space with a base (e_i) ; $K \subset X$ the standard cone w.r. to (e_i) ; $v,w \in X$ and $v \leq w$; $T: [v,w] \to X$ continuous . Suppose also that

(i) $T_i(x_1,\dots,x_{i-1},v_i,x_{i+1},\dots) \geq v_i$ and $T_i(x_1,\dots,x_{i-1},\dots,w_i,x_{i+1},\dots) \leq w_i$ for every $i \geq 1$

(ii) Either $[v,w]$ or $T([v,w])$ is bounded

(iii) λT is β-condensing where $\lambda = \sup_n |P_n|$.

Then T has a fixed point.

Proof. Let $R_n v = v-P_n v$ and consider the IVP

$$z' = P_n T(z+R_n v)-z \ , \ z(0) = x \in [P_n v,P_n w] \ .$$

By Müller's comparison theorem for finite systems of ODEs , see Müller [17] and Deimling/Lakshmikantham [11] for extensions to countable systems, this IVP has a solution in $[P_n v,P_n w]$. Therefore, $\tilde{T}: [P_n v,P_n w] \to X_n$, defined by $\tilde{T}z = P_n T(z+R_n v)$ satisfies (2) with $D = [P_n v,P_n w]$. By Prop. 2 , $\tilde{T}$ has a fixed point z_n . Therefore $P_n T x_n = P_n x_n$ for $x_n = z_n+R_n v \in [v,w]$. Now, we apply Prop. 7.2 in [6] to obtain

$$\beta\{z_n : n \geq 1\} \leq \lambda\beta(T\{x_n : n \geq 1\}) = \beta(\lambda T\{x_n : n \geq 1\}) \ .$$

Since λT is β-condensing and $R_n v \to 0$ as $n \to \infty$, this estimate implies that $\{z_n : n \geq 1\}$ is relatively compact. Thus, we have $x_{n_k} \to x$ for some subsequence and some $x \in [v,w]$.

q.e.d.

Remarks. Since we had the same convergence problem in the proof to Corollary 1 , it is obvious, that we may replace (iii) by the extra condition on T in Corollary 1 . If $v \in K$ and K is normal , i.e. if there exists an M such that $0 \leq x \leq y$ implies $|x| \leq M|y|$, then $[v,w]$ is compact and therefore (iii) is automatically satisfied. If $T:[v,w] \to X$ is quasimonotone w.r. to K , then (i) is equivalent to "$Tv \geq v$ and $Tw \leq w$" , as follows immediately from the consideration of the orthogonal functionals e_i^* , i.e. the functionals satisfying $e_i^*(e_j) = \delta_{ij}$. Evidently, the equivalent form of (i) in terms of functionals is given by

$$x \in \partial[v,w] \ , \ x^* \in K^* \text{ and } x^*(v-x) = 0 \Rightarrow x^*(Tx-x) \geq 0$$

and

$$x \in \partial[v,w] \ , \ x^* \in K^* \text{ and } x^*(w-x) = 0 \Rightarrow x^*(Tx-x) \leq 0 \ ,$$

which allows to consider cones other than those defined by a base, but then we do not know how to prove a fixed point theorem corresponding to Theorem 3 .

5. Applications to integral equations

Let us indicate some problems connected with integral equations where our results might be applied. Details will be given elsewhere.

5.1 Volterra equations of the first kind

Consider the equation

$$\int_0^t k(t,s,u(s))ds = f(t) \quad \text{for} \quad t \in J = [0,a] , \tag{6}$$

and let us assume for simplicity that $f \in C^1(J)$ and that k is continuously differentiable w.r. to t . By differentiation we then obtain the equation

$$k(t,t,u(t)) + \int_0^t k_t(t,s,u(s))ds = f'(t) \tag{7}$$

which may be written as $u = T_1u+T_2u$ with

$$(T_1u)(t) = \int_0^t k_t(t,s,u(s))ds - f'(t)$$

and

$$(T_2u)(t) = u(t) + k(t,t,u(t)) .$$

This fixed point problem can be studied e.g. in $C(J)$ or $L^p(J)$ with $p > 1$ by means of Prop. 3 (i) or Prop. 4 , also in the corresponding standard cones K , since in the first case we have $\overset{\circ}{K} \neq \emptyset$ and for $L^p(J)$ we have X^* uniformly convex and the existence of the continuous metric projection $P: X \to D$. Some conditions for existence in $C(J)$ have been given in Schöneberg [25] .

5.2 Perturbed Volterra equations of the second kind

For equations of the type

$$u(t) = h(t,u(t)) + \int_0^t k(t,s,u(s))ds \quad \text{for } t \geq 0 \tag{8}$$

the operator defined by h may play the role of T_2 and the integral operator that one of T_1 . The same approach may be taken for stochastic equations

$$u(t,w) = h(t,u(t,w),w) + \int_0^t k(t,s,u(s,w);w)ds$$

which can be reduced to (8) for example, if (8) is regarded as an equation in $L^2(\Omega,\mathfrak{A},P)$, where $(\Omega,\mathfrak{A},P)$ is a probability measure space. The usual assumption on h has been a global Lipschitz-condition, see e.g. Milton/Tsokos [16] , while we only need the essentially weaker condition $(h(t,x,w)-h(t,y,w))\operatorname{sgn}(x-y) \le \lambda|x-y|$ and less restrictive growth conditions, and we can obtain information about existence of nonnegative solutions, etc. Clearly, equations (8) with Fredholm - instead of Volterra - operators may be studied similarly.

5.3 Mixed Volterra-Fredholm equations

Among others, Miller/Nohel/Wong [15], Corduneanu [5] and Mamedov [14] have considered equations like

$$u(t) = f(t) + \int_0^t k_1(t,s,u(s))ds + \int_0^\infty k_2(t,s,u(s))ds \quad \text{for } t \ge 0.$$

If, for example, one of the operators defines a strict contraction while the other is compact, then the right hand side defines an α-condensing map in the space under consideration and then Theorem 1 may be applied to obtain solutions, nonnegative solutions, etc.

5.4 Functional integral equations

We remember to have seen several papers dealing with "implicite" integral equations, for example

$$(9) \qquad f(t,y(t) + \int_0^t k(t,s,x(s))ds, x(t)) = 0 \quad ,$$

or with a Fredholm integral operator instead of x(t) in the third argument, etc. However, the conditions on F have always been such that it would have been much easier to reduce (9) to an equivalent, more or less standard type of equations and to apply known results. Recently, Schöneberg [26] has considered (9) in the framework of his degree for semicondensing maps, under weak conditions on f , namely, $f: J\times\mathbb{R}^n\times\mathbb{R}^n \to \mathbb{R}^n$ continuous ; $|f(t,u,0)| \le M(1+|u|)$;

$$(10) \qquad (f(t,u,v)-f(t,u,\bar{v}),v-\bar{v}) \ge c|v-\bar{v}|^2 \quad \text{for all } t\in J \quad ,$$

and $u,v,\bar{v}\in\mathbb{R}^n$, and some $c > 0$; k continuous and $|k(t,s,u)| \le M(1+|u|)$. However, the existence of solutions in C(J) can be shown much easier as follows: For fixed $u\in X = C(J)$, the operator $F_u: X \to X$, defined by $(F_u v)(t) = f(t,u(t),v(t))$ is continuous and satisfies

$$(F_u v - F_u\bar{v}, v-\bar{v})_- \ge c|v-\bar{v}|^2 \quad .$$

Therefore F_u is a homeomorphism onto X , by Corollary 3 in [8] . In particular, F_u has a unique zero Tu ; thus, $f(t,u(t),(Tu)(t)) = 0$. By (10) ,

$$|Tu| \leq \frac{1}{c}|F_u(0)| \leq \frac{M}{c}(1+|u|) \quad ,$$

and since f is uniformly continuous on bounded sets, this implies that $T: X \to X$ is continuous. Therefore, (9) is equivalent to

$$(11) \qquad u(t) = y(t) + \int_0^t k(t,s,(Tu)(s))ds \quad \text{for} \quad t \in J \quad ,$$

where the integral defines a compact operator. By the estimates on k and T , a sufficiently large ball is mapped into itself, and therefore we are done.

Suppose we look for solutions of (9) in some Banach space of continuous functions on $\mathbb{R}^+$. If we know more about f w.r. to u such that T turns out to be α-Lipschitz with constant λ and if k is such that the corresponding Volterra-operator is α-Lipschitz with constant μ , then the operator defined by the right hand side of (11) is α-Lipschitz with constant $\lambda\mu$, and if $\lambda\mu < 1$ then we are in good position to apply Theorem 1 .

5.5 Equilibria for integro-differential equations

There are several models which lead to integro-differential equations of the type

$$(12) \qquad u_t(t,\xi) = h(\xi,u(t,\xi)) + g(\xi,u(t,\xi))\int_\Omega k(\xi,\eta,u(t,\eta))d\eta$$

where one is especially interested in equilibria, i.e. solutions which are independent of the time t . A simple example is a kinetic model for vehicular traffic

$$u_t(t,\xi) = -\lambda(u-u_o) + \mu u \int_{v_1}^{v_2} (\xi-\eta)u(t,\xi)d\xi \quad ,$$

where u_o , λ and μ are constants and ξ is a velocity ; see e.g. Prigogine/Herman [21] , Paveri-Fontana [19] , Barone/Belleni-Morante [1] . The equilibria are the solutions of

$$(13) \qquad v(\xi) = v(\xi)+\rho h(\xi,v(\xi))+\rho g(\xi,v(\xi))\int_\Omega k(\xi,\eta,v(\eta))d\eta$$

with $\rho > 0$ fixed. Here again, $I+\rho H$ may be a candidate for T_2 . Suppose we have a bounded set M of continuous functions v , suppose that the integral operator K is compact on M and that g is Lipschitz with constant L . Then the operator T_1 , defined by

$$(T_1 v)(\xi) = g(\xi, v(\xi)) \int_\Omega \rho k(\xi, \eta, v(\eta)) d\eta$$

is α-Lipschitz with constant $\lambda = \rho L \sup\{|Kv| : v \in M\}$, since

$$|T_1 v - T_1 \bar{v}| \leq L\rho(\sup_M |Kv|)|v-\bar{v}| + \rho(\sup_M |g(\cdot,v)|)|Kv - K\bar{v}| .$$

Since ρ may be choosen arbitrarily small, we then only have to find a subset D of M such that $T = T_1 + T_2$ satisfies the boundary condition (2) .

We hope that these rather heuristic indications are enough stimulating that somebody will work hard to prove the conjecture in section 3.1 .

References

[1] Barone,E. and Belleni-Morante,A.: A nonlinear initial value problem arising from kinetic theory of vehicular traffic (to appear)

[2] Caristi,J.: The fixed point theory for mappings satisfying inwardness conditions. Ph. D. Thesis, University of Iowa, 1975

[3] - Fixed point theorems for mappings satisfying inwardness conditions. Trans. Amer. Math. Soc. 215 , 241-251 (1976)

[4] Clarkson,J.A.: Uniformly convex spaces. Trans. Amer. Math. Soc. 40, 396-414 (1936)

[5] Corduneanu,C.: Nonlinear perturbed integral equations. Rev. Roumaine Math. Pures Appl. 13 , 1279-1284 (1968)

[6] Deimling,K.: Ordinary Differential equations in Banach spaces. Lect. Notes in Math. Vol. 596, Springer-Verlag 1977

[7] - Nichtlineare Gleichungen und Abbildungsgrade. Hochschultext, Springer-Verlag 1974

[8] - Zeros of accretive operators. Manuscripta Math. 13 , 365-374 (1974)

[9] - Cone-valued periodic solutions of ordinary differential equations. Proc. Conf. "Applied Nonlinear Analysis" , Acad. Press (to appear)

[10] - Open problems for ordinary differential equations in Banach spaces. Communications del Conv. Equadiff. 78

(pp. 127-137) Il Centro 2P , Firenze (1978)

[11] Deimling,K. and Lakshmikantham,V.: Existence and comparison theorems for differential equations in Banach spaces (to appear)

[12] Fenske,C. and Eisenack,G.: Fixpunkttheorie. BI-Wissenschaftsverlag Zürich 1978

[13] Halpern,B. and Bergman,G.M.: A fixed point theorem for inward and outward maps. Trans. Amer. Math. Soc. 130 , 353-358 (1968)

[14] Mamedov,J.D. and Asirov,S.A.: Nonlinear Volterra equations. Izdat. Ылым , Aschabad 1977 (Russian)

[15] Miller,R.K., Nohel,J.A. and Wong,J.S.W.: A stability theorem for nonlinear mixed integral equations. J. Math. Anal. Appl. 25 , 446-449 (1969)

[16] Milton,J.S. and Tsokos,C.P.: On a nonlinear perturbed stochastic integral equation of Volterra type. Bull. Austral. Math. Soc. 9 , 227-237 (1973)

[17] Müller,M.: Über das Fundamentaltheorem in der Theorie der gewöhnlichen Differentialgleichungen. Math. Z. 26 , 619-645 (1926)

[18] Nußbaum,R.: The fixed point index for local condensing maps. Annali di Mat. Pura Appl. 89 , 217-258 (1971)

[19] Paveri-Fontana,S.L.: On Boltzmann-like treatments for traffic flow. Transpn. Res. 9 , 225-235 (1975)

[20] Petryshyn,W.V.: On the approximation solvability of equations involving A-proper and pseudo A-proper mappings. Bull. Amer. Math. Soc. 18 , 223-312 (1975)

[21] Prigogine,I. and Herman,R.: Kinetic theory of vehicular traffic. Amer. Elsevier, New York 1971

[22] Reich,S.: Fixed points of condensing functions. J. Math. Anal. Appl. 41 , 460-467 (1973)

[23] Rouche,N. and Mawhin,J.: Equations différentielles ordinaires. Vol. 2. Masson, Paris 1973

[24] Sadovskii,B.N.: Limit compact and condensing operators. Russ. Math. Surveys 27 , 85-155 (1972)

[25] Schöneberg,R.: Fixpunktsätze für einige Klassen kontraktions-

artiger Operatoren in Banachräumen über einen Fixpunktindex, eine Zentrumsmethode und die Fixpunkttheorie nichtexpansiver Abbildungen. Ph.D. Thesis, RWTH Aachen 1977

[26] Schöneberg,R.: A degree theory for semicondensing vector fields in infinite dimensional Banach spaces. Preprint No.193 Sonderforschungs-Bereich 72, Universität Bonn 1978

Klaus Deimling
Fachbereich 17 der GH
Warburger Str. 100
D-4790 Paderborn

WELL-POSEDNESS AND APPROXIMATIONS OF LINEAR VOLTERRA INTEGRODIFFERENTIAL EQUATIONS IN BANACH SPACES

Goong Chen and Ronald Grimmer
Department of Mathematics
Southern Illinois University
Carbondale, Illinois 62901

In this paper we shall be concerned with the integrodifferential equation

$$x'(t) = Ax(t) + \int_0^t B(t-u)x(u)du + f(t), \; x(0) = x_0, \; (t \geq 0) \tag{VE}$$

in a Banach space or Hilbert space. Here $A: D(A) \to X$ will always be the infinitesimal generator of a C_0 semi-group on X. We wish to determine conditions which ensure continuity with respect to initial data but our main goal will be to determine conditions which guarantee that the solutions of the equations

$$x_n'(t) = A_n x_n(t) + \int_0^t B_n(t-u)x_n(u)du + f(t) \qquad (VE)_n$$

satisfy $x_n(t) \to x(t)$. Our work depends greatly on earlier work on the continuity with respect to the parameters x_0 and f done by Miller [5]. Also, we should mention that results concerning this type of problem have been obtained in [1] and that other related work appears in [2, 4, 6, 8].

PRELIMINARIES

Let X be a Banach space with norm $\| \;\|$, $A: D(A) \to X$ a closed linear map with dense domain which generates a C_0 semi-group on X and let R^+ be the interval $[0, \infty)$. Let $B(t)$ be a linear map defined at least on the domain of A with $B(t)x$ a bounded uniformly continuous X valued function for each $x \in D(A)$. Also, we assume $B(\cdot)x(s)$ is continuous in BU when $x(s)$ and $Ax(s)$ are continuous. Let BU denote $\{f \in C(R^+, X)$: f is bounded and uniformly continuous$\}$ and make BU a Banach space by endowing it with the norm $\|f\| = \sup\{\|f(t)\|: t \geq 0\}$. The space $Z = X \times BU$ will be considered with norm given by $\|(x, f)\|^2 = \|x\|^2 + \|f\|^2$. On Z we will be particularly interested in the operators

$$E_\alpha = \begin{bmatrix} A-\alpha I, & \delta_0 \\ B, & D_s-\alpha I \end{bmatrix}$$

where δ_0 is the delta function, B is the map defined by $(Bx)(t) = B(t)x$ in BU, and D_s is the derivative operator. The importance of these operators is illustrated by the following fundamental result proved by Miller under the hypothesis that $\|B(t)x\| \leq \beta(t)\|x\|_A$ for all $x \in D(A)$ where $\beta(t) \in L^1(0, \infty)$ and $\|x\|_A = \|Ax\| + \|x\|$. It is also valid in our setting.

Theorem 1. [5] If $z(t) = (x(t), F(t, \cdot))$ is a solution of $z' = E_0 z$, then $x(t)$ solves (VE) with $x_0 = x(0)$ and $f(t) = F(0, t)$ for all $t \geq 0$.

Here by a solution of $z' = E_0 z$, $z(0) = z_0$, we mean a function $z: R^+ \to D(E_0)$ such that z, z', and $E_0 z$ are continuous and $z' = E_0 z$. A solution of (VE) is a function $x: R^+ \to D(A)$ such that x, x' and Ax are continuous, $x(0) = x_0$, and (VE) is satisfied on R^+.

Recall that the equation $z' = E_0 z$ is called uniformly well posed if for each $z_0 \in D(E_0)$ the initial value problem $z(0) = z_0$ has a unique solution $z(t, z_0)$ and for any $T > 0$ there is a $K > 0$ such that $\|z(t, z_0)\| \leq K\|z_0\|$ for all $z_0 \in D(E_0)$. Equation (VE) is defined to be uniformly well posed if for each pair $(x_0, f) \in D(E_0)$, there is a unique solution $x(t, x_0, f)$ and for any $T > 0$ there is an $M > 0$ such that $\|x(t, x_0, f)\| \leq M(\|x_0\| + \|f\|)$.

From the above discussion it is clear that if (VE) has unique solutions the well posedness of $z' = E_0 z$ implies that of (VE). For a discussion of uniqueness the reader is referred to [5]. We shall assume uniqueness of solutions of (VE) in what follows and concern ourselves with the related differential equation

$$z' = E_0 z, \quad z(0) = z_0. \tag{DE}$$

WELL POSEDNESS

We wish to determine conditions which ensure that E_0 generates a C_0 semi-group. Any such conditions will guarantee that (DE) is uniformly well posed and so, also, (VE) will be uniformly well posed.

First we note that $\{T(t)\}$ defined by $T(t)f(s) = f(s + t)$ is a C_0 semi-group of contractions on BU generated by D_s. Thus, the operator

$$F = \begin{bmatrix} A, & 0 \\ 0, & D_s \end{bmatrix}$$

generates a C_0 semi-group on $Z = X \times BU$. As an easy preliminary result we have

Theorem 2. If $B(t)x \in BU$ for each $x \in X$ and $B: X \to BU$ is a bounded operator then E_0 generates a C_0 semi-group.

Proof. This follows as the operator

$$G = \begin{bmatrix} 0, & \delta_0 \\ B, & 0 \end{bmatrix}$$

is a bounded operator and as F generates a C_0 semi-group so does $E_0 = F + G$, [7, p. 80]. Q.E.D.

A case of some interest is when B is taken to be zero in the above result and A generates a C_0 semi-group $\{S(t)\}$ with $\|S(t)\| \leq e^{\omega t}$. As F generates a contraction semi-group and $\|G\| = 1$ the semi-group generated by E_0 satisfies $\|T(t)\| \leq e^{(\omega+1)t}$ and $E_\alpha = E_0 - \alpha I$ will generate a contraction semi-group $\{e^{-\alpha t}T(t)\}$ if $\alpha \geq \omega + 1$. We can ow perturb with a dissipative operator that is "small" and still have a generator oi a C_0 semi-group. In particular, letting j_1 be a duality map on X and j_2 a duality ıap on BU we have

Theorem 3. Suppose A generates a semi-group $\{T(t)\}$ with $\|T(t)\| \leq e^{\omega t}$. Suppose A_1 and B are linear operators defined on D(A) into X and BU, respectively, with A_1 lissipative and $\|Bx\|^2 \leq -2N\mathrm{Re}\langle A_1x, j_1x\rangle$ for all $x \in D(A)$. Also, assume $\|A_1x\|^2 \leq$ $\|Ax\|^2$ with $b < 1$. Then the operator

$$\begin{bmatrix} A+A_1 , & \delta_0 \\ B , & D_s \end{bmatrix}$$

enerates a C_0 semi-group.

Proof. We show first that the operator

$$H = \begin{bmatrix} A_1 , & 0 \\ B & -\beta I \end{bmatrix}$$

n Z is dissipative with respect to the duality map $j = (j_1, j_2)$ if $\beta > 0$ is chosen ufficiently large. Letting $z = (x, y)$, we have

$$\begin{aligned} \mathrm{Re}\langle Hz, jz\rangle &= \mathrm{Re}\langle A_1x, j_1x\rangle + \mathrm{Re}\langle Bx, j_2y\rangle - \beta\|y\|^2 \\ &\leq \mathrm{Re}\langle A_1x, j_1x\rangle + \frac{\|Bx\|^2}{2N} + \frac{N\|y\|^2}{2} - \beta\|y\|^2 \\ &\leq 0 \end{aligned}$$

f $\beta > N/2$.

Further,
$$\begin{aligned} \|Hz\|^2 &= \|A_1x\|^2 + \|Bx - \beta y\|^2 \\ &\leq \|A_1x\|^2 + \|Bx\|^2 + \beta\|y\|^2 \\ &\leq \|A_1x\|^2 + N|\langle A_1x, j_1x\rangle| + \beta\|y\|^2 \\ &\leq (1 + \gamma)\|A_1x\|^2 + M\|y\|^2 \\ &\leq b(1 + \gamma)\|Ax\|^2 + M\|y\|^2 \end{aligned}$$

here $\gamma > 0$ may be chosen so that $b(1 + \gamma) < 1$. This is sufficient to guarantee that

$$\begin{bmatrix} A+A_1-\alpha I , & \delta_0 \\ B , & D_s-(\alpha+\beta)I \end{bmatrix}$$

enerates a semi-group [7, p. 84] and, thus,

$$\begin{bmatrix} A+A_1 , & \delta_0 \\ B & D_s \end{bmatrix}$$

enerates a semi-group. Q.E.D.

Rather than continue along these lines we turn to the case where X is a Hilbert pace and obtain a result of the same nature.

Before proceeding, however, it will be useful to introduce some notation. If $h \in BU$, h_s will denote the function $h_s(t) = h(t + s)$, $s \geq 0$, while $h^*(\lambda)$ will denote the Laplace transform of h. Also, $\rho(\lambda)$ will denote the operator $[\lambda I - A - B^*(\lambda)]^{-1}$ when it exists.

Theorem 4. Let X be a real Hilbert space and -A a positive self adjoint operator with dense domain D(A). Let A_1 be the positive square root of -A and assume $\|B(t)x\|^2 \leq 2K\|A_1x\|^2$ for $x \in D(A)$, for some $K > 0$. Assume that $\rho(\lambda + \alpha)$ exists as a bounded operator on X for some $\lambda > 0$ and α with $2\alpha > K + 1$. Then E_0 generates a C_0 semi-group.

Proof. We consider the operator E_α where $\alpha > 0$ is chosen so that $2\alpha > K + 1$. In this case, for $z = (x, y)$,

$$\langle E_\alpha z, jz\rangle = \langle (A - \alpha I)x, x\rangle + \langle \delta_0 y, x\rangle + \langle Bx, j_2 y\rangle$$
$$+ \langle D_s y, j_2 y\rangle - \alpha\|y\|^2.$$

As D_s generates a contraction semi-group on BU it follows from the theorem of Lumer-Phillips [7 , p. 16] that D_s is dissipative and so

$$\begin{aligned}\langle E_\alpha z, jz\rangle &\leq \langle (A - \alpha I)x, x\rangle + \tfrac{1}{2}(\|y\|^2 + \|x\|^2) \\ &\quad + (\|Bx\|^2 + K^2\|y\|^2)/2K - \alpha\|y\|^2 \\ &\leq (\tfrac{1}{2} - \alpha)\|x\|^2 + ((K + 1)/2 - \alpha)\|y\|^2 \\ &\leq 0.\end{aligned}$$

Checking the range $R(\lambda - E_\alpha)$ of $\lambda - E_\alpha$, following the calculations in [5] we see that the resolvent of E_α exists for $\lambda > 0$ and $(\lambda - E_\alpha)z = z_1$ may be solved for $z = (x, y)$ by

$$y(s) = (y_{1,s} + B_s x)^*(\lambda + \alpha)$$

and

$$\begin{aligned}x &= [(\lambda + \alpha)I - A - B^*(\lambda + \alpha)]^{-1}(x_1 + y_1^*(\lambda + \alpha)) \\ &= \rho(\lambda + \alpha)(x_1 + y_1^*(\lambda + \alpha)).\end{aligned}$$

It now follows from the Lumer-Phillips theorem that E_α generates a contraction semi-group and E_0 generates a semi-group. Q.E.D.

We note that in each of the preceding results it is easy to calculate the rate of growth of the semi-group if the same information is known about the growth of the semi-group generated by A. In particular, we will say that a linear operator A belongs to $G(M, \beta)$ if A is the generator of a C_0 semi-group $\{T(t)\}$ with $\|T(t)\| \leq Me^{\beta t}$.

Related to the integral equation $(VE)_n$ is the differential equation

$$z_n' = E_0^n z_n \qquad n = 1, 2, \ldots$$

where

$$E_0^n = \begin{bmatrix} A_n & \delta_0 \\ B_n & D_s \end{bmatrix}.$$

Theorem 5. Suppose that E_0 and E_0^n are in $G(M, \beta)$ and that $\rho_n(\lambda) = (\lambda I - A_n - B_n*(\lambda))^{-1}$ exists as a bounded linear operator for $\operatorname{Re} \lambda > \beta$. If $\rho_n(\lambda)x \to \rho(\lambda)x$ and $(B_n)_s*(\lambda)\rho_n(\lambda)x \to B_s*(\lambda)\rho(\lambda)x$ for every $x \in X$ and λ with $\operatorname{Re} \lambda > \beta$, then for any $(x_0, f) \in D(E_0) \cap D(E_0^n)$ for all n we have $x_n(t) \to x(t)$ uniformly on compact intervals where $x_n(t)$ and $x(t)$ are the solutions of $(VE)_n$ and (VE) respectively with initial data (x_0, f).

Proof. One calculates that $R(\lambda, E_0)(x, y)^T = (x_1, y_1)^T$ is given by $x_1 = \rho(\lambda)(x + y*(\lambda))$ and $y_1 = (y_s* + B_s*(\lambda)\rho(\lambda)(x + y*(\lambda)))$. Similar formulas hold for $R(\lambda, E_0^n)$. Our hypotheses thus imply $R(\lambda, E_0^n)z \to R(\lambda, E_0)z$ for every $z \in Z$ and λ with $\operatorname{Re} \lambda > \beta$. The result now follows from the Trotter approximation theorem [7, p. 87] or [3, p. 504]. Q.E.D.

BIBLIOGRAPHY

1. M. G. Crandall and J. A. Nohel, An Abstract Functional Differential Equation and a Related Nonlinear Volterra Equation, Mathematics Research Center, University of Wisconsin Tech. Summary Report #1765,1977.

2. G. Gripenberg, An Existence Result for a Nonlinear Volterra Integral Equation in Hilbert Space. SIAM J. Math. Anal., to appear.

3. T. Kato, "Perturbation Theory for Linear Operators," 2nd ed. Springer-Verlag, New York, 1976.

4. S. O. Londen, On an Integral Equation in Hilbert Space, SIAM J. Math. Anal., to appear.

5. R. K. Miller, Volterra Integral Equation in a Banach Space, Funkcial. Ekvac. 18 (1975), 163-193.

6. R. K. Miller and R. L. Wheeler, Well-posedness and Stability of Linear Volterra Integrodifferential Equations in Abstract Spaces, Funkcial. Ekvac., to appear.

7. A. Pazy, Semi-groups of Linear Operators and Applications to Partial Differential Equations. University of Maryland Lecture Note #10, 1974.

8. G. F. Webb, An Abstract Semilinear Volterra Integrodifferential Equation, Proc. Amer. Math. Soc. 69(1978), 255-260.

Some Notes On The Resolvents Of Volterra Integral Equations

Stanley I. Grossman
Department of Mathematics
University of Montana
Missoula, Montana 59801/USA

I. Introduction

In this paper we consider the linear Volterra integral equation

$$x(t) = f(t) + \int_0^t a(t - s)x(s)ds. \tag{1}$$

This equation has been studied extensively and many results are known concerning existence, uniqueness, the asymptotic behavior of solutions and other important properties (see [4], for example). Many of the facts regarding solutions to (1) depend upon a knowledge of the resolvent kernel, $r(t) = \text{res } a(t)$, defined by

$$r(t) = -a(t) + \int_0^t a(t - s)r(s)ds. \tag{r}$$

If $a(t)$ is locally integrable (i.e. $\int_0^T |a(t)|dt < \infty$ for all $T > 0$) then it is easily established that the resolvent equation has a unique solution $r(t)$ which is locally integrable on $[0, \infty)$. [see [4]].

With the resolvent defined above, we can easily verify that the unique solution to (1) is given by

$$x(t) = f(t) - \int_0^t r(t - s)f(s)ds. \tag{2}$$

Suppose that $r(\cdot) \in L^1[0, \infty)$. Let X be any of the spaces $BC[0, \infty) = \{x$ continuous and bounded on $R^+\}$, $BC_\ell[0, \infty) = \{x \in BC[0, \infty): x(t) \to \ell$ as $t \to \infty$, $|\ell| < \infty\}$, or $L^p[0, \infty)$, $1 \leq p < \infty$. Then since the convolution of an L^1 function with a function in X is again in X, we can prove that the solution $x(t)$ to (1), given by (2), will lie in X whenever $f(t) \in X$.

We can think of the resolvent being in L^1 in other terms. If $r \in L^1[0, \infty)$ and f is bounded on $[0, \infty)$, then the solution $x(t)$ given by (2) will be bounded. That is, $r \in L^1$ implies "bounded input-bounded output" stability. Thus, in a certain sense the integrability of the resolvent is intrinsically related to the stability of the equation (1). This is also related to the admissibility (as defined by Corduneanu) of the linear operator defined by the right side of (1) (see [1] for details).

The above facts suggest that it is quite important to know when the resolvent will be in $L^1[0, \infty)$. The first result in this direction was proved by Paley and Wiener in 1931 [6].

Theorem 1.1 If $a(t) \in L^1[0, \infty)$, then $r(t) \in L^1[0, \infty)$ if and only if $\hat{a}(s) \neq 1$ for Re $s \geq 0$; here $\hat{a}(s) = \int_0^\infty e^{-st} a(t)dt$ is the Laplace transform of $a(\cdot)$.

Other results have been few and far apart. Shea and Wainger [7] showed that if $-a(t) = b(t) + \beta(t)$ where $b(t)$ is positive, decreasing and convex on $(0, \infty)$, $b(t) \in L^1(0, 1)$, $\beta(t) \in L^1[0, \infty)$, $t\beta(t) \in L^1[0, \infty)$, and $\hat{b}(s) \neq 1$ for Re $s \geq 0$, then $r(t) \in L^1[0, \infty)$. In [5] Miller showed that $r(t) \in L^1[0, \infty)$ if $-a(t)$ is completely monotonic ($-a(t) \in C^\infty[0, \infty)$ and $(-1)^{k+1}a^{(k)}(t) \geq 0$ for $k = 0, 1, 2, \ldots$). In [2], this author showed that $r(t) \in L^1[0, \infty)$ if $a(0) = 0$, $a(\infty)$ exists and is finite, $a'(t) \leq 0$ for $t \geq 0$ and is not constant, $ta'(t) \in L^1[0, \infty)$, $a''(t) \geq 0$ and $a''(t)$ is nonincreasing. Finally, in [3], Levin proved the related result that if $a(t) \leq 0$ and $a'(t) \geq 0$ on $0 < t < \infty$ and $a(0) < \infty$, then $r(t)$ satisfies $a(0) \leq r(t) \leq -a(0)$ for $0 \leq t < \infty$, $0 \leq \int_0^t r(\tau)d\tau \leq 1$ for $0 \leq t < \infty$ and $\lim_{t \to \infty} \int_0^t r(\tau)d\tau = R$ exists. Furthermore, if $A = \int_0^\infty a(\tau)d\tau$, then $R = A(1 + A)^{-1}$ if $A < \infty$ and $R = 1$ if $A = -\infty$.

In this paper we will be concerned with finding necessary conditions in order that $r(t)$ be in $L^1[0, \infty)$. These conditions will provide us with numerous examples of functions which do not have integrable resolvents.

Before proceeding, we need an additional definition.

Definition 1. Let $a(t)$ be defined on R^+. Then a is said to be of subexponential order on R^+ if for every $\epsilon > 0$, there exists a constant $M > 0$ such that

$$|a(t)| \leq Me^{\epsilon t}, \quad t \geq 0. \tag{3}$$

Note: If a is of subexponential order on R^+, then $\hat{a}(s)$ exists for Re $s > 0$.

II. The Main Results

We will proceed to the principal results by first proving two lemmas. In this section $r(t)$ will denote res $a(t)$.

Lemma 1. Let $a(t)$ be of subexponential order and let $a_1(t) = e^{-\alpha t}a(t)$. Then

$$r_1(t) = \text{res } a_1 = e^{-\alpha t}r(t).$$

Proof. Since $a(t)$ is of subexponential order, $\hat{a}(s)$ exists for Re $s > 0$. Then, from (1) and the convolution theorem for Laplace transforms, we easily calculate

$$\hat{r}(s) = \frac{-\hat{a}(s)}{1 - \hat{a}(s)}. \tag{4}$$

But $\hat{a}_1(s) = \widehat{e^{-\alpha t}a(t)} = \hat{a}(s + \alpha)$. Thus

$$\hat{r}_1(s) = \frac{-\hat{a}(s+\alpha)}{1-\hat{a}(s+\alpha)} = \hat{r}(s+\alpha) = \widehat{e^{-\alpha t}r(t)}$$

which, by the uniqueness of the inverse transform, proves the lemma.

Lemma 2. If $a(t)$ is of subexponential order and if $r(t) \in L^1[0, \infty)$, then $\hat{a}(s) \neq 1$ for $\operatorname{Re} s > 0$.

Proof: Suppose there is an s^* with $\operatorname{Re} s^* > 0$ such that $\hat{a}(s^*) = 1$. Let α be a real number in $(0, \operatorname{Re} s^*)$. Let $a_1(t) = e^{-\alpha t}a(t)$. Clearly, $a_1(t) \in L^1[0, \infty)$ by hypothesis. Also, $\operatorname{res} a_1 = e^{-\alpha t}r(t)$ by Lemma 1 and, since $r(t) \in L^1[0, \infty)$, we have $\operatorname{res} a_1 \in L^1[0, \infty)$. Thus, by the Paley-Wiener theorem $\hat{a}_1(s) \neq 1$ for $\operatorname{Re} s \geq 0$. But $\hat{a}_1(s) = \hat{a}(s+\alpha)$ and, for $s' = s^* - \alpha$, we have $\operatorname{Re} s' > 0$ and $\hat{a}_1(s') = \hat{a}(s'+\alpha) = 1$. This contradiction establishes the lemma.

We can now state the main results.

Theorem 1. Let $a(t)$ be of subexponential order and suppose that $r(t) = \operatorname{res} a(t) \in L^1[0, \infty)$. Then

(i) $\hat{a}(s) \neq 1$ for $\operatorname{Re} s > 0$

and

(ii) there exists a positive constant $M < \infty$

such that

$$\lim_{\epsilon \to 0^+} \left| \frac{\hat{a}(\epsilon + i\beta)}{1-\hat{a}(\epsilon + i\beta)} \right| \leq M \tag{5}$$

uniformly for $\beta \in \mathbb{R}$.

Proof. (i) is a consequence of Lemma 2.

(ii) Since $r \in L^1[0, \infty)$, $\hat{r}(s)$ exists for $\operatorname{Re} s \geq 0$.

Thus

$$\left| \frac{\hat{a}(\epsilon + i\beta)}{1-\hat{a}(\epsilon + i\beta)} \right| = |\hat{r}(\epsilon + i\beta)| \leq \|r\|_{L_1} \tag{6}$$

The result now follows from taking limits.

Theorem 2. Suppose that $a(t)$ is of subexponential order, $a(t) > 0$ for all $t \geq 0$ and $a \notin L^1[0, \infty)$. Then $r \notin L^1[0, \infty)$.

Proof. Since $\int_0^\infty a(t)dt = \infty$, we have $\lim_{s \to 0^+} \hat{a}(s) = \infty$. But, since $a_1(t) = e^{-\epsilon t}a(t) \in L^1[0, \infty)$ for $\epsilon > 0$, $\hat{a}_1(s) = \hat{a}(s+\epsilon) \to 0$ as $s \to \infty$. Clearly $\hat{a}(s)$ is continuous in $(0, \infty)$ [since, if $s_2 > s_1 > 0$, $|\hat{a}(s_1) - \hat{a}(s_2)| \leq \int_0^\infty e^{-s_1 t} a(t)[1 - e^{(s_1-s_2)t}]dt$ which $\to 0$ as $s_2 \to s_1$ by dominated convergence and the fact that $e^{-s_1 t}a(t) \in L^1(0, \infty)$]. Thus, by the intermediate value theorem, there is a real, positive number s^* such that $\hat{a}(s^*) = 1$ and, by Theorem 1 (i) the theorem is proved.

It is easy to find examples of functions of subexponential order that do not have L^1 resolvents. Any positive function will do. The function $-t$ has the transform $\frac{-1}{s^2}$ and then $\frac{-\hat{a}(s)}{1-\hat{a}(s)} = \frac{1}{s^2+1}$ which is unbounded along the imaginary axis implying that $\operatorname{res}(-t) \notin L^1$. This is easily verified as $\operatorname{res}(-t) = \sin t$.

On the other hand, the function $a(t) = -1$ has the L^1 resolvent e^{-t} and $\frac{-\hat{a}}{1-\hat{a}} = \frac{1}{1+s}$ which does remain bounded on the imaginary axis. This suggests (together with other examples) that the conditions in Theorem 1 are both necessary and sufficient for $r(t)$ to be integrable. One possible approach to proving this fact is the following:

Since all functions of concern are defined only on $[0, \infty)$, it is no restriction to think of $r(t)$ to be defined on $(-\infty, \infty)$ and identically 0 in $(-\infty, 0)$. Suppose that conditions (i) and (ii) of Theorem 1 hold. Then, for $\epsilon > 0$

$$\hat{r}(\epsilon + i\beta) = \frac{-\hat{a}(\epsilon + i\beta)}{1 - \hat{a}(\epsilon + i\beta)}$$

is finite by (i) and, by (ii),

$$\int_{-\infty}^{\infty} e^{-i\beta t} r(t)dt \leq M < \infty$$

That is, $r(t)$ is a subexponential function whose Fourier transform is bounded on the reals! Unfortunately this is not sufficient to conclude that $r \in L^1$ since it is possible to construct a pathological example of a function not in L^1 with a uniformly bounded Fourier transform. However it seems that with the added structure here, such examples can be ruled out.

References

1. C. Corduneanu, "Problemes globaux dans la theorie des equations integrales de Volterra, Ann. Mat. Pura. Appl., 67(1965), 349-363.
2. S. I. Grossman, "Integrability of resolvents of certain Volterra integral equations", J. Math. Anal. Appl. 48(1974), 785-793.
3. J. J. Levin, "Resolvents and bounds for linear and nonlinear Volterra equations, Trans. Amer. Math. Soc., 228 (1977), 207-222.
4. R. K. Miller, Nonlinear Volterra Integral Equations, Benjamin, Menlo Park, Calif., 1971.
5. _______, "On Volterra integral equations with nonnegative integrable resolvents", J. Math. Anal. Appl. 22(1968), 319-340.
6. R. E. A. C. Paley and N. Wiener, Fourier Transforms in the Complex Domain, Amer. Math. Soc. Colloq. Publ., Vol. 19, Amer. Math. Soc., Providence, R.I., 1934.
7. D. F. Shea and S. Wainger, "Variants of the Wiener-Levy theorem with applications to stability problems for some Volterra integral equations", Amer. J. Math. 97(1975), 312-343.

AN INTEGRODIFFERENTIAL EQUATION WITH PARAMETER

Kenneth B. Hannsgen
Virginia Polytechnic Institute and State University
Blacksburg, VA 24061/USA

1. Introduction. Consider a nonegative function $a(t)$ on $[0,\infty)$ $(a(0) = \delta > 0)$ which is nonincreasing, convex, and piecewise linear, with changes of slope only at the integers $t = 1,2,3,\ldots$. Let $c \geq 0$ and for $\lambda > 0$ let $u(t,\lambda)$ be the solution of the linear integrodifferential equation

$$x'(t) + \lambda\int_0^t [c + a(t-s)]x(s)ds = 0, \quad x(0) = 1. \tag{1.1}$$

It is known [4] that $\lim_{t\to\infty} u(t,\lambda) = 0$ unless

$$\lambda = \lambda_j = (2\pi j)^2/(\delta + c) \tag{1.2}$$

for some positive integer j; indeed [3,7]

$$\int_0^\infty |u(t,\lambda)|dt < \infty \quad (0 < \lambda \neq \lambda_j). \tag{1.3}$$

On the other hand,

$$\lim_{t\to\infty} [u(t,\lambda_j) - \frac{2}{\gamma_0} \cos \omega_j t] = 0, \tag{1.4}$$

where $\omega_j = 2\pi j$ and $\gamma_0 = (3\delta + 2c)/(\delta + c)$.

The following result describes more precisely the behavior of $u(t,\lambda)$ for large t as $\lambda \to \lambda_j$.

THEOREM 1. *Let $a(t)$ and c be as above, and let j be a fixed positive integer. Assume that*

$$\int_0^\infty a(t)dt < \infty. \tag{1.5}$$

Then there exist finite positive constants ε and B and continuously differentiable complex-valued functions $\Gamma(\lambda) = \gamma(\lambda) + i\beta(\lambda)$ and $\Omega(\lambda) = \nu(\lambda) + i\rho(\lambda)$, defined for $0 < |\lambda - \lambda_j| < \varepsilon$ such that

$$\int_0^\infty \left|u(t,\lambda) - 2\mathrm{Re}\,\frac{e^{i\Omega(\lambda)t}}{\lambda\Gamma(\lambda)}\right|dt \leq B \quad (0 < |\lambda - \lambda_j| < \varepsilon).$$

Moreover,

(1.6) $$\rho(\lambda) > 0, \quad \lim_{\lambda \to \lambda_j} \Gamma(\lambda) = \gamma_0, \quad \lim_{\lambda \to \lambda_j} \Omega(\lambda) = \omega_j.$$

Note that

$$2 \text{ Re } \frac{e^{i\Omega t}}{\lambda \Gamma} = \frac{2e^{-\rho t}}{\lambda |\Gamma|} [\gamma \cos \nu t - \beta \sin \nu t].$$

In the proof we exhibit formulas for Γ and Ω which enable one to verify (1.6) directly. One also sees, for example that $\rho/(\nu - \omega_j) \to 0 \quad (\lambda \to \lambda_j)$.

With some modifications [6], our proof also shows that

$$\int_0^\infty \left| u(t,\lambda_j) - \frac{2}{\gamma_0} \cos \omega_j t \right| dt < \infty.$$

In the next two sections, we sketch a proof of Theorem 1. Some aspects of the proof are adapted from studies (e.g. [1]) of (1.1) with a(t) nonincreasing and convex (but not of the special piecewise linear form considered here), where estimates such as

$$\int_0^\infty \sup_{1 \le \lambda < \infty} |u(t,\lambda)| dt < \infty$$

are obtained. In the sketch below, we shall emphasize those estimates arising specifically in the piecewise linear case.

2. Preliminaries. In the course of the proof, several *a priori* restrictions are placed on the number ϵ. We take these restrictions to be in force throughout the proof; thus ϵ is a fixed positive number and (3.1), (3.2), (3.11), and (3.14) below hold. We assume throughout that $0 < |\lambda - \lambda_j| < \epsilon$. The symbol M denotes a finite positive constant, independent of λ; its value may change from line to line. B is then any number greater than all the various values of M. In some formulas, λ is suppressed.

By (1.3), the Fourier transform

$$\hat{u}(\tau,\lambda) = \int_0^\infty e^{-i\tau t} u(t,\lambda) dt$$

is analytic in the lower half-plane $\Pi^- = \{\text{Im } \tau < 0\}$ and continuous for $\text{Im } \tau \le 0$. Moreover,

(2.1) $$\hat{u}(-\tau^*) = [\hat{u}(\tau)]^*$$

(* = complex conjugate). The same remarks apply to the Fourier transform of a(t), which we write as

(2.2) $$\hat{a}(\tau) = \varphi(\tau) - i\tau\theta(\tau) = \tau^{-2} \sum_{k=1}^\infty \frac{\delta_k}{k}(1 - ik\tau - e^{-ik\tau}).$$

Here δ_k/k is the change of slope of a(t) at t = k; note that

(2.3) $$\sum_{k=1}^\infty \delta_k = \delta, \quad \sum_{k=1}^\infty k\delta_k = 2\int_0^\infty a(t)dt.$$

Thus the last expression in (2.2) can be differentiated twice term-by-term, and $\hat{a} \in C^2(R^1\backslash\{0\})$.

Now take Fourier transforms in (1.1). By (2.2),

$$\lambda\hat{u}(\tau,\lambda) = \frac{1}{p(\tau,\lambda)}, \tag{2.4}$$

where

$$p(\tau,\lambda) = \varphi(\tau) + i\tau[\lambda^{-1} - c\tau^{-2} - \theta(\tau)]. \tag{2.5}$$

Using (2.2) and (1.5) one can establish the following facts (see [7, Lemma 1], [5, Lemma 2.2] and [6, (4.2)]):

(2.6) (i) $|\hat{a}(\tau)| \le \frac{4\delta}{|\tau|}$, $|\hat{a}'(\tau)| \le \frac{40\delta}{\tau^2}$ (τ real),

(ii) $\int_{-1}^{1} |\hat{a}'(\tau)| d\tau < \infty$,

(iii) $\varphi(2k\pi) = 0$ $(k=1,2,3\ldots)$, $\varphi(\tau) > 0$ for all other $\tau > 0$,

(iv) $\theta(\tau) > 0$, $\theta'(\tau) < 0$ $(\tau > 0)$,

(v) For $\tau > 0$, $\mathrm{Im}\, p(\tau,\lambda) = 0$ iff $\tau = \omega(\lambda)$, where $\omega \in C^1$, $\omega(\lambda_j) = \lambda_j$, and $\omega'(\lambda_j) > 0$.

$\Gamma(\lambda)$ and $\Omega(\lambda)$ are defined as follows. Let

$$\begin{aligned} p_1(\tau,\lambda) &= p(\omega(\lambda),\lambda) + \frac{\partial p}{\partial \tau}(\omega(\lambda),\lambda)(\tau-\omega(\lambda)) \\ &= \varphi(\omega(\lambda)) + [\varphi'(\omega(\lambda)) + i\omega(2c\omega^{-3}(\lambda)-\theta'(\omega(\lambda))](\tau-\omega(\lambda)) \\ &\equiv \Gamma(\lambda)\, i(\tau - \Omega(\lambda)). \end{aligned} \tag{2.7}$$

Setting $\Gamma = \gamma + i\beta$, $\Omega = \nu + i\rho$, we find after a straightforward computation that for each λ,

$$\begin{aligned} \gamma &= \omega(2c\omega^{-3} - \theta'(\omega)) \\ \beta &= -\varphi'(\omega) \\ \rho &= \omega\varphi(\omega)[2c\omega^{-3} - \theta'(\omega)]/|\Gamma|^2 \\ \nu &= \{-\varphi'(\omega)[\varphi(\omega) - \omega\varphi'(\omega)] + \omega^3[2c\omega^{-3} - \theta'(\omega)]^2\}/|\Gamma|^2. \end{aligned}$$

Now (1.6) follows from (2.6).

Note that p_1^{-1} is the Fourier transform of $e^{i\Omega t}/\Gamma$. Thus we need to show that

$$\int_0^\infty |u_1(t,\lambda)|\,dt \le M, \tag{2.8}$$

where

$$\hat{u}_1(\tau,\lambda) = \frac{1}{p(\tau,\lambda)} - \frac{1}{p_1(\tau,\lambda)} - \frac{1}{p_1^*(-\tau^*,\lambda)}. \tag{2.9}$$

Since $\rho > 0$, $\hat{u}_1(\tau,\lambda)$ is continuous and bounded for $\mathrm{Im}\ \tau \le 0$ and continuously differentiable for real $\tau \neq 0$. We shall show that

$$(2.10) \qquad 2\int_0^\infty \left|\frac{\partial}{\partial\tau}\hat{u}_1(\tau,\lambda)\right|d\tau = \int_{-\infty}^\infty \left|\frac{\partial}{\partial\tau}\hat{u}_1(\tau,\lambda)\right|d\tau \le M.$$

An argument involving the Poisson integral representation for $\hat{u}_1$ (see [7, p. 323] for details) shows that the function $f(\tau) \equiv (\partial/\partial\tau)\ \hat{u}_1(\tau,\lambda)$ belongs to the Hardy space $H^1(\Pi^-)$; moreover, by (2.4) and integration by parts (note that $\hat{u}_1(\tau,\lambda) \to 0$ as $\tau \to \pm\infty$),

$$\begin{aligned}\frac{i}{t}\check{f}(t) &\equiv \frac{i}{2\pi t}\int_{-\infty}^{\infty} f(\tau)e^{i\tau t}d\tau \\ &= \lim_{N\to\infty}\frac{1}{2\pi}\int_{-N}^{N}\hat{u}(\tau,\lambda)e^{i\tau t}d\tau \\ &= \begin{cases} u_1(t,\lambda) & (t > 0) \\ 0 & (t < 0). \end{cases}\end{aligned}$$

(Since $\rho > 0$ and (1.1) and (1.3) hold, $u_1(\cdot,\lambda)$ belongs to $L^1 \cap C^1(0,\infty)$; this justifies the last step above.) Then an inequality of Hardy and Littlewood (half-plane version [2, p. 198]) yields (2.8).

3. Proof of (2.10). Since ω is continuous with $\omega(\lambda_j) = \omega_j = 2\pi j$, we may require that

$$(3.1) \qquad \omega(\lambda_j - \epsilon) - \epsilon \ge 1.$$

Clearly,

$$(3.2) \qquad \int_0^\infty \left|\frac{\partial}{\partial\tau}\frac{1}{p_1^*(-\tau^*,\lambda)}\right| d\tau \le \frac{1}{|\Gamma|}\int_0^\infty \frac{d\tau}{(\nu+\tau)^2} \le M.$$

By (1.6), we may choose ϵ so that

$$(3.3) \qquad |\nu(\lambda) - \omega(\lambda)| < \epsilon/2 \quad (|\lambda-\lambda_j| < \epsilon),$$

so

$$(3.4) \qquad \begin{aligned}&\left[\int_0^{\omega-\epsilon} + \int_{\omega+\epsilon}^\infty\right]\left|\frac{\partial}{\partial\tau}\frac{1}{p_1(\tau,\lambda)}\right|d\tau \\ &\le \frac{1}{|\Gamma|}\left[\int + \int\right]\frac{d\tau}{(\nu-\tau)^2} \le 4/\epsilon\ |\Gamma| \le M.\end{aligned}$$

By (2.5) and (2.6 iv,v),

$$(3.5) \qquad |\mathrm{Im}\ p(\tau,\lambda)| \ge \tau/M \quad (|\tau-\omega| \ge \epsilon);$$

since

$$(3.6) \qquad -\frac{\partial}{\partial\tau}\frac{1}{p(\tau,\lambda)} = \frac{i/\lambda + ic/\tau^2 + \hat{a}'(\tau)}{[p(\tau,\lambda)]^2},$$

(2.6i) yields

$$(3.7) \qquad \int_{\omega+\epsilon}^\infty \left|\frac{\partial}{\partial\tau}\frac{1}{p(\tau,\lambda)}\right| d\tau \le M.$$

Since $p(0+,\lambda) = \varphi(0) = \int_0^\infty a(t)dt > 0$, we see from (2.5), (2.6iii) and (3.5) that $|p(\tau,\lambda)| \geq M$ $(0 < \tau \leq \omega(\lambda) - \epsilon)$; moreover, $|p(\tau,\lambda)| \geq c/M\tau$ on the same interval. Now (3.6) and (2.6ii) show that

(3.8) $$\int_0^{\omega-\epsilon} \left|\frac{\partial}{\partial\tau}\frac{1}{p(\tau,\lambda)}\right| d\tau \leq M.$$

To complete the proof, we need only establish that

(3.9) $$\int_{\omega-\epsilon}^{\omega+\epsilon} \left|\frac{\partial}{\partial\tau}\left[\frac{1}{p(\tau,\lambda)} - \frac{1}{p_1(\tau,\lambda)}\right]\right| d\tau \leq M.$$

A little rearranging shows that the derivative in (3.9) can be written as

(3.10) $$\frac{p'(\omega)[p(\tau) - p_1(\tau)]^2}{-p^2(\tau)p_1^2(\tau)} + \frac{[p'(\tau) - p'(\omega)]p(\omega)}{p^2(\tau)p_1(\tau)}$$

$$+ p'(\omega)\frac{[p'(\tau)+p'(\omega)](\tau-\omega) - 2[p(\tau)-p(\omega)]}{p^2(\tau)p_1(\tau)}.$$

Roughly speaking, the third term here measures $\|u_1(\cdot,\lambda_j)\|_{L^1}$, while the second measures $\|u_1(\cdot,\lambda) - u_1(\cdot,\lambda_j)\|_{L^1}$ (note that $p(\omega) = \varphi(\omega) = 0$ when $\omega = \omega_j$). The first term is merely a byproduct of the rearrangement and is easy to estimate; we do this first.

Since $\partial^2 p/\partial\tau^2$ is continuous, $\text{Im } p(\omega(\lambda),\lambda) = \text{Im } p_1(\omega(\lambda),\lambda) = 0$, and $\text{Im}(\partial p/\partial\tau)(\omega(\lambda),\lambda) > 0$, we may assume that

(3.11) $$|\text{Im } p(\tau,\lambda)|, |\text{Im } p_1(\tau,\lambda)| \geq |\tau-\omega|/M \quad (|\tau-\omega| \leq \epsilon,\ |\lambda-\lambda_j| \leq \epsilon).$$

By the mean value theorem and continuity

(3.12) $$|p(\tau)-p_1(\tau)| \leq M(\tau-\omega)^2,\ |p'(\tau)-p'(\omega)| \leq M|\tau-\omega|,\ |p(\omega)| + |p'(\omega)| \leq M.$$

Therefore

(3.13) $$\int_{\omega-\epsilon}^{\omega+\epsilon} \left|\frac{p'(\omega)[p(\tau)-p_1(\tau)]^2}{-p^2(\tau)p_1^2(\tau)}\right| d\tau \leq M.$$

Since $\varphi \in C^2$, $\varphi(\omega_j) = \varphi'(\omega_j) = 0$, and $\varphi''(\omega_j) > 0$, we may assume that

(3.14) $$2\varphi''(\omega_j)(\omega-\omega_j)^2 \geq \varphi(\omega) \geq \varphi''(\omega_j)(\omega-\omega_j)^2/4,$$

$$|\varphi'(\omega)| \leq 2\varphi''(\omega_j)|\omega-\omega_j|,\ (\tau-\omega_j)\varphi'(\tau) \geq 0$$

for $|\lambda-\lambda_j| \leq \epsilon$, $|\tau-\omega| \leq \epsilon$. Then

(3.15) $$\varphi(\tau) \geq \varphi(\omega) \quad \text{if} \quad \frac{\tau-\omega_j}{\omega-\omega_j} \geq 1.$$

For τ between ω and ω_j note that

(3.16) $$\varphi(\tau) = \varphi(\omega) + \varphi'(\omega)(\tau-\omega) + \tfrac{1}{2}\varphi''(\tilde{\tau})\ (\tau-\omega)^2$$

with $\tilde{\tau}$ between ω and τ. Let

$$M_1 = \max_{\omega(\lambda_j-\epsilon) \le \tau \le \omega(\lambda_j+\epsilon)} |\varphi''(\tau)|,$$

$$M_2 = \max\{\frac{1}{32}, \frac{4\varphi''(\omega_j)}{M_1}\}.$$

Then if $|\tau-\omega| \le M_2|\omega-\omega_j|$, (3.16) and (3.14) show that

$$\begin{aligned}|\varphi(\tau) - \varphi(\omega)| &\le 4\varphi''(\omega_j)\ |\omega-\omega_j|\ |\tau-\omega| \\ &\le \varphi''(\omega_j)(\omega-\omega_j)^2/8 \\ &\le \tfrac{1}{2}\varphi(\omega).\end{aligned}$$

Combining this with (3.15), we see that

(3.17) $$\varphi(\tau) \ge \tfrac{1}{2}\varphi(\omega) \quad (\tau \in E_\lambda),$$

where $E_\lambda = \{\omega-\epsilon \le \tau \le \omega + M_2(\omega_j-\omega)\}$ if $\omega < \omega_j$ and $E_\lambda = \{\omega-M_2(\omega-\omega_j) \le \tau \le \omega + \epsilon\}$ if $\omega > \omega_j$. Let $F_\lambda = [\omega-\epsilon,\omega+\epsilon]\backslash E_\lambda$. Then by (3.11), (3.12), (3.14), and (3.17),

(3.18) $$\int_{\omega-\epsilon}^{\omega+\epsilon} \left|\frac{p(\omega)[p'(\tau)-p'(\omega)]}{p^2(\tau)p_1(\tau)}\right| d\tau$$

$$\le M\varphi(\omega)\left[\int_{E_\lambda} \frac{d\tau}{\varphi^2(\omega) + (\tau-\omega)^2} + \int_{F_\lambda} \frac{d\tau}{(\tau-\omega)^2}\right]$$

$$\le M\left[\int_{-\infty}^{\infty} \frac{\varphi(\omega)d\sigma}{\varphi^2(\omega)+\sigma^2} + 2\varphi''(\omega_j)(\omega-\omega_j)^2 \int_{M_2|\omega-\omega_j|}^{\infty} \frac{d\sigma}{\sigma^2}\right]$$

$$\le M.$$

By (3.11), the third term in (3.10) is bounded by

$$\frac{M|p'(\tau)+p'(\omega) - 2[p(\tau)-p(\omega)]/(\tau-\omega)|}{(\tau-\omega)^2}$$

and we may use the argument of [6, Lemma 3.1]. By (2.2),

(3.19) $$p(\tau,\lambda) = h(\tau,\lambda) + \frac{1}{\omega^2}\sum_{k=1}^{\infty} \frac{\delta_k}{k}(1-ik\tau-e^{-ik\tau}),$$

where $\partial^2h/\partial\tau^2$ is Lipschitz continuous in τ, uniformly in λ. Using Taylor's formula one easily shows that

(3.20) $$\left|\frac{h'(\tau)+h'(\omega)-2[h(\tau)-h(\omega)]/(\tau-\omega)}{(\tau-\omega)^2}\right| \le M$$

for $|\tau-\omega| \le \epsilon$. Let $b(\tau) = b(\tau,\lambda)$ denote the infinite sum in (3.19). By direct computation,

$$|b'(\tau) + b'(\omega) - 2[b(\tau)-b(\omega)]/(\tau-\omega)|$$

$$\le \sum_{k=1}^{\infty} \delta_k J[(\tau-\omega)k]$$

where $J \in C^2$ with $J(0) = J'(0) = 0, J''(0) = -\frac{1}{3}$, and $|J(x)| \le M$. Choose $\mu > 0$ such that $|J(x)| \le x^2$ $(|x| \le \mu)$. Then (recall (2.3))

$$\int_{\omega-\epsilon}^{\omega+\epsilon} (\tau-\omega)^{-2}|b'(\tau) + b'(\omega) - 2(\tau-\omega)^{-1}[b(\tau)-b(\omega)]|d\tau$$

$$\le \int_{\omega-\epsilon}^{\omega+\epsilon} (\tau-\omega)^{-2}[\sum_{k|\tau-\omega|\le\mu} \delta_k k^2(\tau-\omega)^2 + M \sum_{k|\tau-\omega|>\mu} \delta_k]d\tau$$

$$\le \sum_{k=1}^{\infty} \delta_k[k^2 \int_{\substack{k|\tau-\omega|\le\mu \\ |\tau-\omega|\le\epsilon}} d\tau + M \int_{\substack{k|\tau-\omega|>\mu \\ |\tau-\omega|\le\epsilon}} (\tau-\omega)^{-2}d\tau$$

$$\le \sum_{k=1}^{\infty} \delta_k[2\mu k + 2Mk/\mu] \le M$$

Combining this with (3.20), (3.18), and (3.13), we get (3.9). This completes the proof of Theorem 1.

REFERENCES

1. R. W. Carr and K. B. Hannsgen, A nonhomogeneous integrodifferential equation in Hilbert space, SIAM J. on Math. Anal., to appear.

2. P. L. Duren, <u>Theory of H^p Spaces</u>, Academic Press, New York, 1970.

3. S. I. Grossman and R. K. Miller, Nonlinear Volterra integrodifferential systems with L^1-kernels, J. Differential Equations 13(1973), 551-566.

4. K. B. Hannsgen, Indirect abelian theorems and a linear Volterra equation, Trans. Amer. Math. Soc., 142(1969), 539-555.

5. ____________, Uniform L^1 behavior for an integrodifferential equation with parameter, SIAM J. on Math. Anal. 8 (1977), 626-639.

6. ____________, An L^1 remainder theorem for an integrodifferential equation with asymptotically periodic solution, Proc. Amer. Math. Soc., to appear.

7. D. F. Shea and S. Wainger, Variants of the Wiener-Lévy theorem, with applications to stability problems for some Volterra integral equations, Amer. J. Math. 97 (1975), 312-343.

FUNCTIONAL DIFFERENTIAL EQUATIONS WITH DISCONTINUOUS RIGHT HAND SIDE

T. L. Herdman and J. A. Burns*
Department of Mathematics
Virginia Polytechnic Institute and State University
Blacksburg, Virginia 24061/U.S.A.

*This research was supported in part by the Army Research Office under grant DAAG-29-78-G-0125.

1. Introduction.

Suppose that $T > 0$ and $x:(-\infty,\sigma + T) \to R^n$ is a given function. For $t \in [\sigma,\sigma+T)$, we define the function $x_t:(-\infty,0) \to R^n$ by $x_t(s) = x(t+s)$. In this paper we consider some fundamental questions regarding the functional differential equation with infinite delay

$$\dot{x}(t) = F(t,x(t),x_t). \tag{1.1}$$

Particular cases of equation (1.1) are wide classes of differential-difference equations and integro-differential equations of Volterra type. The principal objective of the present work is to obtain existence, uniqueness and continuous dependence for (1.1) under minimal assumptions on the functional F. In particular, if z denotes (η,φ), then we wish to study (1.1) without the standard continuity assumptions on the map $z \to F(t,z) = F(t,\eta,\varphi)$. The results are applicable to a large class of differential-difference equations and certain integro-differential equations with "singular" kernels.

In order to discuss the initial value problem associated with (1.1) it is necessary to select a space of initial conditions, i.e. a phase space. In many applications the appropriate choice for the phase space is usually dictated by the form of the equation and the particular objectives of the study (i.e. numerical approximation, control, qualitative theory, etc.). Generally speaking, the phase spaces most frequently used are either spaces of continuous functions (see [7] [8], [9], [14]) or a $\mathcal{L}_p$-type space with an appropriate measure μ on $(-\infty,0]$ (see [1], [2], [3], [4], [5], [6]). Recently, the theory of functional differential equations has been studied using a phase space which satisfy some very general axioms (see [9], [15]).

Regardless of which phase space used, a minimal continuity assumption on F has always been one of two types; i) the classical hypothesis that the map $(t,z) \to F(t,z)$ is continuous or ii) the Carathéodory type assumptions (see [6]) which imply that for almost all t the map $z \to F(t,z)$ is continuous. Any continuity requirement of this type (while natural for ordinary differential equations) can eliminate a large class of functional differential equations.

In Section 2 we treat the initial value problem for (1.1) by using the history spaces of Coleman and Mizel [3] as phase spaces. We restrict ourselves to these spaces because it allows us to present the basic ideas in a concrete (yet general) setting without getting bogged down in details that would be necessitated by the abstract problem. Existence, uniqueness and continuous dependence are obtained for (1.1) without assuming continunity of the map $z \to F(t,z)$.

Examples and concluding remarks are given in Section 3 and outlines of proofs of the theorems are presented in Section 4.

2. The Initial Value Problem In History Spaces.

History spaces and spaces with fading memory were introduced in 1966 by Coleman and Mizel [3], and have been used in numerous studies of functional differential equations. The interested reader is refered to the paper by Coleman and Owen [5] and the recent dissertation of Lima [14] for a complete discussion of these spaces.

Suppose that $g: I \to R$ is a locally integrable function defined on some interval $I \subseteq (-\infty,+\infty)$ and $g(s) > 0$ a.e. on I. Denote by $\mathcal{L}_p^n(I;g)$ the usual Lebesgue space of equivalence classes of functions $x: I \to R^n$ such that $\int_I |x(s)|^p g(s)ds < +\infty$. If μ is a measure defined on I, then the space of μ-measurable "functions" satisfying $\int_I |x(s)|^p d\mu(s) < +\infty$ will be denoted by $\mathcal{L}_p^n(I;d\mu)$.

A locally integrable function $g:(-\infty,0) \to R^n$ is called an <u>influence function</u> if g satisfies;

(I) $g(s) > 0$ a.e. on $(-\infty,0)$,

(II) $G(t) = \operatorname*{ess.\,sup}_{-\infty < s < 0} \dfrac{g(s-t)}{g(s)}$ is finite for each $t \geq 0$,

(III) $H(t) = \operatorname*{ess.\,sup}_{-\infty < s < -t} \dfrac{g(s+t)}{g(s)}$ is finite for each $t \geq 0$.

In all that follows we assume that g is an influence function on $(-\infty,0)$ and define $\bar{g}:(-\infty,+\infty) \to R$ by

$$\bar{g}(s) = \begin{cases} g(s) & , \quad -\infty < s < 0, \\ 1 & , \quad 0 \leq s < +\infty. \end{cases}$$

<u>Remark 2.1</u>. If g is an influence function on $(-\infty,0)$, then it is known (see [4] [5]) that there are positive constants a,b,c,d such that

$$ae^{bs} < g(s) < ce^{-ds} \quad \text{a.e. on } (-\infty,0).$$

Moreover, if g is integrable on $(-\infty,0)$, then there is a constant M such that $g(s) < M$ a.e. on $(-\infty,0)$ and $\lim_{s \to -\infty} sg(s) = 0$. We shall see later that because of these bounds on g, the requirement that the map $z \to F(t,z)$ be continuous greatly reduces the class of admissible functions F.

We take as our phase space the space

$$Z_p^n(g) = R^n \times \mathcal{L}_p^n((-\infty,0);g)$$

with norm

$$\|(\eta,\varphi)\|^p = |\eta|^p + \int_{-\infty}^{0} |\varphi(s)|^p g(s)ds.$$

If μ is the measure defined on $(-\infty,0]$ by

$$\mu(E) = \begin{cases} \int_E g(s)ds & , \quad E \subsetneq (-\infty,0) \text{ measurable}, \\ 1 & , \quad E = \{0\}, \end{cases}$$

then it is clear that the product space $Z_p^n(g)$ is equivalent to the Lebesgue space $\mathcal{L}_p^n((-\infty,0];d\mu)$. Consequently, if $x:(-\infty,\sigma + T) \to R^n$ is given and $x_t \in \mathcal{L}_p^n((-\infty,0],d\mu)$, then x_t as an element of $\mathcal{L}_p^n((-\infty,0];d\mu)$ may be identified with the element $z(t) = (x(t),x_t)$ in the space $Z_p^n(g)$. This representation can prove to be very useful (see for example [2]).

Suppose that $\Omega = [\sigma,+\infty) \times \mathcal{D}$ where $\mathcal{D}$ is a dense subset of $Z_p^n(g)$. We shall be interested in functions $F:\Omega \to R^n$ that satisfy the following hypothesis.

($\mathcal{H}$) 1. _If_ $x \in \mathcal{L}_p^n((-\infty,\sigma + T); \bar{g})$ _is such that_ x _is continuous on_ $[\sigma,\sigma + T)$, _then the map_

$$q(t) = F(t,x(t),x_t)$$

is well defined a.e. _on_ $[\sigma,\sigma + T)$, _depends only on the equivalence class of_ x _and_ $q(t)$ _is integrable on_ $[\sigma,\sigma + T)$.

2. _Given_ $\beta > 0$ _there is a locally bounded measurable function_ $\Gamma_\beta:[\sigma,+\infty) \to R$ _such that for any_ $T > 0$, _the following condition holds:_

_If_ $x,y \in \mathcal{L}_p^n((-\infty,\sigma + T);\bar{g})$ _are continuous on_ $[\sigma,\sigma +T)$ _and_ $\|x\| \leq \beta$, $\|y\| \leq \beta$, _then_

$$\int_\sigma^t |F(s,x(s),x_s) - F(s,y(s),y_s)|ds \leq \Gamma_\beta(t)[\int_{-\infty}^t |x(s)-y(s)|^p \bar{g}(s)ds]^{1/p},$$

for all $t \in [\sigma,\sigma + T)$.

Remark 2.2. It is important to note that hypothesis ($\mathcal{H}$) _does not imply_ that the map $z \to F(t,z)$ is continuous. In fact, $F(t,z)$ may not be defined for all $z \in Z_p^n(g)$. Consequently, hypothesis ($\mathcal{H}$) is a much weaker condition than the usual Lipschitzian and Carathéodory type assumption. Hypothesis ($\mathcal{H}$) was first suggested by Borisovic and Turbabin [1] for linear finite delay equations. The nonlinear finite delay case was recently treated by Kappel and Schappacher [10]. A form of hypothesis ($\mathcal{H}$) was employed by Burns and Herdman [2] in their study of certain semi-groups generated by linear infinite delay equations.

Given a function $F:\Omega \to R^n$, a locally integrable function h, an initial time σ and an initial pair $z = (\eta,\varphi) \in Z_p^n(g)$, consider the initial value problem defined by the equation

$$(2.1)\qquad \dot{x}(t) = F(t,x(t),x_t) + h(t) \quad , \quad t \geq \sigma$$

with initial condition

$$(2.2)\qquad x(\sigma) = \eta \quad , \quad x_\sigma = \varphi .$$

A solution to (2.1) - (2.2) is a function $x \in \mathcal{L}_p^n((-\infty,\sigma + T);\bar{g})$ such that x is absolutely continuous on $[\sigma,\sigma + T)$, $x(\sigma) = \eta$, $x_\sigma(s) = \varphi(s)$ a.e. on $(-\infty,0)$ and x satisfies (2.1) a.e. on $[\sigma,\sigma + T)$.

The main results of this paper are summarized in the following theorems.

THEOREM (A). *Suppose hypothesis* ($\mathcal{H}$) *holds for* F *and* $(\sigma,z) \in \Omega$ *is given. If* h *is a locally integrable function, then there is a* $T > 0$ *such that the initial value problem* (2.1) - (2.2) *has a unique solution* $x(t;z,h)$ *existing on the interval* $(-\infty,\sigma + T)$.

Remark 2.3. If in hypothesis ($\mathcal{H}$) $\Gamma_\beta(t)$ is independent of β (i.e. $\Gamma_\beta(t) = \Gamma(t)$ for all $\beta > 0$), then the solutions of (2.1) - (2.2) exist globally.

THEOREM (B). *Suppose hypothesis* ($\mathcal{H}$) *holds for* F *and* $x(t) = x(t;z,h)$, $y(t) = y(t;\hat{z},\hat{h})$ *are two solutions existing on the same interval* $(-\infty,\sigma+T)$, *then there are constants* k_0, k_1, λ *such that for all* $t \in [\sigma,\sigma + T)$

$$|x(t)-y(t)|^p \leq (k_0\|z - \hat{z}\| + k_1 \|h-\hat{h}\|^p)e^{\lambda t}.$$

Remark 2.4. In [10] Kappel and Schappacher make an interesting observation regarding hypothesis ($\mathcal{H}$) - 2. They note that if ($\mathcal{H}$) - 2 holds only for those functions x,y satisfying $x(t) = y(t)$ on $(-\infty,\sigma)$, then the proof of Theorem A remains valid and one obtains existence and uniqueness. However, this weakened form of ($\mathcal{H}$) - 2 is not sufficient to imply continuous dependence. They cite the example

$$\dot{x}(t) = \mathrm{sgn}[x(t-1)]$$

which does satisfy the weak version of ($\mathcal{H}$) - 2, but does not satisfy ($\mathcal{H}$) - 2.

3. An Illustrative Example.

Consider the Volterra integro-differential equation

$$(3.1)\qquad \dot{x}(t) = ax(t) + bx(t-1) + \int_{-\infty}^{0} k(s)x(t+s)ds,$$

with initial condition

$$(3.2)\qquad x(0) = \eta \quad , \quad x_0 = \varphi .$$

We assume that the function k(s) is integrable on $(-\infty,0)$. It is desirable to discuss this equation in a phase space setting and allow initial pairs (η,φ) where φ is not necessarily continuous. In particular, we would like to find a $p \in [1,+\infty)$ and an influence function g such that (3.1) - (3.2) can be studied in the phase space $\mathcal{Z}_p(g) = R \times \mathcal{L}_p((-\infty,0);g)$. If we define F by

$$F(t,\eta,\varphi) = a\eta + b\varphi(-1) + \int_{-\infty}^{0} k(s)\varphi(s)ds, \tag{3.3}$$

then equation (3.1) is equivalent to the functional differential equation

$$\dot{x}(t) = F(t,x(t),x_t).$$

Since (3.1) - (3.2) is a special case of the system (2.1) - (2.2), one would hope that existence theory developed for (2.1) - (2.2) would apply to (3.1) - (3.2). It is easy to check that F defined by (3.3) satisfies hypothesis ($\mathcal{H}$) for $g(s) \equiv 1$ and any $p \in [1,+\infty)$, consequently existence uniqueness and continuous dependence are guaranteed by Theorems A and B.

On the other hand, suppose that we require that the map $z = (\eta,\varphi) \to F(t,z)$ be continuous for almost all $t \geq 0$, and attempt to formulate (3.1) - (3.2) as an initial value problem in some $\mathcal{Z}_p(g)$ space. It is obvious that such a requirement can not be satisfied for any space $\mathcal{Z}_p(g)$, since the function $F(t,\eta,\varphi)$ involves point evaluation of φ at -1. This problem can be eliminated in two ways; either change state spaces and use some appropriate space of continuous functions or let $b = 0$ and not consider equations with discrete delays. By allowing discontinuous initial functions, we are forced to restrict ourselves to equations of the form

$$\dot{x}(t) = ax(t) + \int_{-\infty}^{0} k(s)x(t+s)ds. \tag{3.4}$$

However, there is another possible problem.

Let k(s) be defined by

$$k(s) = \begin{cases} \frac{1}{\sqrt{-s}} & -1 \leq s < 0, \\ e^{s+1} & -\infty < s \leq -1, \end{cases} \tag{3.5}$$

and suppose that it is desired to consider the equation

$$\dot{x}(t) = \int_{-\infty}^{0} k(s)x(t+s)ds,$$

with initial pair (η,φ) in a Hilbert space of the form $\mathcal{Z}_2(g)$. There will exist no influence function g such that the function $F(t,\eta,\varphi) = F(\varphi) = \int_{-\infty}^{0} k(s)\varphi(s)ds$ is continuous on $\mathcal{Z}_2(g)$!

In order to see how restrictive the continuity assumption can be, consider the function

$$k(s) = \begin{cases} (\frac{1}{s})[\ell n(2/-s)]^{-2} & -1 \leq s < 0 \\ -e^{s+1}/[\ell n(2)]^2 & -\infty < s \leq -1. \end{cases} \tag{3.6}$$

It is easy to show that k is integrable, and yet for any $q > 1, \int_{-1}^{0} |k(s)|^q ds = +\infty$.

If the map $F(\varphi) = \int_{-\infty}^{0} k(s)\varphi(s)ds$ is to be continuous on some space $Z_p(g)$, then the Riesz representation theorem implies that $k(t) \in \mathcal{L}_q((-\infty,0);g)$, where $q > 1$ satisfies $1/p + 1/q = 1$. Since g is an influence function there are positive constants a,b such that $ae^{bs} < g(s)$ a.e. on $[-1,0)$. It follows that

$$\int_{-\infty}^{0} |k(s)|^q g(s)ds \geq ae^{-b} \int_{-1}^{0} |k(s)|^q ds = +\infty,$$

and hence k(t) can not belong to $\mathcal{L}_q((-\infty,0);g)$ for any $q > 1$ and influence function g. Consequently, for the function k(t) defined in (3.6) it is not possible to find any history space $Z_p(g)$ such that the initial value problem

$$\dot{x}(t) = \int_{-\infty}^{0} k(s)x(t+s)ds$$

$$x(0) = \eta \quad , \quad x_0 = \varphi$$

will have the right hand side continuous.

Equation (3.1) is a special case of certain nonlinear equations considered among others by Levin [12] and Levin and Shea [13]. Therefore it is clear from the linear case above that if these equations are formulated as functional differential equations in history spaces, then the resulting right hand side F may not have the map $z \to F(t,z)$ continuous (even in the Carathéodory sense). Consequently, any existence result that assumes this continuity condition on F may not be applicable to such equations.

Finally we mention that Leitman and Mizel [11], have considered a problem that is related to this paper in that they consider functions defined on certain fading memory spaces that in some circumstances are not continuous.

4. Proofs of Theorems.

Proof of Theorem A. Let $T_0 > 0$ be given and for $(\eta,\varphi) \in R^n \times \mathcal{L}_p^n((-\infty,\sigma);g)$ define

$$\bar{\eta}(s) = \begin{cases} \varphi(s-\sigma) & \text{for } -\infty < s < \sigma, \\ \eta & \text{for } \sigma \leq s < T_0 + \sigma, \end{cases}$$

$$S_1(\sigma,\sigma + T_0) = \{\psi \in C([\sigma,\sigma + T_0];R^n) \,|\, \|\psi-\eta\|_\infty \leq 1\}$$

and for $\psi \in S_1(\sigma,\sigma + T_0)$

$$\bar{\psi}(s) = \begin{cases} \varphi(s-\sigma) & \text{for } -\infty < s < \sigma, \\ \psi(s) & \text{for } \sigma \leq s \leq \sigma + T_0. \end{cases}$$

For $\psi \in S_1(\sigma,\sigma + T_0)$ we define the operator V by

$$[V\psi](t) = \eta + \int_\sigma^t F(s,\psi(s),\overline{\psi}_s)\ ds + \int_\sigma^t h(s)ds,$$

where $t \in [\sigma,\sigma + T_0]$. In view of hypothesis (H), we have that $V\psi$ exists and is absolutely continuous on $[\sigma,\sigma + T_0]$.

For $\psi \in S_1(\sigma,\sigma + T_0)$ the function $\overline{\psi}$ is in $\mathcal{L}_p^n((-\infty,\sigma + T_0);\overline{g})$ and we have

$$\begin{aligned}\|\overline{\psi}\|^p &= \int_{-\infty}^{\sigma+T_0} |\overline{\psi}(s)|^p\overline{g}(s)ds \\ &\le \|\varphi\|^p + T_0\ \|\psi\|_\infty^p \\ &\le \|\varphi\|^p + (|\eta| + 1)^p T_0 \\ &\le \|\varphi\|^p + 2^p(|\eta|^p + 1)T_0,\end{aligned}$$

thus $\|\overline{\psi}\| \le \beta$ where $\beta = (\|\varphi\|^p + 2^p(|\eta|^p + 1)T_0)^{1/p}$. Indeed, for any $T_1 \in (0,T_0]$ it follows that for $t \in [\sigma,\sigma + T_1]$

$$\begin{aligned}|[V\psi](t)-\eta| &\le \left|\int_\sigma^t F(s,\psi(s),\overline{\psi}_s) - F(s,\eta,\overline{\eta}_s)ds\right| \\ &\quad + \int_\sigma^t f(s,\eta,\overline{\eta}_s)ds + \int_\sigma^t h(s)ds\Big| \\ &\le \sup_{t\in[\sigma,\sigma+T_1]} \{\Gamma_\beta(t)\}(T_1)^{1/p} \\ &\quad + \int_\sigma^{\sigma+T_1} |f(s,\eta,\overline{\eta}_s|ds + \int_\sigma^{\sigma+T_1} |h(s)|ds.\end{aligned}$$

Also, for any $T_1 \in (0,T_0]$ and $\psi,\chi \in S_1(\sigma,\sigma + T_0)$ we have that for $t \in [\sigma,\sigma + T_1]$

$$|[V\psi](t) - [V\chi](t)| \le [\sup_{t\in[\sigma,\sigma+T_1]} \{\Gamma_\beta(t)\}(T_1)^{1/p}][\sup_{t\in[\sigma,\sigma+T_1]} \{|\psi(t)-\chi(t)|\}].$$

To obtain the desired result we pick $T_1 \in (0,T_0]$ sufficiently small so that $V:S_1(\sigma,\sigma + T_1) \to S_1(\sigma,\sigma + T_1)$ is a contraction and thus has a unique fixed point.

<u>Proof of Theorem B</u>. For the solutions $x(t)$ and $y(t)$ we define $\beta = \max\{\|x\|,\|y\|\}$. Hypothesis (H) together with some elementary inequalities implies that for $t \in [\sigma,\sigma+T]$

$$\begin{aligned}|x(t)-y(t)| \le |\eta-\hat{\eta}| &+ 2^{1/p} \sup_{t\in[\sigma,\sigma+T]} \{\Gamma_\beta(t)\}[\|\varphi-\hat{\varphi}\| + (\int_\sigma^t |x(s)-y(s)|^p ds)^{1/p}] \\ &+ T^{\frac{p-1}{p}} \|h-\hat{h}\|.\end{aligned}$$

We now define $m = \max\left\{2^{1/p} \sup_{t\in[\sigma,\sigma+T]} \{\Gamma_\beta(t)\},1\right\}$ and note that

$$\begin{aligned}|x(t)-y(t)|^p \le\ & 2^p[m(|\eta-\hat{\eta}| + \|\varphi-\hat{\varphi}\| + T^{\frac{p-1}{p}}\|h-\hat{h}\|]^p \\ & + 2^p m^p \int_\sigma^t |x(s)-y(s)|^p ds\end{aligned}$$

for $t \in [\sigma, \sigma + T]$. An application of Gronwalls inequality gives the desired result where

$$k_0 = 2^{3p} m^p e^{-2^p m^p \sigma}$$

$$k_1 = 2^{2p} (T)^{p-1} e^{-2^p m^p \sigma}$$

and

$$\lambda = 2^p m^p.$$

References

[1] J. G. Borisovič and A. S. Turbabin, On the Cauchy problem for linear non-homogeneous differential equations with retarded argument, Soviet Math. Doklady, 10 (1969), 401-405.

[2] J. A. Burns and T. L. Herdman, Adjoint semigroup theory for a class of functional differential equations, SIAM J. Math. Anal., 7 (1976), 729-745.

[3] B. D. Coleman and V. J. Mizel, Norms and semigroups in the theory of fading memory, Arch. Rational Mech. Anal., 23 (1966), 87-123.

[4] B. D. Coleman and V. J. Mizel, On the general theory of fading memory, Arch. Rational Mech. Anal., 29 (1968), 18-31.

[5] B. D. Coleman and D. R. Owen, On the initial value problem for a class of functional-differential equations, Arch. Rational Mech. Anal., 55 (1974), 275-299.

[6] M. C. Delfour and S. K. Mitter, Hereditary differential systems with constant delays. I. General Case, J. Diff. Eqs., 12 (1972), 213-235.

[7] J. K. Hale, Functional differential equations with infinite delays, J. Math. Anal. Appl., 48 (1974), 276-283.

[8] J. K. Hale and C. Imaz, Existence, continuity and continuation of solutions of retarded differential equations, Bol. Soc. Mat. Mex. (1967), 29-37.

[9] J. K. Hale and J. Kato, Phase space for retarded equations with infinite delay, (1978), preprint.

[10] F. Kappel and W. Schappacher, Autonomous nonlinear functional differential equations and averaging approximations, (1977), preprint.

[11] M. J. Leitman and V. J. Mizel, On fading memory spaces and hereditary integral equations, Arch. Rational Mech. Anal., 55 (1974), 18-51.

[12] J. J. Levin, Boundedness and oscillation of some Volterra and delay equations, J. Diff. Eqs., 5 (1969), 369-398.

[13] J. J. Levin and D. F. Shea, On the asymptotic behavior of the bounded solutions of some integral equations, I, II, III, J. Math. Anal. Appl., 37 (1972), 42-82, 288-326, 537-575.

[14] P. F. Lima, Hopf bifurcation in equations with infinite delays, Ph.D. Thesis, Brown University, Providence, R. I., June 1977.

[15] T. Naito, On linear autonomous retarded equations with an abstract phase space for infinite delay, (1978), preprint.

SOME NONLINEAR SINGULARLY PERTURBED VOLTERRA INTEGRODIFFERENTIAL EQUATIONS

G.S. Jordan
University of Tennessee

Knoxville, TN 37916/USA

1. Introduction

In a recent paper [7] Lodge, McLeod, and Nohel study the nonlinear Volterra integrodifferential equation

$$(1.1)\qquad \begin{cases} -\mu y'(t) = \int_{-\infty}^{t} a(t-s)F(y(t), y(s))ds & (t > 0;\ ' = d/dt), \\ y(t) = g(t) & (-\infty < t \leq 0), \end{cases}$$

where μ is a small positive parameter, a is a given real kernel, and F, g are given real functions. They discuss qualitative properties of the solutions of (1.1) and of the reduced equation, obtained by setting $\mu = 0$, and the relation between those solutions as $\mu \to 0+$, both for large t and for t near zero where a boundary layer occurs.

For particular choices of the functions a, F, and g, the initial value problem (1.1) models the elongation ratio of a homogeneous polyethylene filament which is stretched on the time interval $(-\infty, 0]$, released, and allowed to undergo elastic recovery.

Our purpose here is to describe some results analogous to those of [7] for the corresponding nonconvolution problem

$$(1.2)\qquad \begin{cases} -\mu y'(t) = \int_{-\infty}^{t} b(t, s)F(y(t), y(s))ds & (t > 0), \\ y(t) = g(t) & (-\infty < t \leq 0), \end{cases}$$

where $b(t, s)$ is a given real kernel and μ, F, and g are as above. Most of our assumptions concerning $b(t, s)$ are natural generalizations of those of [7] concerning $a(t)$; in fact, if $a(t)$ satisfies the hypotheses of [7], then $b(t, s) \equiv a(t - s)$ satisfies our hypotheses. The assumptions on F and g in [7] are left unchanged.

Many of the arguments of [7] do not depend on the convolution nature of problem (1.1); thus, once the essential properties of $a(t)$ are identified and suitably generalized to $b(t, s)$, these arguments may be extended to problem (1.2). However, some of the arguments of [7] do depend on the convolution nature of (1.1) and, hence, must be replaced with new arguments for the nonconvolution problem.

Other results for Volterra integral and integrodifferential equations of non-convolution type may be found in, for example, [1], [3], [4], [5], [6], [8], [9].

In Section 2 we state and discuss several results for equation (1.2) and the associated reduced equation. Section 3 contains some examples, and in Section 4 a typical new argument needed for the nonconvolution problem is presented. Additional results and an expanded discussion will appear in [2].

2. Statement and discussion of results

Let $\mathbb{R}$ and $\mathbb{R}^+$ denote the real numbers and the positive real numbers, respectively, and let C^k be the set of k times continuously differentiable functions. The hypotheses on F and g are those of [7]:

$$\text{H}(F) \qquad \begin{cases} F: \mathbb{R}^+ \times \mathbb{R}^+ \to \mathbb{R}\,; \; F(x, x) = 0 \quad \text{for every} \quad x > 0, \\ F \in C^1(\mathbb{R}^+ \times \mathbb{R}^+) \quad \text{and} \quad F_1(y, z) > 0\,, \quad F_2(y, z) < 0 \; (y, z \in \mathbb{R}^+)\,, \\ \text{where the subscripts denote partial differentiation;} \end{cases}$$

and

$$\text{H}(g) \qquad \begin{cases} g: (-\infty, 0] \to \mathbb{R}^+\,; \; g(-\infty) = 1\,, \; g(0) > 1\,; \\ g \in C(-\infty, 0] \quad \text{and} \quad g \quad \text{is nondecreasing.} \end{cases}$$

The condition $H(g)$ is the only one required of g throughout the paper; however, a stronger smoothness assumption will be imposed on F in Theorem 6.

For purposes of comparison we include the basic assumption on the kernel $a(t)$ in [7]:

$$\text{H}(a) \qquad \begin{cases} a \in C^1[0, \infty)\,; \; a(t) > 0\,, \; a'(t) < 0 \qquad (0 \le t < \infty)\,; \\ a \in L^1(0, \infty); \; \log a(t) \quad \text{is convex} \quad (0 \le t < \infty), \text{ i.e.,} \\ a'(t)/a(t) \quad \text{is nondecreasing.} \end{cases}$$

The hypothesis $H(a)$ is sufficient for most of the results in [7]; however, a moment condition on the kernel is needed in some of the theorems.

Now let $D = \{(t, s) \mid 0 \le t < \infty\,, \; -\infty < s \le t\}$. We replace $H(a)$ in part with the underlying hypothesis

$$H_1(b) \quad \begin{cases} b(t, s),\ b_1(t, s) \in C(D)\ ;\ 0 < b(t, s) \le B\ ,\ b_1(t, s) < 0\ , \\ \text{and } b_1(t, s)/b(t, s) \ge -\sigma_0\ ((t, s) \in D);\ \text{for each } t \ge 0 \\ b(t, \cdot) \in L^1(-\infty, t)\ ,\ \int_{-\infty}^{t} b(t, s)\, ds \ge b > 0\ ,\ \text{and} \\ b_1(t, s)/b(t, s) \text{ is nonincreasing in } s\ ;\ \text{either} \\ \text{(i) } b_1(t, s)/b(t, s) \text{ is not constant in } s \text{ for any } t \ge 0\ , \text{ or} \\ \text{(ii) } b_1(t, s)/b(t, s) \text{ is constant in } s \text{ for each } t \ge 0\ . \end{cases}$$

Here, B, σ_0, and b are finite positive constants.

The basic hypotheses $H_1(b)$, H(F) , and H(g) are sufficient to deduce several useful properties of the solution of (1.2).

Theorem 1. *Let $H_1(b)$, H(F), H(g) be satisfied. Then for each $\mu > 0$, the initial value problem (1.2) has a unique solution $\phi(t, \mu)$ on $[0, \infty)$ having the following properties:*

$$(2.1) \qquad \phi'(t, \mu) < 0 \ \text{and}\ 1 < \phi(t, \mu) \le g(0) \quad (0 \le t < \infty);$$

$$(2.2) \qquad \begin{cases} \text{if } g_1, g_2 \text{ satisfy } H(g) \text{ and if } g_1(t) \ge g_2(t) \quad (-\infty < t \le 0)\ , \\ \text{then the corresponding solutions } \phi_1(t, \mu),\ \phi_2(t, \mu) \text{ of (1.2)} \\ \text{satisfy } \phi_1(t, \mu) \ge \phi_2(t, \mu) \quad (0 \le t < \infty)\ ; \end{cases}$$

$$(2.3) \qquad \begin{cases} \text{if } \mu_1 > \mu_2\ , \text{ then the corresponding solutions of (1.2) satisfy} \\ \phi(t, \mu_1) > \phi(t, \mu_2) \quad (0 < t < \infty)\ . \end{cases}$$

Applying the classical Picard successive approximations to the integrated form of (1.2) yields a unique local solution $\phi(t, \mu)$ in C^1 . By establishing (2.1) on any interval of existence of $\phi(t, \mu)$, one may continue the solution uniquely to the interval $[0, \infty)$. To prove (2.2), subtract equations (1.2) for ϕ_1 and ϕ_2 , apply the mean value theorem to F , and then use H(F) to show that there is no last point $t = t_0 \ge 0$ for which $\phi_1(t, \mu) \ge \phi_2(t, \mu)$. Finally, (2.3) may be established in a similar manner by showing that $z(t) \equiv \phi(t, \mu_1) - \phi(t, \mu_2)$ satisfies $z(0) = 0$, but that there cannot exist $T > 0$ such that $z(T) = 0$. For details, see the proofs for the convolution case in [7].

A simple but important consequence of (2.1) and (2.3) is

Corollary 1.1. *Let $H_1(b)$, H(F), H(g) be satisfied. Then for each fixed* $\mu > 0$

$$(2.4) \qquad \alpha(\mu) = \lim_{t\to\infty} \phi(t, \mu) \ \text{exists and}\ \alpha(\mu) \ge 1\ ;$$

moreover, if $\mu_1 > \mu_2$, *then* $\alpha(\mu_1) \ge \alpha(\mu_2) \ge 1$.

If alternative (ii) holds in $H_1(b)$, then $b(t, s) = u(t)v(s)$ where $v \in L^1(-\infty, t)$ for each $t \geq 0$ and $v(s) > 0 (-\infty < s < \infty)$. It follows from $H(F)$, $H(g)$, and the intermediate value theorem that there exists a unique number y_0 in the interval $(1, g(0))$ such that $\int_{-\infty}^{0} v(s) F(y_0, g(s))\, ds = 0$. As a consequence of the proof of Theorem 1, in this special case conclusions (2.1) and (2.4) may be strengthened to the following:

(2.1') $$\phi'(t, \mu) < 0 \text{ and } 1 < y_0 < \phi(t, \mu) \leq g(0) \quad (0 \leq t < \infty),$$

(2.4') $$\alpha(\mu) = \lim_{t\to\infty} \phi(t, \mu) \text{ exists and } \alpha(\mu) \geq y_0 > 1 .$$

In Theorem 4 we shall establish under additional assumptions a result which implies that $\alpha(\mu) > 1$ when alternative (i) holds in $H_1(b)$.

An analysis of the equation for ϕ'' obtained by differentiating in (1.2) provides useful estimates on ϕ' and ϕ.

Theorem 2. Let $H_1(b)$, $H(F)$, $H(g)$ be satisfied. Then there exists a constant $K_1 > 0$ (independent of μ) and constants $\mu_0 > 0$ and $\tilde{K} = \tilde{K}(\mu_0) > 0$ such that the solution $\phi(t, \mu)$ of (1.2) satisfies

(2.5) $$0 < -\phi'(t, \mu) \leq \frac{\tilde{K}}{\mu} \exp(-K_1 t/\mu) + \tilde{K} \int_{-\infty}^{0} b(t, s)ds \quad (0 \leq t < \infty; 0 < \mu \leq \mu_0) .$$

Integrating the inequality (2.5) yields the first conclusion of

Corollary 2.1. Let $H_1(b)$, $H(F)$, $H(g)$ be satisfied and assume

$H_2(b)$ $$\int_0^{\infty} \int_{-\infty}^{0} b(t, s)\, ds\, dt < \infty .$$

Then there exist constants $\mu_1 > 0$ and $\bar{K} = \bar{K}(\mu_1) > 0$ such that

(2.6) $$0 < \phi(t, \mu) - \alpha(\mu) \leq \bar{K} \exp(-K_1 t/\mu) + \bar{K} \int_t^{\infty} \int_{-\infty}^{0} b(u, s)ds\, du \quad (0 \leq t < \infty; 0 < \mu \leq \mu_1).$$

If, in addition,

$H_3(b)$ $$\int_{-\infty}^{0} b(t, s)ds \geq \alpha \exp(-\beta t) \quad (\alpha > 0, \beta > 0; 0 \leq t < \infty) ,$$

then there exist constants $\mu_0 > 0$ and $K = K(\mu_0) > 0$ such that

(2.7) $$0 < \phi(t, \mu) - \alpha(\mu) \leq K \int_t^{\infty} \int_{-\infty}^{0} b(u, s)ds\,du \quad (0 \leq t < \infty; 0 < \mu \leq \mu_0).$$

The hypothesis $H_2(b)$ may be interpreted as a generalized moment condition, for in the convolution case $b(t, s) = a(t - s)$ a change of variable and an interchange in the order of integration show that the integral in $H_2(b)$ is simply $\int_0^\infty ta(t)dt$. Thus, $H_2(b)$ corresponds to assumption (2.6) in [7]. In addition, in the convolution case the logarithmic convexity of $a(t)$ assumed in $H(a)$ implies that $H_3(b)$ holds. Finally, we remark that in the convolution case the conclusions (2.5), (2.7) are precisely the estimates (2.5), (2.7) of [7].

For small $\mu > 0$ the initial value problem (1.2) may be regarded as a singular perturbation of the reduced equation

$$(2.8) \qquad \begin{cases} 0 = \int_{-\infty}^{t} b(t, s)F(y(t), y(s))ds & (t > 0) \\ y(t) = g(t) & (-\infty < t < 0) . \end{cases}$$

The next theorem describes several useful properties of the solution of (2.8).

Theorem 3. Let $H_1(b)$, $H(F)$, $H(g)$ be satisfied. Then (2.8) has a unique (continuous) solution ϕ_0 on $[0, \infty)$ having the following properties:

(2.9) if $y_0 = \phi_0(0)$, then $1 < y_0 < g(0)$;

if alternative (i) of $H_1(b)$ holds, then

(2.10) $\phi_0 \in C^1[0, \infty)$, $\phi_0'(t) < 0$ and $1 < \phi_0(t) \leq y_0$ $(0 \leq t < \infty)$;

if alternative (ii) of $H_1(b)$ holds, then $\phi_0(t) \equiv y_0$ $(0 \leq t < \infty)$;

(2.11) if g_1, g_2 satisfy $H(g)$ and if $g_1(t) \geq g_2(t)$ $(-\infty < t \leq 0)$, then the corresponding solutions $\phi_0^{(1)}$, $\phi_0^{(2)}$ of (2.8) satisfy $\phi_0^{(1)}(t) \geq \phi_0^{(2)}(t)$ $(0 \leq t < \infty)$

(2.12) if $\phi(t, \mu)$ is the solution of (1.2) for a fixed $\mu > 0$, and if $\phi_0(t)$ is the solution of (2.8), then $\phi_0(t) < \phi(t, \mu) (0 \leq t < \infty)$

A solution ϕ_0 of (2.8) may be obtained as a limit of solutions $\phi(t, \mu_n)$ of equation (1.2) where $\{\mu_n\}$ is an arbitrary positive sequence with $\mu_{n+1} < \mu_n$ and $\lim_{n\to\infty} \mu_n = 0$. Uniqueness of the solution may be proved with Gronwall's inequality. The conclusion (2.9) follows from $H(F)$, $H(g)$ and the intermediate value theorem applied to the function $\int_{-\infty}^{0} b(0, s)\, F(y, g(s))\, ds$ on the interval $1 \le y \le g(0)$. The remaining conclusions may be obtained from the differentiated form of (2.8) by arguments somewhat similar to the proofs of (2.1)-(2.3). For details in the convolution case see [7].

As an immediate consequence of Theorem 3 we have the first part of

Corollary 3.1. *Let* $H_1(b)$, $H(F)$, $H(g)$ *be satisfied. Then*

$$\alpha_0 = \lim_{t\to\infty} \phi_0(t) \text{ exists and } 1 \le \alpha_0 \le \alpha(\mu) \text{ for } \mu > 0 ; \tag{2.13}$$

if alternative (ii) *of* $H_1(b)$ *holds, then, in fact,* $\alpha_0 = y_0 > 1$. *If also* $H_2(b)$ *and* $H_3(b)$ *hold, then*

$$\lim_{\mu\to 0+} \alpha(\mu) = \alpha_0 \tag{2.14}$$

and

$$0 < \phi(t, \mu) - \phi_0(t) \le \alpha(\mu) - \alpha_0 + K \int_t^{\infty}\int_{-\infty}^{0} b(u, s)\,ds\,du \quad (0 \le t < \infty;\ 0 < \mu \le \mu_0). \tag{2.15}$$

Under additional assumptions a result more precise than (2.14), (2.15) is established in Theorem 6. Estimates similar to (2.5), (2.7) for the solution ϕ_0 of (2.8) also may be obtained.

By (2.4) $\alpha(\mu) \ge 1$ for $\mu > 0$. The next theorem gives conditions which imply that $\alpha(\mu) > 1$; in fact, they yield $\alpha_0 > 1$ so that, by (2.13), $\alpha(\mu) \ge \alpha_0 > 1$ for $\mu > 0$. When alternative (ii) in $H_1(b)$ is satisfied, this result follows from (2.9), the statement below (2.10), and (2.12). If alternative (i) in $H_1(b)$ holds and $b(t, s) = a(t - s)$, where $a(t)$ satisfies the hypotheses of [7], then the result may be deduced by means of a tauberian argument for Laplace transforms (cf. [7]). We shall deduce the result for general kernels $b(t, s)$ of convolution or nonconvolution type by means of an abelian argument; however, we shall need an additional hypothesis concerning $b(t, s)$ and the functions

$$C(t, u) = \int_{-\infty}^{u} b(u, v)\,dv - \int_{u}^{t} b(s, u)\,ds \quad (0 \le u \le t < \infty), \tag{2.16}$$

$$c(u) = \int_{-\infty}^{u} b(u, v)\,dv - \int_{u}^{\infty} b(s, u)\,ds \quad (0 \le u < \infty). \tag{2.17}$$

The existence of the second integral in the definition of $c(u)$ is assured by the first part of

$$
H_4(b) \quad \begin{cases} b(\cdot, s) \in L^1(s, \infty) \quad (0 \le s < \infty); \text{ for some constant } C > 0 \\ \int_0^t |C(t, u)|du \le C \ (0 \le t < \infty) \ ; \ c(u) \le 0 \quad (0 \le u < \infty) \ ; \\ \lim_{t\to\infty} \int_0^T (C(t, u) - c(u))du = 0 \quad \text{for each } T > 0 \ ; \\ \gamma \equiv \lim_{t\to\infty} \int_0^t C(t, u)du \ \text{ exists and } \ \gamma > \int_0^\infty c(u)du \ . \end{cases}
$$

Note that the positivity of $b(t, s)$ implies that $C(t, u) - c(u) > 0$ $(0 \le u \le t < \infty)$. In addition, a simple argument using the first three lines of $H_4(b)$ shows that

$$
\int_0^\infty |c(u)|du \le C \ . \tag{2.18}
$$

Finally, in the convolution case $b(t, s) = a(t - s)$, we have $C(t, u) = \int_{t-u}^\infty a(s)ds$ and $c(u) \equiv 0$; hence, $H_4(b)$ holds if $a(t)$ satisfies $H(a)$ and the moment condition (2.6) of [7]: $\int_0^\infty ta(t)dt < \infty$.

<u>Theorem 4</u>. <u>Let</u> $H_1(b)$, $H_2(b)$, $H_4(b)$, $H(F)$, $H(g)$ <u>be satisfied</u>. <u>Then</u>

$$
\alpha_0 > 1 \ .
$$

The hypothesis $H_2(b)$ may not be omitted in Theorem 4. A Laplace transform argument is used in [7] to show that if $H(a)$, $H(g)$ are satisfied, $F(y, z) = y - z$, and $\int_0^\infty ta(t)dt = \infty$, then the solution ϕ_0 of (2.8) with $b(t, s) = a(t - s)$ satisfies $\lim_{t\to\infty} \phi_0(t) = 1$.

We remarked earlier that the estimate (2.7) holds with ϕ and $\alpha(\mu)$ replaced by ϕ_0 and α_0, respectively. One proof may be obtained by modifying the proofs of Theorems 2 and 3. It may be of interest to note that another proof which is independent of Theorem 2 may be obtained by using an abelian argument similar to the one used to establish Theorem 4. In fact, the estimate (2.7) may be proved in the same manner. These abelian arguments will not be given here (cf. [2]).

The next goal is to establish the existence of a boundary layer in a neighborhood of $t = 0$ as $\mu \to 0+$. For small $t > 0$ we may approximate the problem (1.2) by the equation

$$
-\mu v'(t) = \int_{-\infty}^0 b(0, s)F(v(t), g(s))ds \qquad (t > 0; \ v(0) = g(0)) \ . \tag{2.19}
$$

Setting $t = \mu\tau$, $w(\tau) = v(t)$ transforms (2.19) to

$$-\frac{dw}{d\tau} = \int_{-\infty}^{0} b(0, s)\, F(w(\tau), g(s))ds \qquad (\tau > 0;\ w(0) = g(0)) . \tag{2.20}$$

Theorem 5. Let $H_1(b)$, $H(F)$, $H(g)$ be satisfied. Then the initial value problem (2.20) has a unique solution $w = \xi(\tau)$ existing on $0 \le \tau < \infty$ and having the following properties:

$$\lim_{\tau\to\infty} \xi(\tau) = y_0 = \phi_0(0) \ ;\ 0 < \xi(\tau) - y_0 \le (g(0) - y_0)e^{-K\tau} \ (0 \le \tau < \infty) \tag{2.21}$$

where ϕ_0 is the solution of (2.8) and K is some positive constant.

Moreover, if $\phi(t, \mu)$ is the unique solution of (1.2) and if $\xi(t/\mu)$ is the unique solution of (2.19) for $\mu > 0$, then for any $t_0 > 0$ there exists a constant $\overline{K} > 0$ (independent of μ) such that

$$|\phi(t, \mu) - \xi(t/\mu)| \le \overline{K}t \ (0 \le t \le t_0;\ \mu > 0) . \tag{2.22}$$

The estimate (2.22) establishes the existence of a boundary layer in a positive neighborhood of $t = 0$.

The existence of a unique solution $\xi(\tau)$ of (2.20) satisfying $\xi'(\tau) < 0$, $y_0 < \xi(\tau) \le g(0)$ $(0 \le \tau < \infty)$ may be proved in a manner similar to the proof of the first part of Theorem 1. To establish (2.21) first subtract from equation (2.20) the defining equation for y_0, $\int_{-\infty}^{0} b(0, s)F(y_0, g(s))ds = 0$; an application of the mean value theorem and a simple estimate then yield a differential inequality which may be integrated to obtain the second part of (2.21), from which the first part follows immediately. A similar argument using equations (1.2) and (2.20) yields (2.22).

By Corollary 3.1 the solutions $\phi(t, \mu)$ of (1.2) and $\phi_0(t)$ of (2.8) are close for large t and small $\mu > 0$. The final result gives an estimate of their difference under additional assumptions.

$$H_5(b) \qquad \begin{cases} b_2(t, s) \in C(D);\ b_2(t, s) \ge 0 & \text{and} \\ b_2(t, s)/b(t, s) \le \tau_0 < \infty & ((t, s) \in D) . \end{cases}$$

Theorem 6. Let $H_i(b)$ $(i = 1,\dots,5)$, $H(F)$, $H(g)$ be satisfied. In addition, assume that $F \in C^2(\mathbb{R}^+ \times \mathbb{R}^+)$. Then there exist constants $K > 0$, $\mu_0 > 0$ and a positive, bounded, nondecreasing function $\gamma \in C^1[0, \infty)$ such that

$$\phi_0(t) < \phi(t,\mu) < \phi_0(t) + (g(0) - \phi_0(0))\exp(-Kt/\mu) + \gamma(t)\mu|\log\mu| \tag{2.23}$$

$$(0 \le t < \infty\ ;\ \ 0 < \mu < \mu_0)\ .$$

In particular, as an immediate consequence of (2.23), *there exists a constant* $\tilde{K} > 0$ *such that*

$$0 < \phi(t,\mu) - \phi_0(t) = O(\mu|\log\mu|) \quad (\mu \to 0+\ ;\ \tilde{K}\mu|\log\mu| \le t < \infty)\ . \tag{2.24}$$

The estimates given in Theorem 6 are precisely those obtained in [7]. The proof uses upper and lower solutions of (1.2); details for the convolution case are in [7] and a short discussion of the nonconvolution case is in [2].

3. Examples

If $a(t)$ satisfies $H(a)$ and

$$\int_0^\infty ta(t)\,dt < \infty\ , \tag{3.1}$$

then $b(t,s) \equiv a(t-s)$ satisfies $H_i(b)$ $(i = 1,\dots,5)$. Thus, the kernels of [7] satisfy the hypotheses of Theorems 1-6.

A larger class of kernels to which our theorems apply may be obtained by perturbing convolution kernels. For example, let

$$b(t,s) = \alpha(s)\,\beta(t)\,a(t-s) \quad ((t,s) \in D)\ , \tag{3.2}$$

where $a(t)$ satisfies $H(a)$ and (3.1), and $\alpha(s)$ and $\beta(t)$ satisfy

$$\begin{cases} \alpha \in C^1(-\infty,\infty);\ \ \alpha'(s) \ge 0\ ,\ \alpha'(s)/\alpha(s) \le K < \infty \quad (-\infty < s < \infty)\ ; \\ 0 < \alpha(-\infty) \le \alpha(\infty) < \infty\ ;\ \displaystyle\int_0^\infty (\alpha(\infty) - \alpha(s))ds < \infty\ ; \end{cases} \tag{3.3}$$

$$\begin{cases} \beta \in C^1[0,\infty)\ ;\ \beta'(t) \le 0\ ,\ \beta'(t)/\beta(t) \ge -\overline{K} > -\infty \quad (0 \le t < \infty)\ ; \\ \beta(\infty) > 0\ ;\ \displaystyle\int_0^\infty (\beta(t) - \beta(\infty))dt < \infty\ ; \end{cases} \tag{3.4}$$

$$\int_0^\infty a(w)\{\beta(u)\,\alpha(u-w) - \alpha(u)\beta(u+w)\}\,dw \le 0 \quad (0 \le \mu < \infty)\ . \tag{3.5}$$

Simple calculations show that $b(t, s)$ satisfies $H_i(b)$ $(i = 1,\ldots,5)$. In particular, the function $c(u)$ in $H_4(b)$ is given by the integral in (3.5), and
$\gamma \equiv \lim_{t\to\infty} \int_0^t C(t, u)du = \int_0^\infty c(u)du + \alpha(\infty)\ \beta(\infty) \int_0^\infty ta(t)\ dt > \int_0^\infty c(u)du$.

In the convolution case $c(u) \equiv 0$. If in (3.2) $\alpha(s)$ satisfies $\alpha(-\infty) < \alpha(\infty)$ and $\beta(t)$ is a constant function, then $c(u) < 0$ for u sufficiently large. For an example in which $c(u) < 0$ for all $u \geq 0$, let $\alpha(s) = 1 + e^s$ $(s \leq 0)$. $\alpha(s) = 3 - e^{-s}(s \geq 0)$, $\beta(t) = 4 + e^{-t}(t \geq 0)$.

4. Proof of Theorem 4

The monotonicity property (2.11) of solutions of (2.8) permits us to prove Theorem 4 only for a function g of the form

$$(4.1) \qquad g(t) = \begin{cases} 1 + \delta & (-\eta \leq t \leq 0) \\ 1 & (-\infty < t < -\eta) \end{cases}$$

where $\delta > 0$, $\eta > 0$. Since this function is not continuous, it does not satisfy $H(g)$; however, it is easy to alter the proof given below if the function in (4.1) is replaced by a continuous function which sufficiently approximates it. We also assume without loss of generality that

$$(4.2) \qquad F_1(1, 1) = 1,\ F_2(1, 1) = -1$$

(note that $F_1(x, x) = -F_2(x, x)$ since $F(x, x) = 0$ for $x > 0)$.

For $t > 0$ equation (2.8) with g defined in (4.1) may be written as

$$(4.3) \qquad F(\phi_0(t), 1) \int_{-\infty}^{-\eta} b(t, s)ds + F(\phi_0(t), 1 + \delta) \int_{-\eta}^{0} b(t, s)ds$$
$$+ \int_0^t b(t, s) F(\phi_0(t), \phi_0(s))ds = 0 .$$

Since $1 \leq g(t) \leq 1 + \delta$ $(-\infty < t \leq 0)$, it follows from Theorem 3 that $1 < \phi_0(t) < 1 + \delta$ $(0 \leq t < \infty)$. Thus, by (4.2), $F \in C^1(\mathbb{R}^+ \times \mathbb{R}^+)$, and several applications of the mean value theorem, we have

$$(\phi_0(t) - 1) \int_{-\infty}^{-\eta} b(t, s)ds + (\phi_0(t) - 1 - \delta) \int_{-\eta}^{0} b(t, s)ds$$
$$+ o(\delta) \int_{-\infty}^{0} b(t, s)ds + \int_0^t b(t, s)(\phi_0(t) - \phi_0(s))ds +$$

$$+ \int_0^t b(t, s)\,[o(\phi_0(t) - \phi_0(s))]ds = 0 ,$$

or, equivalently,

(4.4)
$$(\phi_0(t) - 1) \int_{-\infty}^t b(t, s)ds - \int_0^t b(t,s)(\phi_0(s) - 1)ds = \delta \int_{-\eta}^0 b(t, s)ds$$
$$+ o(\delta) \int_{-\infty}^0 b(t, s)ds + \int_0^t b(t, s)[o(\phi_0(t) - \phi_0(s)]ds .$$

The notation $w(t) = o(\delta) \int_{-\infty}^0 b(t, s)ds$ means that for each $\varepsilon > 0$ one has $|w(t)| \leq \varepsilon\delta \int_{-\infty}^0 b(t, s)ds$ for $t \geq 0$ and for $\delta > 0$ sufficiently small.

Since $1 < \phi_0(t) < 1 + \delta$ $(0 \leq t < \infty)$, it follows from H(F) that the first term in (4.3) is positive and the other two terms are negative. Thus, the absolute value of the third term in (4.3) is less than the first term. Consequently, the last term in (4.4) is $o(\delta) \int_{-\infty}^{-\eta} b(t, s)\, ds$. Hence, putting $z(t) = \phi_0(t) - 1$ in (4.4) now yields

(4.5)
$$z(t) \int_{-\infty}^t b(t, s)ds - \int_0^t b(t, s)\, z(s)ds =$$
$$\delta \int_{-\eta}^0 b(t, s)ds + o(\delta) \int_{-\infty}^0 b(t, s)ds \quad (0 \leq t < \infty) .$$

The proof to this point is a straight-forward modification of the proof of Theorem 4 of [7]. Now, let

$$\psi(t) = \int_{-\eta}^0 b(t, s)ds , \quad \omega(t) = \int_{-\infty}^0 b(t, s)ds .$$

Then integrating (4.5) and interchanging the order of integration in the second term on the left-hand side yield

$$\int_0^t z(s) \int_{-\infty}^s b(s, u)duds - \int_0^t z(u) \int_u^t b(s, u)dsdu =$$
$$\delta \int_0^t \psi(s)ds + o(\delta) \int_0^t \omega(s)ds$$

or, by (2.16)

(4.6)
$$\int_0^t z(u)\, C(t, u)du = \delta \int_0^t \psi(s)ds + o(\delta) \int_0^t \omega(s)ds \quad (0 \leq t < \infty) .$$

By Theorem 3 $z(t)$ is positive and decreasing. Thus

$$z(\infty) = \lim_{t\to\infty} z(t) \text{ exists and } z(\infty) \geq 0 \quad . \tag{4.7}$$

We next show that Theorem 3, $H_4(b)$, and (4.7) imply

$$\lim_{t\to\infty} \int_0^t z(u)\ C(t,\ u)du = \int_0^\infty z(u)\ c(u)du + z(\infty)(\gamma - \int_0^\infty c(u)du) \ , \tag{4.8}$$

where $c(u)$ and γ are defined in (2.17) and $H_4(b)$, respectively.

First, let $\varepsilon > 0$ and choose T by (4.7) such that

$$z(T) - z(\infty) < \varepsilon/(8C) \ , \tag{4.9}$$

where the constant C occurs in $H_4(b)$. Next, by Theorem 3, $H_4(b)$ and (2.18), choose $t_0 > T$ such that the following inequalities hold for $t \geq t_0$:

$$\int_0^T (C(t,\ u) - c(u))du < \varepsilon/(4z(0)) \ , \tag{4.10}$$

$$\left| \int_0^t C(t,\ u)du - \gamma \right| < \varepsilon/(4z(0)) \ , \tag{4.11}$$

$$\int_t^\infty |c(u)|du < \varepsilon/(4z(0)) \ . \tag{4.12}$$

Finally, let L denote the right-hand side of (4.8) and use (4.9)-(4.12), $H_4(b)$, (2.18) and the fact $z(t)$ is decreasing to obtain for $t \geq t_0$

$$\begin{aligned}
\left| \int_0^t z(u)C(t,\ u)du - L \right| &\leq \left(\int_0^T + \int_T^t\right)(z(u) - z(\infty))(C(t,\ u) - c(u))du \\
&\quad + z(\infty)\left|\int_0^t C(t,\ u)du - \gamma\right| + \int_t^\infty (z(\infty) - z(u))c(u)du \\
&\leq z(0) \int_0^T (C(t,\ u) - c(u))du + (z(T) - z(\infty))\left(\int_0^t |C(t,u)|du\right. \\
&\quad \left. + \int_0^t |c(u)|du\right) + z(\infty)\ \varepsilon/(4z(0)) + z(0) \int_t^\infty |c(u)|du \\
&< \varepsilon \ .
\end{aligned}$$

Thus, (4.8) holds.

Now let $t \to \infty$ in (4.6) and use (4.8) to obtain

$$\int_0^\infty z(u)c(u)du + z(\infty)(\gamma - \int_0^\infty c(u)du) = \delta\int_0^\infty \psi(s)ds + o(\delta)\int_0^\infty \omega(s)ds \ , \tag{4.13}$$

where the last two integrals exist by $H_2(b)$. For small $\delta > 0$ the first term on the right-hand side of (4.13) dominates the second term and, therefore, for δ sufficiently small there exists a $K(\delta) > 0$ such that

$$z(\infty) > (K(\delta)\int_0^\infty \psi(s)ds - \int_0^\infty z(u)c(u)du)/(\gamma - \int_0^\infty c(u)du) . \tag{4.14}$$

The denominator in (4.14) is positive by $H_4(b)$, $\psi(s) > 0$ by $H_1(b)$, $z(u) > 0$ by Theorem 3, and $c(u) \leq 0$ by $H_4(b)$. Thus, $z(\infty) > 0$ or, equivalently, $\alpha_0 = \lim_{t\to\infty} \phi_0(t) > 1$. The proof of Theorem 4 is now complete.

REFERENCES

1. G. Gripenberg, On Volterra equations with nonconvolution kernels, Report - HTKK-MAT-A118(1978), Helsinki Univ. of Tech., Inst. of Math.

2. G.S. Jordan, A nonlinear singularly perturbed Volterra integrodifferential equation of nonconvolution type, Proc. Roy. Soc. Edinburgh Sect. A, to appear.

3. T.R. Kiffe, On nonlinear Volterra equations of nonconvolution type, J. Differential Equations 22(1976), 349-367.

4. __________, A Volterra equation with a nonconvolution kernel, SIAM J. Math. Anal. 8(1977), 938-949.

5. __________, The asymptotic behavior of bounded solutions of a nonconvolution Volterra equation, J. Differential Equations, to appear.

6. J.J. Levin, A nonlinear Volterra equation not of convolution type, J. Differential Equations 4(1968), 176-186.

7. A.S. Lodge, J.B. McLeod, and J.A. Nohel, A nonlinear singularly perturbed Volterra integrodifferential equation occurring in polymer rheology, Proc. Roy. Soc. Edinburgh Sect. A., to appear.

8. C.L. Rennolet, Abstract nonlinear Volterra integrodifferential equations of nonconvolution type, Thesis, Univ. of Wisconsin-Madison, 1977.

9. M.C. Smith, On a nonlinear Volterra equation of nonconvolution type, submitted.

EXISTENCE AND COMPARISON RESULTS FOR VOLTERRA INTEGRAL EQUATIONS IN A BANACH SPACE

V. Lakshmikantham
Department of Mathematics
University of Texas at Arlington
Arlington, Texas 76019/USA

I. PRELIMINARIES

Let E be a real Banach space with a norm $\|\cdot\|$. Let $K \subset E$ be a proper solid cone, that is,

(a) $K + K \subseteq K$,

(b) $\lambda K \subseteq K$ for $\lambda \geq 0$,

(c) $K \cap (-K) = \{0\}$,

(d) the interior of K, i.e. K^0 is nonempty,

(e) $\bar{K} = K$.

If $u,v \in E$, we say that

$$u \leq v \quad \text{if} \quad v - u \in K,$$

$$\text{and} \quad u < v \quad \text{if} \quad v - u \in K^0.$$

Also let K^* and K_0^* denote the following sets of functionals:

$$K^* = \{\phi: \ \phi(x) \geq 0 \ \text{ if } \ x \in K\},$$

and

$$K_0^* = \{\phi: \ \phi(x) > 0 \ \text{ if } \ x \in K^0\}.$$

In this paper, we always assume the cone under consideration is a proper solid cone.

We consider the Volterra integral equation of the form

$$x(t) = x_0(t) + \int_{t_0}^{t} K(t,s,x(s))ds, \tag{1.1}$$

where $x_0 \in C[J,\bar{\Omega}]$, $K \in C[J \times J \times \Omega, \bar{\Omega}]$, $J = [t_0, t_0+a] \subset R$ and Ω is an open subset of E.

We need the following notions and known results. Let us begin by defining Kurotowski's measure of noncompactness α. The measure α is defined by

$$\alpha(A) = \inf\{\varepsilon > 0: \text{ A can be covered by a finite number of sets each having diameter } \leq \varepsilon\},$$

where A is a bounded subset of the Banach space E.

Theorem 1.1 (Darbo). Let E be a Banach space and A is a closed, bounded, convex, nonvoid subset of E. If $T \in C[\bar{A},\bar{A}]$ is such that $\alpha(T(B)) \leq K\alpha(B)$ where $K < 1$ for each bounded subset B of A, then T has a fixed point.

Theorem 1.2 (Mazur). Let E be a Banach space and let $K \subseteq E$ be a cone. Let $x \in \partial K$, the boundary of K. Then $\exists$ a continuous linear functional ϕ: $E \to R$ with $\phi(x) = 0$; further $\phi(y) > 0$ for every y in the interior of K.

Theorem 1.3. Let $g \in C[J \times J \times R^+, R^+]$, $g(t,s,x)$ monotone nondecreasing in x for each $(t,s) \in J \times J$, and

$$x(t) \leq \int_{t_0}^{t} g(t,s,x(s))ds \quad \text{for} \quad t \geq t_0,$$

where $x \in C[J,R^+]$. Then $x(t) \leq r(t)$, where $r(t)$ is the maximal solution of

$$y(t) = \int_{t_0}^{t} g(t,s,y(s))ds.$$

For the properties of α and the proofs of Theorems 1.1 to 1.3 see [1,2,3,5,6,8].

For convenience, we list below some assumptions:

(H_1) $||K(t,s,x)|| \leq M$ for all $(t,s,x) \in J \times J \times \Omega$.

(H_2) $\limsup_{t\to\tau} \left\{\int_I ||K(t,s,x(s)) - K(\tau,s,x(s))||ds: x \in C[I,B]\right\} = 0$, for every bounded set $B \subseteq \Omega$ and every interval $I \subseteq J$.

(H_3) $\alpha(K(J \times J \times B)) \leq \beta\alpha(B)$, where $\beta > 0$ for each bounded set $B \subseteq \Omega$.

(H_4) $\alpha(K(t,s,B)) \leq g(t,s,\alpha(B))$ for every bounded $B \subseteq \Omega$, where $g \in C(J \times J \times R^+, R^+)$, $g(t,s,u)$ is monotone nondecreasing in u for each $(t,s) \in J \times J$, and the equation $u(t) = \int_{t_0}^{t} g(t,s,u(s))ds$ has a unique solution $u(t) \equiv 0$.

II. EXISTENCE

We are now in a position to prove the existence of solutions of (1.1). We shall give two types of results.

Theorem 2.1. Let $K \in C[J \times J \times \Omega, \Omega]$, $x_0 \in C[J, \Omega]$ and suppose (H_1), (H_2), and (H_3) hold. Then $\exists$ a solution $x(t)$ for the problem (1.1) on $[t_0, t_0+\gamma]$ for some $\gamma > 0$.

Indication of Proof: Let $\eta = \sup\{\varepsilon > 0 \mid B_\varepsilon(x_0(t_0)) \subseteq \Omega\}$. Since x_0 is uniformly continuous on J, $\exists$ $\delta_1 > 0$ such that

$$|t - t_0| < \delta_1 \Rightarrow ||x_0(t) - x_0(t_0)|| \leq \eta/2.$$

Choose $\gamma = \min\{a, \delta_1, \eta/2M, 1/2\beta\}$ and let $J_0 = [t_0, t_0+\gamma]$. Define $A \subseteq C[J_0, \Omega]$ by $A = \{\phi \in C[J_0, \Omega],\ \sup_{t \in J_0} ||\phi(t) - x_0(t) \leq \eta/2\}$. Clearly A is closed, bounded and convex. Define a map $T\colon C(J_0, \Omega) \to C(J_0, \Omega)$ by

$$(T\phi)(t) = x_0(t) + \int_{t_0}^{t} K(t, s, \phi(s))ds.$$

To show T is continuous, we utilize the continuity of K, hypothesis (H_1) and the bounded convergence theorem. Next we show that TA is bounded. For this purpose if $t, \tau \in J_0$ and $\phi \in A$, we get

$$||T\phi(t) - T\phi(\tau)|| \leq ||x_0(t) - x_0(\tau)|| + M(t-\tau) + \int_{t_0}^{t} ||K(t, s, \phi(s)) - K(\tau, s, \phi(s))||ds$$

$$= I_1 + I_2 + I_3 \quad \text{(say)}.$$

Now each of I_1, I_2, I_3 can be made less than $\varepsilon/3$ by the uniform continuity of x_0, by the choice of $\delta_2^* < \varepsilon/3M$ where $|t - \tau| < \delta_2^*$, and by assumption (H_2) respectively.

Let $B \subset A$. Then using the properties of α, we can see that $\alpha(TB(t)) \leq \frac{1}{2}\alpha(B)$. Thus $\alpha(TB) = \sup_{t \in J_0} (TB(t)) \leq \frac{1}{2}\alpha(B)$. Hence the conclusion follows from Darbo fixed point Theorem.

If we replace (H_3) by a weaker assumption (H_4), then we need the uniform continuity of K. The next theorem deals with this situation where proof differs from that of Theorem 2.1.

Theorem 2.2. Assume (H_1), (H_2) and (H_4) hold. Suppose further that K is uniformly continuous. Then the conclusion of Theorem 2.1 is valid.

Indication of Proof: We choose η and δ, as in the proof of Theorem 2.1 and let $\gamma = \min(a, \delta_1, \eta/2M)$. Define $\{x_n\}_{n=1}^{\infty} \subseteq C[J_0, \Omega]$ by

$$x_n(t) = x_0(t_0), \quad t_0 - \gamma < t < t_0,$$

$$x_n(t) = x_0(t) + \int_{t_0}^{t} K(t,s,x_n(s - \gamma/n))ds, \quad t_0 \le t \le t_0 + \gamma.$$

It can be observed that x_n is equicontinuous and uniformly bounded. The sequence $\{K(t,s,x_n(s - \gamma/n)\}$ can be seen to be equicontinuous since K is uniformly continuous and $\{x_n\}$ is equicontinuous.

Using the properties of α and (H_4) we arrive at

$$\alpha(\{x_n(t)\}) \le \int_{t_0}^{t} g(t,s,\alpha\{x_n(s)\})ds.$$

Consequently by Theorem 1.3, $\alpha(\{x_n(t)\}) \le r(t)$, $t \in J_0$, where $r(t)$ is the maximal solution of $v(t) = \int_{t_0}^{t} g(t,s,v(s))ds$. Hence $\alpha(\{x_n(t)\}) \equiv 0$ in view of (H_4).

Thus $\{x_n\}$ contains a uniformly convergent subsequence $\{x_{n_k}\}$. If $x_{n_k} \to x$, then by the bounded convergence theorem $\int_{t_0}^{t} K(t,s,x_{n_k}(s))ds \to \int_{t_0}^{t} K(t,s,x(s))ds$. Thus x is a solution of (1.1).

III. EXISTENCE OF EXTREMAL SOLUTIONS

In this section we prove the existence of maximal solution. The existence of minimal solution can be proved similarly.

We prove the following result on integral inequalities before we proceed further.

Theorem 3.1. Let $K \in C[J \times J \times E, E]$, $x_0, u, v \in C[J,E]$ with $K(t,s,u)$ monotone nondecreasing in u for each $(t,s) \in J \times J$. Then

$$u(t) \le x_0(t) + \int_{t_0}^{t} K(t,s,u(s))ds,$$

$$v(t) \ge x_0(t) + \int_{t_0}^{t} K(t,s,v(s))ds, \quad t > t_0$$

with one of the inequalities strict implies $u(t) < v(t)$ $\forall$ $t \ge t_0$ provided $u(t_0) < v(t_0)$.

Indication of Proof: Suppose the theorem is false. Then $\exists$ $t_1 > t_0$ so that $v(t_1) - u(t_1) \in \partial K$, the boundary of K, $v(t) - u(t) \in K^0$ for $t \in [t_0, t_1)$. Then by Theorem 1.2 $\exists$ $\phi \in K_0^*$ with $\phi(u(t_1)) = \phi(v(t_1))$. By monotonicity of K, $\phi(u(t_1)) < \phi(v(t_1))$ which contradicts $\phi(u(t_1)) = \phi(v(t_1))$. Hence the theorem.

We can now prove the existence of extremal solutions.

<u>Theorem 3.2.</u> Assume the hypothesis of Theorem 2.1. Suppose further K is monotone nondecreasing in u for each $(t,s) \in J \times J$. Then $\exists\ \gamma > 0$ so that the maximal and minimal solutions to (1.1) exist on $[t_0, t_0+\gamma]$.

<u>Indication of Proof:</u> Since the proof of minimal solution is very similar, we shall only prove the existence of maximal solution. Let η and δ_1 be as in Theorem 2.1. Let $\gamma = \min\{a, \delta_1, \eta/_{4M}, 1/_{2\beta}\}$.

Let $A = \left\{\phi \in C\left[[t_0, t_0+\gamma], \Omega\right] \;\middle|\; ||\phi - x_0|| \leq \eta/_2\right\}$. Define $T: C\left[[t_0,t_0+\gamma],\Omega\right] \to C\left[[t_0,t_0+\gamma],\Omega\right]$ by

$$(T\phi)(t) = x_0(t) + \int_{t_0}^{t} K(t,s,\phi(s))ds.$$

Let $y_0 \in K^0$ with $||y_0|| \leq \eta/_4$ and $y_n = \frac{1}{n} y_0$, $n = 1,2,\ldots$.

Define T_n: $C\left[[t_0,t_0+\gamma],\Omega\right] \to C\left[[t_0,t_0+\gamma],\Omega\right]$ by

$$T_n\phi = T\phi + y_n, \quad n = 1,2,\ldots .$$

T_n is continuous, bounded and equicontinuous since T is so. Hence T_n has a fixed point ϕ_n.

$y_n > y_m \Rightarrow \phi_n(t) > \phi_m(t) \quad \forall \quad t \in [t_0, t_0+\gamma]$ when $m > n$.

Now consider $\{\phi_n\}$. It can be seen $||\phi_n(t) - x_0(t_0)|| < \eta$. Hence $\{\phi_n\}$ is uniformly bounded.

Also $||\phi_n(t) - \phi_n(\tau)|| = ||(T\phi_n)(t) - (T\phi_n(\tau)||$. Hence T_n is equicontinuous.

We claim $\overline{\{\phi_n\}}$ is compact. If not $\alpha(\{\phi_n\}) > 0$ but

$$\begin{aligned}\alpha(\{\phi_n(t)\}) &\leq \alpha(\{(T\phi_n)(t) + y_n\}) \\ &\leq \alpha(\{(T\phi_n)(t)\}) = \alpha(T\{\phi_n(t)\}) \\ &\leq \frac{1}{2}\,\alpha\{\phi_n(t)\}, \quad \text{which is impossible.}\end{aligned}$$

Thus $\alpha(\{\phi_n\}) = 0$. Hence $\{\phi_n\}$ is precompact. Hence there exists a uniformly convergent subsequence.

Suppose $\phi_{n_k} \to \phi$. By the assumption (H_1) and the bounded convergence theorem ϕ is a fixed point of T.

If x is any other fixed point of T,

$$x(t) = x_0(t) + \int_{t_0}^{t} K(t,s,x(s))ds.$$

$$\begin{aligned}\phi_{n_k}(t) &= x_0(t) + y_{n_k} + \int_{t_0}^{t} K(t,s,\phi_{n_k}(s))ds \\ &> x_0(t) + \int_{t_0}^{t} K(t,s,\phi_{n_k}(s))ds\end{aligned}$$

but $x(t_0) < \phi_{n_k}(t_0)$. Therefore $x(t) < \phi_n(t)$ for $t \in [t_0, t_0+\gamma]$. This by Theorem 1.3 implies $x(t) \leq \lim_{k\to\infty} \phi_{n_k}(t) = \phi(t)$ for $t \in [t_0, t_0+\gamma]$. Thus ϕ is the maximal solution of (1.1).

Remark. Obviously one could prove existence of extremal solutions under the hypothesis of Theorem 2.2 provided K is monotone nondecreasing in u.

IV. COMPARISON THEOREMS

We can now prove a general comparison result in this set-up.

Theorem 4.1. Suppose the hypothesis of Theorem 2.1 are satisfied. Let $m \in C[J,\Omega]$ and $m(t) \leq x_0(t) + \int_{t_0}^{t} K(t,s,m(s))ds$ on J. Then $m(t) \leq r(t)$, $t \in [t_0, t_0+\gamma]$, where $r(t)$ is the maximal solution of (1.1).

Proof: Let $x_n(t)$ be a solution of

$$x(t) = x_0(t) + \int_{t_0}^{t} K(t,s,x(s))ds + y_n, \quad \text{on} \quad [t_0, t_0+\gamma]$$

Since $\lim_{n\to\infty} x_n(t) = r(t)$, it is enough to prove $m(t) < x_n(t)$ for $t \in [t_0, t_0+\gamma]$. Since $m(t_0) < x_n(t_0) = x_0(t_0) + y_n$ and $x_n(t) > x_0(t) + \int_{t_0}^{t} K(t,s,x_n(s))ds$, Theorem 3.1 yields $m(t) < x_n(t)$ for $t \in [t_0, t_0+\gamma]$ for each n. Hence the proof.

To obtain upper and lower bounds on solutions of (1.1) when K is not monotone, the following result will be useful.

Theorem 4.2. Let K satisfy the hypothesis of Theorem 2.1 or 2.2. Let the following integral inequalities hold:

$$\phi(y(t)) < \phi\left\{x_0(t) + \int_{t_0}^{t} K(t,s,\xi(s))ds\right\} \quad \text{for all} \quad \xi(s)$$

$$\text{whenever} \quad y(s) \leq \xi(s) \leq z(s), \quad t_0 \leq s \leq t_1$$

$$\text{and} \quad \phi(y(t_1)) = \phi(\xi(t_1)) \quad \forall \quad t_1 \in (t_0, t_0+a),$$

$$\phi(z(t)) > \phi\left\{x_0(t) + \int_{t_0}^{t} K(t,s,\xi(s))ds\right\} \quad \text{for all} \quad \xi(s)$$

$$\text{whenever} \quad y(s) \leq \xi(s) \leq z(s) \quad \text{for} \quad t_0 \leq s \leq t_1$$

$$\text{and} \quad \phi(\xi(t_1)) = \phi(z(t_1)) \quad \forall \quad t_1 \in (t_0, t_0+a),$$

where $\phi \in K_0^*$.

Then $y(t) < x(t) < z(t)$ provided $y(t_0) < x(t_0) < z(t_0)$ where $x(t)$ is any solution to (1.1).

Indication of Proof: By Theorem 2.1 or 2.2, there exists a solution $x(t)$ of (1.1). If the conclusion is false, $\exists$ a $\hat{t}$ and a $\phi \in K_0^*$ such that $z(t) > x(t) > y(t)$, $t \in [t_0, \hat{t})$ and either $\phi(z(\hat{t})) = \phi(x(\hat{t}))$ or $\phi(x(\hat{t})) = \phi(y(\hat{t}))$.

Suppose that the first case holds. Then using the hypothesis, we arrive at $\phi(z(\hat{t})) > \phi(x(\hat{t}))$ which is a contradiction. Similar contradiction results in the other case. Hence the proof is complete.

Remark. Most of the results presented here are adapted from the work of [9]. For corresponding results in finite dimension see [1,4,7,8].

REFERENCES

1. Corduneanu, C., Integral Equations and Stability of Feedback Systems, Academic Press, New York, NY, 1973.

2. Darbo, G., "Punti uniti in transformaziani a codomino noncompatto", Rend. Sem. Mat. Univ. Padova, 24 (1955), 84-92.

3. Deimling, K., Ordinary Differential Equations in Banach Spaces, Springer Verlag, 1977.

4. Lakshmikantham, V., and Leela, S., Differential and Integral Inequalities, Vol. 1, Academic Press, New York, NY, 1969.

5. Lakshmikantham, V., and Leela, S., An Introduction to Nonlinear Differential Equations in Abstract Spaces, (to be published).

6. Martin, R. H., Jr., Nonlinear Operators and Differential Equations in Banach Space, J. Wilen and Sons, New York, NY, 1976.

7. Miller, R. K., Nonlinear Volterra Integral Equations, W. A. Benjamin, Inc., Menlo Park, 1971.

8. Nohel, J. A., "Some problems in nonlinear Volterra integral equations", Bull. Amer. Math. Soc. 68 (1962), 323-329.

9. Vaughn, R. L., "Existence and comparison results for nonlinear Volterra integral equations in a Banach space", (to appear in Applicable Analysis).

On Plane Waves Propagating Into A Random Fluid: Asymptotic Behavior

M. J. Leitman*

BERAN and McCOY [1], [2], [3] have developed a mathematical model for the propagation of acoustic waves in water which incorporates the scattering effect of microscopic variations in density (sound speed) into the classical model of geometric optics. This note is concerned with the dispersal and consequent loss in resolution of the wave as a function of distance from the source. Computational aspects of this problem and a short treatment of the asymptotics may be found in a paper by BERAN, LEITMAN, and SCHWARTZ [4].

If we let the function $\mu \to x_t(\mu)$, $-\infty < \mu < \infty$, denote the acoustic intensity spectral density at a distance $t \geq 0$ from the source, then BERAN and McCOY [1] show, under suitable physical assumptions, that x_t satisfies an initial value problem of the following form:

$$\frac{d}{dt}x_t(\mu) = \int_{-\infty}^{\infty} \Psi(\mu,\nu)(x_t(\nu) - x_t(\mu))d\nu, \quad t \geq 0, \qquad -\infty < \mu < \infty$$
$$x_0(\mu) = \mathring{x}(\mu),$$

where Ψ is a kernel determined by the physics of the problem and $\mathring{x}$ is a prescribed initial state.

*Case Western Reserve University, Cleveland, Ohio, 44106.

We require the kernel Ψ to satisfy the following standing hypotheses:

(A1) $\Psi(\mu,\nu) > 0$ (positivity);

(A2) $\Psi(\mu,\nu) = \Psi(\nu,\mu)$ (symmetry);

(A3) $\Psi(\mu,\nu) \le \overline{\Psi} < \infty$ (boundedness);

(A4) $\varphi(\mu) \overset{\text{def}}{=} \int_{-\infty}^{\infty} \Psi(\mu,\nu)d\nu \le \overline{\varphi} < \infty$ (sectional integrability and boundedness);

(A5) $\lim\limits_{M\to\infty} \frac{1}{M} \int\limits_{|\nu|\ge M} \int\limits_{|\mu|\le M} \Psi(\mu,\nu)d\mu d\nu = 0.$

Properties (A1-4) arrise more or less naturally from the physical model; but (A5) is a technical assumption satisfied in the cases of interest. From the point of view of existence, uniqueness, and asymptotic stability of solutions to the initial value problem, some of these hypotheses are overly restrictive. Satisfactory results obtain under weaker assumptions.

It is important throughout to bear in mind two specific examples. The <u>optical model</u>, without scattering, has a kernel of convolution type so that

$$\Psi(\mu,\nu) = f(\mu-\nu).$$

In this case (A1-4) are equivalent to $f \ge 0$, $f \in L^1 \cap L^\infty$, and f even; moreover, (A5) is a consequence of (A1-4). The <u>acoustic model</u>, with scattering, has a kernel of non-convolution form typified by

$$\Psi(\mu,\nu) = [1 + (\mu^2 - \nu^2)^2]^{-1}.$$

Henceforth we suppose that $\mu \in (-\infty,\infty)$, $t \in [0,\infty)$, and write L^p for $L^p((-\infty,\infty))$, $1 \le p \le \infty$. If $x \in L^p$ and $y \in L^q$, where p and q are conjugate, we write

$$\langle x,y\rangle = \int_{-\infty}^{\infty} x(\mu)y(\mu)d\mu.$$

It is also convenient to define an operator A in the class of functions on $(-\infty,\infty)$ by

$$(Ax)(\mu) = \int_{-\infty}^{\infty} \Psi(\mu,\nu)(x(\nu) - x(\mu))d\nu.$$

In terms of the operator A, the initial value problem has the form of a Cauchy problem:

$$\left.\begin{aligned} \frac{d}{dt}x_t &= Ax_t, \quad t \geq 0 \\ x_0 &= \overset{\circ}{x} \end{aligned}\right\} (*)$$

Regarding the solutions of this problem we assert the following:

(T1) Each initial state $\overset{\circ}{x} \in L^p$ determines a unique solution $x_t = T_t\overset{\circ}{x}$, where $T_t \equiv \exp tA$, $t \geq 0$, is an analytic semi-group of linear operators in L^p, $1 \leq p \leq \infty$;

(T2) For each $t \geq 0$, T_t is a positive linear operator in L^p, $1 \leq p \leq \infty$, so that

$$x \geq 0 \Rightarrow T_t x \geq 0;$$

(T3) For each $t \geq 0$, $\|T_t\|_p = 1$, $1 \leq p \leq \infty$, and, more specifically,

(i) if $p=1$, $x \geq 0$, $x \not\equiv 1 \Rightarrow 0 < \|T_t x\|_1 = \|x\|_1$,

(ii) if $p=2$, $x \not\equiv 0 \Rightarrow 0 < \|T_t x\|_2 < \|x\|_2$,

(iii) if $p=\infty$, $x \equiv 1 \Rightarrow T_t 1 = 1$;

(T4) $\lim\limits_{t\to\infty} \|T_t x\|_2 = 0$, $\forall x \in L^2$; and

(T5) There is at least one $\overset{\circ}{x} \in L^2$, $\overset{\circ}{x} \geq 0$, such that

$$\lim_{t\to\infty} \frac{d}{dt} \ln \|T_t\overset{\circ}{x}\|_2 = 0. \qquad (**)$$

We thus see that positive initial profiles produce positive solution profiles (T2), and that intensity (area under positive solution profiles) is conserved (T3(i)). However, profiles flatten out and resolution is lost in the sense that $x_t = T_t\overset{\circ}{x} \to 0$ in the mean square as $t \to \infty$ (T2 (ii),4). The most significant assertion (T5) implies that the rate at which resolution is lost is expected to be very slow. For if (**) holds, then every positive initial state lies arbitrarily close to one for which the rate of decay, in the mean square, is slower than any exponential.

Before we outline the proofs of these assertions we make some additional remarks. If no scattering is assumed, so that A has a convolution kernel, then (**) holds for <u>every</u> positive initial state $\overset{\circ}{x} \in L^2$. This is a straightforward application of Plancherel's Theorem. This observation motivates us to make the following

<u>Conjecture</u>. Under the hypotheses (A1-5), solution to the Cauchy problem (*) satisfy (**) for every positive initial state $\overset{\circ}{x} \geq 0$, $\overset{\circ}{x} \not\equiv 0$ in L^2.

The positivity of the initial state seems to be essential here. We note that if A has a positive eigenfunction $\hat{x} \geq 0$ in L^2 corresponding to the eigenvalue $-\omega$ (A will be shown to be symmetric and negative definite in L^2), then (**) must fail. For in this case

$$x_t = T_t\hat{x} = e^{-\omega t}\hat{x}.$$

But T_t is positive (and symmetric) in L^2 and $\|T_t\|_2 = 1$, from which it follows that $e^{-\omega t} = 1$, $t \geq 0$, and hence $\omega = 0$. (See COFFMAN, DUF-

FIN, and MIZEL [5]). It will be shown that zero is not an eigenvalue of A in L^2, so we conclude that A can have no positive eigenfunction. Of course this fact alone does not validate the conjecture.

Verification of assertions (T1-5) depend upon establishing the following properties of the generator A. We provide comments on the proof of each proposition, omitting full computational detail.

(P1) (A1-4) $\Rightarrow A : L^p \to L^p$, $1 \le p \le \infty$, is a bounded linear operator whose norm satisfies

$$\|A\|_p \le 2\tilde{\varphi}.$$

Proof of (P1). The cases $p = 1$ and $p = \infty$ follow by routine computation. The result for $1 < p < \infty$ is a consequence of Riesz's Convexity Theorem. The specific case $p = 2$ can also be verified directly by using an integral version of an argument due to Shur. $\square$

(P2) (A1-4) $\Rightarrow$ for $A : L^2 \to L^2$,

(i) $\langle Ax, y\rangle = \langle x, Ay\rangle$

(ii) $x \neq 0 \Rightarrow \langle Ax, x\rangle < 0$

(iii) if, in addition, (A5) holds, then

$$\sup_{\|x\|_2 = 1} \langle Ax, x\rangle = 0.$$

Thus A is a symmetric, negative definite linear operator in L^2 containing the point zero in its continuous spectrum whenever (A5) holds as well as (A1-4).

Proof of (P2). This result follows by establishing the formula

$$\langle Ax, y\rangle = -\tfrac{1}{2}\int_{-\infty}^{\infty}\int_{-\infty}^{\infty} \Psi(\mu,\nu)(x(\nu) - x(\mu))(y(\nu) - y(\mu))\,d\mu\,d\nu. \qquad \square$$

(P3) (A1-4) $\Rightarrow$

(i) $x \in L^1$, $Ax = 0 \Rightarrow x = 0$

(ii) $x \in L^1 \Rightarrow \int_{-\infty}^{\infty} (Ax)(\mu)d\mu = 0$

(iii) $x \in L^\infty$, $Ax = 0 \Rightarrow x =$ constant.

Proof of (P3). Assertions (i)-(iii) follow by routine computation. □

(P4) (A1-4) $\Rightarrow J_\lambda \equiv (I - \frac{1}{\lambda}A)^{-1}$, $\lambda > 0$, is a strictly positive contraction in L^p, $1 \le p \le \infty$: for $\lambda > 0$,

$$x \ge 0, \quad x \not\equiv 0 \Rightarrow J_\lambda x > 0,$$

$$\|J_\lambda\|_p \le 1.$$

Moreover:

(i) if $p=1$, $\|J_\lambda\|_1 = 1$, and $x \ge 0 \Rightarrow \|J_\lambda x\|_1 = \|x\|_1$;

(ii) if $p=\infty$, $\|J_\lambda\|_\infty = 1$, and $J_\lambda 1 = 1$; and

(iii) if $p=2$, $x \ne 0 \Rightarrow \|J_\lambda x\|_2 < \|x\|_2$.

If, in addition, (A5) holds, then $\|J_\lambda\|_2 = 1$ as well.

Thus, (A1-5) $\Rightarrow \|J_\lambda\|_p = 1$, $1 \le p \le \infty$, for $\lambda > 0$.

Proof of (P4). First observe that $x = J_\lambda y$ must be a solution of

$$x - \frac{1}{\lambda}Ax = y, \qquad \lambda > 0,$$

or

$$x + \frac{1}{\lambda}\varphi x - \frac{1}{\lambda}\psi x = y, \quad \lambda > 0,$$

where

$$(\varphi x)(\mu) = \varphi(\mu)x(\mu),$$

$$(\psi x)(\mu) = \int_{-\infty}^{\infty} \Psi(\mu,\nu)x(\nu)d\nu.$$

A priori estimates for $\|J_\lambda\|_p$, valid on the range of $(I - \frac{1}{\lambda}A)$ are fairly easy to obtain in case $p = 1,2,$ or ∞. That the range of

$(I - \frac{1}{\lambda}A)$ is all of L^p is established for $p = 1$ and $p = \infty$ by means of the Neumann series, and for $p = 2$ by a continuity argument. For the case $1 < p < \infty$, the Riesz Convexity Theorem can be used. The positivity of J_λ follows in case $p = 1$ or $p = \infty$ from (P3 (ii, (iii)) and then for $1 < p < \infty$ by writing $y \in L^p$ as a sum $y = y_1 + y_\infty$, where $y_1 \in L^1 \cap L^p$ and $y_\infty \in L^\infty \cap L^p$ are each positive. The assertions (i)-(iii) are easy to establish by using the order preserving properties of J_λ. Finally, we note that $\|J_\lambda\|_2 = 1$ iff zero is in the spectrum of A. The condition (A5), in addition to (A1-4), suffices to guarantee that zero is in the continuous spectrum of A. □

Having established the needed properties of A, we can now proceed to verify the assertions (T1-5). Assertions (T1-3) follow at once from (P1-4) and the familiar exponential formulae

$$T_t = \int_{0^-}^{2\bar{\varphi}} e^{-\lambda t} d\,E(\lambda),$$

where $\{E(\lambda) : -\infty < \lambda < \infty\}$ is the spectral resolution of the identity for $(-A)$, normalized to be left continuous. It is a standard result that $\lim_{t\to\infty} T_t = E(0^+)$. But since $\lambda = 0$ is in the continuous spectrum of $(-A)$, it follows that $E(0^+) \equiv 0$. Finally, (T5) follows from the following technical

<u>Lemma</u>. If $\{E(\lambda) : -\infty < \lambda < \infty\}$ is an increasing family of projections in L^2 such that $E(\lambda) \neq 0$, for $\lambda > 0$, then there is at least one positive vector $\overset{\circ}{x} \geq 0$ in L^2 such that $E(\lambda)\overset{\circ}{x} \neq 0$ for $\lambda > 0$.

Using the Lemma and the formula

$$\|T_t \mathring{x}\|_2 = \int_{0^-}^{2\bar{\varphi}} e^{-\lambda t} \, d(\|E(\lambda)\mathring{x}\|_2),$$

we easily verify (**) of (T5). □

In view of the fact that the equation arrises from a multiple scattering problem, we might expect a connection between its solution and stochastic processes. Indeed, for $t \geq 0$, $\mu \in (-\infty, \infty)$, and any Borel set $E \subset (-\infty, \infty)$ define

$$P(t, \mu, E) = (T_t \chi_E)(\mu),$$

where χ_E is the characteristic function of the set E. It is not too hard to show that P is a Markoff process which is temporally homogeneous and spatially inhomogeneous (except in the optical case). Furthermore, P admits a non-trivial invariant measure, namely Lebesgue measure; however the process is dissipative in the sense of stochastic processes. We see that our original Cauchy problem corresponds to the Kolmogorov equation associated with the Markoff process P.

With a view toward approximating solutions to the original problems, we replace the kernel Ψ by $\Psi^{(M)} = \chi_{M\times M}\Psi$, where $\chi_{M\times M}$ is the characteristic function of the square $\{(\mu,\nu) : |\mu|, |\nu| \leq M\}$. The approximate problem thus obtained possesses all the features of the original problem except thatits solution semi-group $\{T_t^{(M)} : t \geq 0\}$ satisfies

$$\lim_{t\to\infty} T_t^{(M)} = P^{(M)}$$

where $P^{(M)}$ is the projection in L^2 given by

$$(P^{(M)}x)(\mu) = \begin{cases} \frac{1}{2M}\int_{-M}^{M} x(\nu)d\nu & : \mu \in [-M,M] \\ x(\mu) & : \mu \notin [-M,M] \end{cases}$$

or $\overset{\circ}{x} \in L^2$ the TROTTER-KATO Theorem guarantees that

$$T_t^{(M)}\overset{\circ}{x} \to T_t\overset{\circ}{x}$$

s $M \to \infty$ uniformly for $t \in [0,\alpha]$, $\alpha > 0$. Of course, $P^{(M)}\overset{\circ}{x} \to 0$ as $\to \infty$.

Now if it happens that $\underline{\Psi}^{(M)} \equiv \inf_{|\nu|,|\mu|\leq M} \Psi^{(M)}(\mu,\nu) > 0$ as it does in our typical xample, then the decay rate is exponential. Indeed, for every $\overset{\circ}{x} \in L^2$,

$$\|T_t^{(M)}\overset{\circ}{x} - P^{(M)}\overset{\circ}{x}\|_2 \leq e^{-2M\underline{\Psi}^{(M)}t}\|\overset{\circ}{x} - P^{(M)}\overset{\circ}{x}\|_2 .$$

urthermore, if $\lim_{M\to\infty} 2M\,\underline{\Psi}^{(M)} = 0$, this exponential rate becomes slower s the degree of approximation improves. Other approximation schemes also xhibit this phenomenon.

We conclude this note by observing that our analysis depended in no ssential way upon the boundedness of A. Moreover, techniques other than pectral theory may be used to get results of a similar nature in case A s not symmetric or normal. What is essential in our analysis is the veraging property, namely

$$\int_{-\infty}^{\infty}(Ax)(\mu)d\mu = 0.$$

eneralization to include non-linear hereditary effects also seems feasible, ay by integrating against the solution semi-group.

REFERENCES

[1] Beran, M.J. and J.J. McCoy, Propagation through an anisotropic random medium, J. Math. Phys. 15, 11(1i74), 1901-1912.

[2] Beran, M.J. and J.J. McCoy, Propagation from a finite beam or source through an anisotropic random medium, J. Accoust. Soc. of Am. 56, 6(1974), 1667-1672.

[3] Beran, M.J. and J.J. McCoy, Propagation through anisotropic random medium. An integro-differential formulation, J. Math. Phys. 17, 7(1976), 1186-1189.

[4] Beran, M.J., M.J. Leitman and N. Schwartz, Scattering in the depth direction for an anisotropic random medium, to appear in the J. Accoust. Soc. of Am.

[5] Private communication of a result of Coffman, Duffin and Mizel.

ON THE ASYMPTOTIC BEHAVIOR OF SOLUTIONS OF INTEGRAL EQUATIONS

J. J. Levin *

University of Wisconsin
Madison, WI 53706, USA

Consider

(1f) $$x'(t) + \int_{-\infty}^{\infty} g(x(t-\xi))\, dA(\xi) = f(t) \qquad (' = \frac{d}{dt},\ -\infty < t < \infty),$$

under the assumptions

(2) $$g \in C(\mathbb{C}^N, \mathbb{C}^N)$$

(3) $$A = (A_{ij}),\ A_{ij} \in BV(R^1, \mathbb{C}^1) \qquad (i, j = 1, \ldots, N)$$

(4) $$f \in L^\infty(R^1, \mathbb{C}^N),\quad \lim_{t\to\infty} f(t) = f(\infty) \text{ exists.}$$

In studies of the asymptotic behavior as $t \to \infty$ of the bounded solutions of (1f) a key role is often played by

($1^* f(\infty)$) $$y'(t) + \int_{-\infty}^{\infty} g(y(t-\xi))\, dA(\xi) = f(\infty) \qquad (-\infty < t < \infty),$$

the associated limit equation. For

(5f) $$x(t) + \int_{-\infty}^{\infty} g(x(t-\xi))\, dA(\xi) = f(t) \qquad (-\infty < t < \infty),$$

under the further assumption

(6) $$f \in \mathcal{B}(R^1, \mathbb{C}^N),$$

a similar role is played by its limit equation

($5^* f(\infty)$) $$y(t) + \int_{-\infty}^{\infty} g(y(t-\xi))\, dA(\xi) = f(\infty) \qquad (-\infty < t < \infty).$$

It is convenient to denote

($1^* \mu$) $$y'(t) + \int_{-\infty}^{\infty} g(y(t-\xi))\, dA(\xi) = \mu \qquad (-\infty < t < \infty)$$

---

* This research was supported by the U. S. Army Research Office.

$$(5^*\mu) \qquad y(t) + \int_{-\infty}^{\infty} g(y(t-\xi))\, dA(\xi) = \mu \qquad (-\infty < t < \infty),$$

where $\mu \in \mathbb{C}^N$. These equations reduce to $(1^* f(\infty))$ and $(5^* f(\infty))$, respectively, when $\mu = f(\infty)$. If $f(t) \equiv \mu$, then $(1f)$, $(1^* f(\infty))$, and $(1^* \mu)$ are, except for x being replaced by y, the same equation. No ambiguity will arise from this redundancy nor from the analogous one involving $(5f)$, $(5^* f(\infty))$, and $(5^* \mu)$.

It is understood that: x is a bounded solution of $(1f)$ if $x \in L^\infty \cap AC_{loc}(\mathbb{R}^1, \mathbb{C}^N)$ satisfies $(1f)$ a.e. on $\mathbb{R}^1$, x is a bounded solution of $(5f)$ if $x \in \mathcal{B} \cap L^\infty(\mathbb{R}^1, \mathbb{C}^N)$ satisfies $(5f)$ on $\mathbb{R}^1$. While only the behavior of the bounded solutions of $(1f)$ and of $(5f)$ are analyzed here, it is not assumed that all solutions of these equations are bounded.

The notation: $\mathbb{R}^1 = (-\infty, \infty)$, $\mathbb{R}^+ = [0, \infty)$, $\mathbb{C}^1 =$ complex-plane, $\mathbb{C}^N = \{(z_1 \ldots z_N) \mid z_k \in \mathbb{C}^1\}$, $|z| = \sum_{k=1}^N |z_k|$ $(z \in \mathbb{C}^N)$ is employed. Also: $x_n \to x (n \to \infty)$ $(x_n, x \in \mathbb{C}^N)$ if $|x_n - x| \to 0 (n \to \infty)$, $\bar{A} =$ closure of A $(A \subset \mathbb{C}^N)$ with respect to the metric associated with $|\cdot|$, $B_r = \{x \in \mathbb{C}^N \mid |x| < r\}$. Further: $C, C_u, C_u^{(k)}$, L^∞, $\mathcal{B}$, BV, and AC respectively denote continuous, uniformly continuous, uniformly continuous k^{th} derivative, essentially bounded, Borel measurable, bounded variation, and absolutely continuous. The subscript loc denotes that the indicated property need only hold on compact subsets of the domain. In statements such as $x_n \to x$ it will, for brevity, be understood that $n \to \infty$.

A special choice of A reduces $(1f)$, $(1^* f(\infty))$ to

$$x'(t) + g(x(t)) = f(t), \quad y'(t) + g(y(t)) = f(\infty) \qquad (-\infty < t < \infty).$$

Another choice of A together with $g(x) = x$ reduces $(5f)$, $(5^* f(\infty))$ to

$$(7f) \qquad \int_{-\infty}^{\infty} x(t-\xi)\, dB(\xi) = f(t) \qquad (-\infty < t < \infty)$$

$$(7^* f(\infty)) \qquad \int_{-\infty}^{\infty} y(t-\xi)\, dB(\xi) = f(\infty) \qquad (-\infty < t < \infty),$$

where $B \in BV(\mathbb{R}^1, \mathbb{C}^{N^2})$. Thus, our considerations are related both to problems concerning nonautonomous perturbations of autonomous nonlinear ordinary differential equations and to problems in tauberian theory.

Certain asymptotic problems for the Volterra equations

$$x'(t) + \int_0^t g(x(t-\xi))\, dA(\xi) = f(t) \qquad (0 \le t < \infty)$$

$$x(t) + \int_0^t g(x(t-\xi))\, dA(\xi) = f(t) \qquad (0 \le t < \infty)$$

are important motivations for the present study. The first equation includes, of course, systems of ordinary differential equations on $\mathbb{R}^+$ as a special case. Although they are defined on $\mathbb{R}^+$, it is an elementary matter (see, e.g., [4]) to transform their bounded solutions into bounded solutions of equations of type (1f) and (5f). It turns out that the appropriate limit equations for these equations are

$$y'(t) + \int_0^\infty g(y(t-\xi))\, dA(\xi) = f(\infty) \qquad (-\infty < t < \infty)$$

$$y(t) + \int_0^\infty g(y(t-\xi))\, dA(\xi) = f(\infty) \qquad (-\infty < t < \infty),$$

which are defined on $\mathbb{R}^1$. Similarly, the asymptotic behavior of the bounded solutions of various delay and differential-delay equations on $\mathbb{R}^+$ can be studied as special cases of (1f) and (5f).

The functions defined by

$$x \in T(\mathbb{R}^1, \mathbb{C}^N) \text{ if } \lim_{t \to \infty, \tau \to 0} |x(t+\tau) - x(t)| = 0,$$

$$x \in \tilde{T}(\mathbb{R}^1, \mathbb{C}^N) \text{ if } x \in T(\mathbb{R}^1, \mathbb{C}^N) \text{ and } \limsup_{t \to \infty} |x(t)| < \infty,$$

play a very important role in our analysis. If $x \in T(\mathbb{R}^1, \mathbb{C}^N)$, x is said to be tauberian.

All solutions of (1f) and (5f) obviously belong to

$$\mathcal{Y} = \{y : \mathbb{R}^1 \to \mathbb{C}^N\}.$$

All elements of $\mathcal{Y}$ will be called curves - whether or not they are, for example, measurable and bounded. Subsets of $\mathcal{Y}$ such as $L^\infty(\mathbb{R}^1, \mathbb{C}^N)$, $T(\mathbb{R}^1, \mathbb{C}^N), \ldots$ will, for brevity, be denoted by $L^\infty, T, \ldots$ Setting

$$d(x, y) = \sum_{n=1}^{\infty} \frac{1}{2^n} \sup_{|t| \le n} \frac{|x(t) - y(t)|}{1 + |x(t) - y(t)|} \qquad (x, y \in \mathcal{Y}),$$

defines a complete metric on $\mathcal{Y}$. It is well known that $d(x_n, x) \to 0$ $(x_n, x \in \mathcal{Y})$ if and only if $x_n \to x$ c.o., where, by definition,

$$x_n \to x \text{ c.o.} \iff \lim_{n \to \infty} [\sup_{t \in I} |x_n(t) - x(t)|] = 0 \quad \forall \text{ compact } I \subset \mathbb{R}^1.$$

Here c.o. denotes the compact open topology on $\mathcal{Y}$. Let

$$y_\tau : t \mapsto y(t+\tau) \qquad (y \in \mathcal{Y},\ \tau \in \mathbb{R}^1).$$

$Y \subset \mathcal{Y}$ is called translation invariant if $y_t \in Y$ for all $y \in Y$ and $t \in \mathbb{R}^1$. T may be characterized by: $x \in T$ if and only if for each $\varepsilon > 0$ there exist $t_0(\varepsilon) \in \mathbb{R}^1$ and $\delta(\varepsilon) > 0$ such that $d(x_s, x_t) < \varepsilon$ whenever $s, t \geq t_0(\varepsilon)$ and $|s-t| \leq \delta(\varepsilon)$.

Here we discuss some extensions and improvements of the papers of Levin and Shea [4] and Levin [3]. By focusing on $\mathcal{Y}$ and, in particular, on $\tilde{T}$ we avoid having to give parallel proofs for (1f) and (5f), as in [4], or having to confine ourselves to one of them, as in [3] for (1f). That is, once a result is obtained for $\tilde{T}$, its consequences for integral equations will be fairly immediate. However, besides thus obtaining analogues of results in [3] for (5f), considerably stronger forms of results of [3] and [4] are obtained. Only, what might be called, the analysis portion of our results are, for brevity, indicated here. That is, an element $x \in \tilde{T}$ is related to the elements of its limit set, $\Gamma(x)$, defined below; correspondingly, bounded tauberian solutions of integral equations are related to solutions of their limit equations. This culminates in formula (9) of Theorems 1, 1a, 1b below. The use of (9) (with $\eta = 0$) to synthesize appropriate solutions of (1μ) into a solution of (1f) with $f(\infty) = \mu$, corresponding problems for (5μ), (5f), as well as more inclusive synthesis problems in the setting of $\mathcal{Y}$ without specific reference to integral equations will be reported on elsewhere. References to some of the many papers concerned with limit equations are given in [3] and [4]. The papers of Hale [1], Miller [7], [8], and (concerning the above Volterra equations) Londen [5], [6] particularly influenced this work. Quite different generalizations of portions of [4] have been obtained by Sell [9] and, for linear equations, by Jordan and Wheeler [2].

For each $x \in \mathcal{Y}$ and $Y \subset \mathcal{Y}$ let

$$R(x) = \text{range of } x, \quad R(Y) = \bigcup_{y \in Y} R(y),$$

$$\Omega(x) = \{\omega \in \mathbb{C}^N \mid x(t_n) \to \omega \quad \text{for some } t_n \to \infty\},$$

$$\Gamma(x) = \{y \in \mathcal{Y} \mid x_{t_n} \to y \ \text{c.o.} \quad \text{for some } t_n \to \infty\}.$$

Some properties of tauberian curves are collected in the first result; several of them hold under less than the stated hypothesis.

<u>Lemma 1.</u> If $x \in \tilde{T}$, <u>then</u>

$$\Omega(x) \subset \bar{B}_{\limsup_{t\to\infty} |x(t)|} \cap \overline{R(x)},$$

$$\Omega(x) = \Omega(x_\tau), \quad \Gamma(x) = \Gamma(x_\tau) \qquad (\tau \in \mathbb{R}^1),$$

$\Gamma(x)$ <u>is translation invariant and equicontinuous</u>,

$\Omega(x)$ <u>is nonempty, compact, and connected in</u> $\mathbb{C}^N$,

$\Gamma(x)$ <u>is nonempty, compact, and connected in the</u> c.o. <u>topology</u>,

$$\Omega(x) = R(\Gamma(x)), \quad \lim_{t\to\infty} d(x_t, \Gamma(x)) = 0.$$

A key ingredient in the proof of Lemma 1 is: If $x \in \tilde{T}$ and $t_n \to \infty$, then there exist a subsequence $\{t_{n_k}\}$ of $\{t_n\}$ and a $y \in \Gamma(x)$ such that $x_{t_{n_k}} \to y$ c.o. If $x \in L^\infty \cap C_u$, it follows immediately from the Ascoli-Arzela lemma; an adaptation of the proof of the latter establishes the general result.

For each $Q \subset \mathbb{C}^N$ let

$$S_Q(1^*\mu) = \{y \in L^\infty \cap C_u^{(1)} \mid y \text{ satisfies } (1^*\mu),\ R(y) \subset Q\}$$

$$S_Q(5^*\mu) = \{y \in L^\infty \cap C_u \mid y \text{ satisfies } (5^*\mu),\ R(y) \subset Q\}$$

denote certain solution subsets of $\mathfrak{Y}$. Trivially: $R(S_Q(1^*\mu)) \subset Q$, $R(S_Q(5^*\mu)) \subset Q$ and $S_Q(1^*\mu)$, $S_Q(5^*\mu)$ are translation invariant. The connection between (1f), (5f), their limit equations, and tauberian curves is given by

<u>Lemma 2a.</u> <u>If</u> (2) - (4) <u>hold and</u> x <u>is a bounded solution of</u> (1f), <u>then</u> $x \in \tilde{T}$ <u>and</u>

$$\Gamma(x) \subset S_{\Omega(x)}(1^* f(\infty)), \quad \Omega(x) = R(S_{\Omega(x)}(1^* f(\infty))).$$

<u>Lemma 2b.</u> <u>If</u> (2) - (4), (6) <u>hold and</u> $x \in T$ <u>is a bounded solution of</u> (5f), <u>then</u> $x \in \tilde{T}$ <u>and</u>

$$\Gamma(x) \subset S_{\Omega(x)}(5^* f(\infty)), \quad \Omega(x) = R(S_{\Omega(x)}(5^* f(\infty))).$$

In Lemma 2a the hypothesis easily implies that $x' \in L^\infty$ and, thus, that $x \in C_u$. Hence, $x \in \tilde{T}$. However, in Lemma 2b $x \in T$ is not implied by the remaining hypotheses. To see this, in (7f) let

$$B(t) = \int_{-\infty}^{t} e^{-\xi} e^{-e^{-\xi}} d\xi, \quad f(t) = \frac{1}{e^t + e^{-t}} \qquad (-\infty < t < \infty).$$

A calculation shows that $x(t) = \sin e^t$ is a solution of the resulting equation. Clearly, $x \notin T$. It should also be noted that here (see [4])

$$\Omega(x) = [-1, 1] \neq R(S_{\Omega(x)}(7^* f(\infty))) = R(\{y(t) \equiv 0\}) = 0 .$$

We need the

Definition. $\{t_m\} \subset \mathbf{R}^1$ is an expanding t-sequence if $t_1 < t_2 < \dots$ and $t_m - t_{m-1} \to \infty$.

Definition. $\psi_m \in C^\infty(\mathbf{R}^1, [0,1])$ $(m = 1, 2, \dots)$ is a ψ-sequence associated with an expanding t-sequence $\{t_m\}$ if

$$\sum_{m=1}^{\infty} \psi_m(t) \equiv 1 (t \in \mathbf{R}^1),$$

$$\lim_{m \to \infty} \|\psi_m^{(j)}\|_\infty = 0, \quad \lim_{t \to \infty} \sum_{m=1}^{\infty} |\psi_m^{(j)}(t)| = 0 \qquad (j = 1, 2, \dots),$$

$$\psi_1(t) \equiv 1 (t \le t_1), \quad \psi_1(t) \equiv 0 (t_2 \le t), \quad \psi_1'(t) \le 0 (t_1 \le t \le t_2),$$

$$\psi_m(t) \equiv 0 (t \le t_{m-1} \text{ and } t \ge t_{m+1}), \quad \psi_m(t_m) = 1,$$

$$\psi_m'(t) \ge 0 (t_{m-1} \le t \le t_m), \; \psi_m'(t) \le 0 (t_m \le t \le t_{m+1}) \qquad (m = 2, 3, \dots).$$

Thus, an expanding t-sequence, $\{t_m\}$, is an increasing sequence of real numbers whose successive differences tend to infinity. An associated ψ-sequence, $\{\psi_m\}$, is a partition of unity for $\mathbf{R}^1$ in which the successive ψ_m vary slower and slower as $m \to \infty$, increasing from 0 to 1 as $t \to t_m-$ and then decreasing from 1 to 0 as $t \to t_{m+1}-$. It is easily seen that every expanding t-sequence has infinitely many associated ψ-sequences.

If $x \in \mathcal{V}$ and $Y \subset \mathcal{V}$, then the condition

$$(8) \qquad \lim_{s \to \infty} \{ \inf_{y \in Y} [\sup_{|t-s| \le d} |x(t) - y(t)|] \} = 0 \;\; \forall d > 0$$

is a way of stating that Y approximates x for large t. It may be thought of as stating that for any window of fixed width $(2d)$, centered at $t = s$ and moving along the t-axis towards ∞, there is a curve in Y which (when viewed only through the window) is arbitrarily close to x. If Y is translation invariant, it can be shown, whether or not $x \in T$, that (8) holds if and only if $d(x_t, Y) \to 0 (t \to \infty)$. The relationship of (8) to ψ-sequences is given by

Lemma 3. If $x \in \mathcal{V}$, $Y \subset \mathcal{V}$, and (8) hold, then there exist a sequence $\{y_m\} \subset Y$ and an expanding t-sequence $\{t_m\}$ such that

$$\lim_{m\to\infty} \{ \sup_{t_{m-2} \le t \le t_{m+2}} |x(t) - y_m(t)| \} = 0 .$$

Moreover, for any ψ-sequence, $\{\psi_m\}$, associated with $\{t_m\}$

(9) $$x = \sum_{m=1} \psi_m y_m + \eta ,$$

where $\eta(t) \to 0 \quad (t \to \infty)$;

Thus, (8) implies that x "drifts" among some sequence $\{y_m\} \subset Y$ which are successively closer on longer and longer intervals.

Summarizing our results thus far, first for tauberian curves and then for bounded solutions of integral equations, we have

Theorem 1. If $x \in \tilde{T}$, then (8), with $Y = \Gamma(x)$, and the conclusions of Lemmas 1 and 3 are satisfied.

Theorem 1a. If (2) - (4) hold and x is a bounded solution of (1f), then the conclusions of Theorem 1, Lemma 2a, and

$$\lim_{m\to\infty} \{ \operatorname*{ess\,sup}_{t_{m-1} \le t \le t_{m+1}} |x'(t) - y_m'(t)| \} = 0,$$

$$\lim_{t\to\infty} \{ \operatorname*{ess\,sup}_{t \le \tau < \infty} |\eta'(\tau)| \} = 0,$$

are satisfied.

Theorem 1b. If (2) - (4), (6) hold and $x \in T$ is a bounded solution of (5f), then the conclusions of Theorem 1 and Lemma 2b are satisfied.

For a bounded solution, x, of an integral equation, the existence or non-existence of $x(\infty)$ is often of interest. We first study this problem for tauberian curves. Some preliminary definitions: the identically constant elements of $\Gamma(x)$ are called the stationary curves of x and denoted by

$$SC(x) = \Gamma(x) \cap \{y(t) \equiv y(0)\} \qquad (x \in \mathcal{Y}).$$

Further denote

$$SC(Y) = \bigcup_{y \in Y} SC(y) \qquad (Y \subset \mathcal{Y}).$$

Then for any $x \in \mathcal{Y}$

$$SC(\Gamma(x)) = \bigcup_{y \in \Gamma(x)} SC(y) = SC(x), \quad R(SC(x)) \subset \Omega(x).$$

If $x(\infty)$ exists, then obviously $\Omega(x) = x(\infty)$ and $\Gamma(x) = SC(x) = \{y(t) \equiv x(\infty)\}$. If $y \in \Gamma(x)$ and either $y(\infty) = c$ or $y(-\infty) = c$ exists, then $w(t) \equiv c \in SC(x)$.

The critical points of $(1^*\mu)$ and $(5^*\mu)$ are defined by

$$CP(1^*\mu) = \{c \in \mathbb{C}^N \mid g(c)A(\infty) = \mu\},$$
$$CP(5^*\mu) = \{c \in \mathbb{C}^N \mid c + g(c)A(\infty) = \mu\},$$

respectively, where A has been normalized so that $A(-\infty) = 0$. Trivially, $y(t) \equiv c \in S_{\mathbb{C}^N}(1^*\mu)$ if and only if $c \in CP(1^*\mu)$ and $y(t) \equiv c \in S_{\mathbb{C}^N}(5^*\mu)$ if and only if $c \in CP(5^*\mu)$. The notions of stationary curves and critical points are related by

<u>Lemma 4a.</u> <u>If</u> (2) - (4) <u>hold and</u> x <u>is a bounded solution of</u> (1f), <u>then</u> $R(SC(x)) \subset CP(1^* f(\infty))$.

<u>Lemma 4b.</u> <u>If</u> (2) - (4), (6) <u>hold and</u> $x \in T$ <u>is a bounded solution of</u> (5f), <u>then</u> $R(SC(x)) \subset CP(5^* f(\infty))$.

For each $Y \subset \mathcal{Y}$ and $\alpha, \beta \in \mathbb{C}^N$ let

$$Y^{(\alpha,\beta)} = Y \cap \{y(-\infty) = \alpha,\ y(\infty) = \beta\}.$$

We will employ the

<u>Definition.</u> $Y \subset \mathcal{Y}$ is of type 1 if either $Y = \emptyset$ or there exist distinct $c_1, \dots, c_n \in \mathbb{C}^N$ $(n \geq 1)$ such that

$$Y = \bigcup_{i=1}^{n} [Y \cap \{y(t) \equiv c_i\}] \cup \bigcup_{(i,j) \in E} Y^{(c_i, c_j)},$$

where $E \subset \{(i,j) \mid i, j = 1, 2, \dots\}$ is either empty or satisfies:

(i) $i \neq j \ \forall (i,j) \in E$,

(ii) there does not exist a sequence $\{(i_k, j_k)\}_{k=1}^{p} \subset E$ for any $p \geq 2$ such that $i_{k+1} = j_k$ $(1 \leq k \leq p-1)$ and $i_1 = j_p$,

(iii) $Y^{(c_i, c_j)} \neq \emptyset \ \forall (i,j) \in E$.

Thus, if $Y \neq \emptyset$, each $y \in Y$ is either in $SC(Y) \subset \{c_1, \dots, c_n\}$ or $Y^{(c_i, c_j)}$ for some $(i,j) \in E$. Moreover, no subset of $\{c_1, \dots, c_n\}$ is "joined" by a closed "loop" of nonconstant elements of Y, where each element of the loop is traversed in the direction of increasing t. Clearly:

Lemma 5. If $Y_1 \subset Y_2 \subset \Psi$ and if Y_2 is of type 1, then Y_1 is of type 1.

The existence of $x(\infty)$, for $x \in \tilde{T}$, is characterized in

Theorem 2. Let $x \in \tilde{T}$. Then $x(\infty)$ exists if and only if $\Gamma(x)$ is of type 1.

The following counterparts for solutions of integral equations are immediate consequences of Theorem 2 and Lemmas 2a, 2b, and 5.

Theorem 2a. Let (2) - (4) hold and let x be a bounded solution of (1f). Then $x(\infty)$ exists if and only if $S_{\Omega(x)}(1^* f(\infty))$ is of type 1.

Theorem 2b. Let (2) - (4), (6) hold and let $x \in T$ be a bounded solution of (5f). Then $x(\infty)$ exists if and only if $S_{\Omega(x)}(5^* f(\infty))$ is of type 1.

A typical application of, for example, Theorem 2a would involve showing that $S_{B_r}(1^* f(\infty))$ is of type 1 for all r. Then Theorem 2a and Lemma 5 imply that $x(\infty)$ exists for all bounded solutions of (1f).

Outline of Proofs:

Lemma 1. Except for the equicontinuity of $\Gamma(x)$, the first three lines of Lemma 1 are true for all $x \in \Psi$ and follow directly from the definitions of $\Omega(x)$ and $\Gamma(x)$. To show that $\Gamma(x)$ is equicontinuous for $x \in T$ it obviously suffices to show that any sequence $\{y^{(1)}, y^{(2)}, \dots\} \subset \Gamma(x)$ is equicontinuous. The definition of $\Gamma(x)$ yields the existence of $t_k^{(n)}$ $(k, n = 1, 2, \dots)$ such that

$$(10) \quad \begin{cases} t_k^{(n)} < t_{k+1}^{(n)}, \quad t_k^{(n)} < t_k^{(n+1)}, \quad t_k^{(n)} \to \infty & (k \to \infty) \\ x_{t_k^{(n)}} \to y^{(n)} \quad (k \to \infty) \quad \text{c.o.} & (n = 1, 2, \dots). \end{cases}$$

Let $\varepsilon > 0$. Then $x \in T$ implies that

$$|x(\tau+\lambda) - x(\lambda)| \le \varepsilon \qquad (\lambda \ge \lambda_0, \quad |\tau| \le \tau_0)$$

for some $\lambda_0(\varepsilon) \in R^1$, $\tau_0(\varepsilon) > 0$. Hence, using (10),

$$|x_{t_k^{(n)}}(t+\tau) - x_{t_k^{(n)}}(t)| \le \varepsilon \qquad (n \ge 1, \quad k \ge k_0, \quad |\tau| \le \tau_0)$$

for some $k_0(\varepsilon, t)$, which with the triangle inequality yields

$$|y^{(n)}(t+\tau) - y^{(n)}(t)| \le \varepsilon + |y^{(n)}(t+\tau) - x_{t_k^{(n)}}(t+\tau)|$$
$$+ |x_{t_k^{(n)}}(t) - y^{(n)}(t)|$$

for $n \ge 1$, $k \ge k_0$, $|\tau| \le \tau_0$. Letting $k \to \infty$ implies

$$|y^{(n)}(t+\tau) - y^{(n)}(t)| \le \varepsilon \qquad (n \ge 1,\ |\tau| \le \tau_0),$$

completing the proof of equicontinuity of $\Gamma(x)$.

$\Omega(x) \ne \emptyset$ since $\{x(n)\} \subset \mathbb{C}^N$ is bounded. If $\omega_n \in \Omega(x)$ and $\omega_n \to \omega$, then there exist $t_k^{(n)}$ satisfying the first line of (10) and $|x(t_k^{(n)}) - \omega_n| \le 1/k$. The triangle inequality now yields $x(t_n^{(n)}) \to \omega$. Thus, $\Omega(x)$ is compact. If $\Omega(x)$ is not connected, there exist nonempty, disjoint, and compact $E, F \subset \mathbb{C}^N$ such that $\Omega(x) = E \cup F$. Then $\text{dist}(E,F) = \rho > 0$. However, $x \in T$ easily yields a sequence $\tau_n \to \infty$ such that $\rho/4 \le \text{dist}(x(\tau_n), E) \le 3\rho/4$. There exist, since $\{x(\tau_n)\}$ is bounded, a subsequence $\{x(\tau_{n_k})\}$ of $\{x(\tau_n)\}$ and an $\omega \in \Omega(x)$ such that $x(\tau_{n_k}) \to \omega$. Hence, $\text{dist}(\omega, \Omega(x)) > 0$, which contradicts $\omega \in \Omega(x)$ and establishes the connectedness of $\Omega(x)$.

If $x \in \tilde{T}$ and $t_n \to \infty$, then the translates x_{t_n} are, essentially, equicontinuous and uniformly bounded on compact sets. Employing this fact in the usual diagonalization proof of the Ascoli-Arzela lemma shows that

$$\text{(11)} \qquad x \in \tilde{T},\ t_n \to \infty \Longrightarrow x_{t_{n_k}} \to y \ \text{ c.o. for some } \{t_{n_k}\} \subset \{t_n\} \text{ and } y \in \Gamma(x).$$

This result (stated in the paragraph following the statement of Lemma 1) and the arguments of the preceding paragraphs, with c.o. convergence in $\mathfrak{Y}$ replacing convergence in $\mathbb{C}^N$, establish the last two lines of Lemma 1.

<u>Lemma 2a</u>. The hypothesis of 2a implies that $x \in \tilde{T}$. Hence, Lemma 1 implies for each $y \in \Gamma(x)$

$$\text{(12)} \qquad x_{t_n} \to y \ \text{ c.o.},\quad y \in C_u,\quad R(y) \subset \Omega(x),\quad \Omega(x) \text{ is compact}$$

for some $t_n \to \infty$. From (1f)

$$x'_{t_n}(t) = -\int_{-\infty}^{\infty} g(x_{t_n}(t-\xi))\, dA(\xi) = f_{t_n}(t).$$

The first assertion of the lemma follows from (12) and the reasoning of (4.5)-(4.7) of [4]. The second assertion follows from the first and Lemma 1.

Lemma 2b. The hypothesis of 2b implies that $x \in \tilde{T}$. Hence, (12) again follows from Lemma 1. From (5f)

$$x_{t_n}(t) + \int_{-\infty}^{\infty} g(x_{t_n}(t-\xi))\, dA(\xi) = f_{t_n}(t) .$$

The first assertion follows from (12) and the reasoning of (5.5)-(5.6) of [4]. Again the second assertion follows from the first and Lemma 1.

Lemma 3. This is Lemma 3.1 of [4].

Lemma 4a. Let $c \in R(SC(x))$ and let $w(t) \equiv c$. Then the definition of $SC(x)$ and Lemma 2a imply that

$$w \in SC(x) \subset \Gamma(x) \subset S_{\Omega(x)}(1^* f(\infty)) \subset S_{\mathbb{C}^N}(1^* f(\infty)) .$$

This together with $(1^* f(\infty))$ shows that $c \in CP(1^* f(\infty))$ and completes the proof.

Lemma 4b. Replacing Lemma 2a and $(1^* f(\infty))$ by Lemma 2b and $(5^* f(\infty))$, respectively, in the proof of Lemma 4a establishes this result.

Lemma 5. This follows immediately from the definition of type 1.

Theorem 2. The basic procedure of the proof of Theorem 2.1 of [3] without, however, appealing to (1f) (as in [3]) but using (11) directly instead establishes this result.

As already noted, Theorems 1, 1a, 1b follow from Lemmas 1, 2a, 2b, 3 and Theorems 2a, 2b follow from Theorem 2 and Lemmas 2a, 2b, 5.

REFERENCES

1. J. K. Hale, Sufficient conditions for stability and instability of autonomous functional-differential equations, J. Differential Equations 1(1965), 452-482.
2. G. S. Jordan and R. L. Wheeler, Linear integral equations with asymptotically almost periodic solutions, J. Math. Anal. Appl. 52(1975), 454-464.
3. J. J. Levin, On some geometric structures for integrodifferential equations, Advances in Math. 22(1976), 146-186.
4. J. J. Levin and D. F. Shea, On the asymptotic behavior of the bounded solutions of some integral equations, I, II, III, J. Math. Anal. Appl. 37 (1972), 42-82, 288-326, 537-575.
5. S-O. Londen, The qualitative behavior of the solutions of a nonlinear Volterra equation, Michigan Math. J. 18(1971), 321-330.
6. S-O. Londen, On the solutions of a nonlinear Volterra equation, J. Math. Anal. Appl. 39 (1972), 564-573.

7. R. K. Miller, Asymptotic behavior of nonlinear delay-differential equations, J. Differential Equations 1 (1965), 293-305.

8. R. K. Miller, Asymptotic behavior of solutions of nonlinear Volterra equations, Bull. Amer. Math. Soc. 72 (1966), 153-156.

9. G. R. Sell, A Tauberian condition and skew flows with applications to integral equations, J. Math. Anal. Appl. 43 (1973), 388-396.

ON THE ASYMPTOTICS OF A NONLINEAR SCALAR VOLTERRA INTEGRODIFFERENTIAL EQUATION

by

STIG-OLOF LONDEN

Institute of Mathematics
Helsinki University of Technology
Otaniemi, Finland

1. INTRODUCTION

We investigate the qualitative behavior of the bounded solutions of the nonlinear scalar Volterra integrodifferential equation

$$(1.1) \qquad x'(t) + \int_0^t g(x(t-\tau))da(\tau) = f(t), \quad t \geq 0,$$

where g,a,f are given real functions and x denotes the solution. Our primary interest is to analyze (1.1) under the following assumption on f:

$$(F_\infty) \qquad f \in L^\infty(R^+), \lim_{t\to\infty} f(t) = 0,$$

and in particular to give conditions on the kernel $a(t)$ which permi one to reduce the case when only (F_∞) is assumed to the case when the essentially stronger condition

$$(F_1) \qquad f \in L^1(R^+)$$

is imposed.

Specifically we have the following

THEOREM. *Suppose*

(1.2) $g \in C(R)$,

(1.3) $a \in NBV(R^+)$,

(1.4) $\int_t^\infty |da(\tau)| \in L^1(R^+)$,

(1.5) $\operatorname{Re} \hat{a}(\omega) \geq 0,\ \omega \in R$,

(1.6) $\hat{a}(\omega) = 0,\ \omega \in S \overset{\text{def}}{=} \{\omega \mid \operatorname{Re} \hat{a}(\omega) = 0\}$,

(1.7) $\hat{a}(0) > 0$,

(1.8) $x \in L^\infty \cap LAC(R^+)$, x *satisfies* (1.1) *a.e. on* R^+, *and assume* (F_∞) *holds. Then either* i) *or* ii) *below is satisfied*.

i)

(1.9) $\lim_{t\to\infty} g(x(t)) = 0$.

ii) *There exist* $\tilde{f},\tilde{x}$ *such that*

(1.10) $\tilde{f} \in L^1(R^+)$,

(1.11) $\tilde{x} \in L^\infty \cap LAC(R^+)$,

(1.12) $\lim_{t\to\infty} t^{-1} V_{\tilde{x}}[0,t] = 0$,

(1.13) $\lim_{t\to\infty} g(\tilde{x}(t))$ *does not exist, and satisfying*

(1.14) $\tilde{x}'(t) + \int_0^t g(\tilde{x}(t-\tau))da(\tau) = \tilde{f}(t)$ on R^+.

From the above result follows that if the assumptions (1.2)-(1.8) together with (F_1) can be shown to imply $\lim_{t\to\infty} g(x(t)) = 0$, then (1.2)-(1.8) combined with (F_∞) imply the same.

Concerning the above hypothesis we note first that only the usual continuity assumption (1.2) is imposed on the nonlinear function g. In particular we do not make any restrictions on the zero set of this function.

It is not surprising that something like (1.4) is needed if the (F_∞)-case is to be reduced to the $(F_1$-case). Interestingly enough a single moment on the kernel suffices. As we are aiming for a more specific result we incorporate the positivity condition (1.5). Concerning the hypothesis on the imaginary part of the Fourier-Stieltjes transform we recall that a result by Staffans [2, Th. 3.3 (ii)] shows this condition to be necessary for the limit set of $x(t)$ to contain constant functions. Observe that this is consistent with our use of (1.6); it is needed only in Lemma 1; where it's use leads to a result analogous to (1.12).

Above we have assumed $\hat{a}(0) > 0$. This of course implies $g(y) = 0$ for any constant limit function y of $x(t)$ and so does have non-trivial consequences for the treatment. The case $\hat{a}(0) = 0$ will be analyzed in another paper.

A recent result, partially overlapping the above, by Staffans [3] establishes that (1.2)-(1.8), (F_∞) and the assumption that S is countable imply i). The question thus arises what happens if (1.2)-(1.8) and (F_1) hold and S is large; i.e. noncountable.

Finally note that the above Theorem improves upon a recent result by Levin [1, Th. 2.5] where stronger conditions are imposed on both a and g.

2. PROOF OF THE THEOREM

The first part of the proof is contained in Lemmas 1-5 to follow. We postpone the proofs of these Lemmas to Sections 3-7.

LEMMA 1. *Assume* (1.2)-(1.6) *hold and let* $y(t)$ *be any solution of*

$$(2.1)\qquad y'(t) + \int_0^\infty g(y(t-s))da(s) = 0, \qquad t \in R,$$

such that

$$(2.2)\qquad y \in L^\infty \cap LAC(R).$$

Then

$$(2.3)\qquad \lim_{t\to\infty} t^{-1} \int_0^t |y'(\tau)|d\tau = 0.$$

This result implies that a bounded solution of the limit equation (2.1) cannot, under the hypotheses (1.2)-(1.6) keep oscillating at a rate bounded away from zero on arbitrarily long intervals. Any existing oscillations must be more and more spread out. As was pointed out in the introduction the proof of Lemma 1 is the only place in the proof of the Theorem where (1.6) is needed.

LEMMA 2. *Let* (1.2)-(1.8) *and* (F_∞) *hold. Define* $\Gamma_c(x)$, G, a, b *by*

(2.4) $\quad \Gamma_c(x) = \{y \in R \mid$ there exist $r_n \to \infty$ such that $x(t+r_n) \to y$ uniformly on compact sets$\}$.

(2.5) $\quad G(y) = \int_0^y g(u)du,\ a = \liminf_{t\to\infty} x(t),\ b = \limsup_{t\to\infty} x(t).$

Then

(2.6) $\quad G(\Gamma_c(x)) = \sup_{a\le y\le b} G(y).$

The function G, when considered on the asymptotic values of $x(t)$, thus attains its maximum on the set of constant limit functions. Observe that by Lemma 1 $\Gamma_c(x)$ is nonempty.

For the subsequent analysis it is convenient to define sets $B(\alpha,\beta)$ by

(2.7) $\quad B(\alpha,\beta) = \{y \mid \alpha \le y \le \beta,\ g(y) = 0,\ G(y) = \sup_{\alpha\le y\le\beta} G(y)\}.$

Take a,b as in Lemma 2 and consider $B(a,b)$. Clearly either $m(B(a,b)) > 0$ or $m(B(a,b)) = 0$. Lemmas 3 and 4 below are needed to handle the former case. Although these Lemmas do hold if $m(B(a,b))=0$ they play no role in this latter case. Also note that the construction (2.12)-(2.18) after Lemma 4 is rather simple in this case. In fact, if $m(B(a,b)) = 0$ then one may proceed directly to (2.23) where one takes Γ_0,λ,μ as in (2.12), (2.14) and observes that by Lemma 2 one has $m(\Gamma_c(x)) \le m(B(a,b))$ and therefore in this case $m(\Gamma_0(x)) = 0$.

LEMMA 3. Let (1.2)-(1.7) hold, let $y(t)$ satisfy (2.1), (2.2) and define α,β by $\alpha = \inf_{t\in R} y(t) \le \sup_{t\in R} y(t) = \beta$. Then

(2.8) $y'(t) = 0, \ t \in \{s \mid y(s) \in B(\alpha,\beta)\}$.

The next Lemma implies that any density point of $B(a,b)$ acts much like a barrier for the solution $x(t)$. For large t and provided (F_∞) is assumed a solution of (1.1) can pass a neighborhood of such a point only very slowly. Notice that using Lemma 4 one can show that if also (F_1) is assumed then a solution of (1.1) can get through a neighborhood of a density point of $B(a,b)$ at most a finite number of times. We include $a(b)$ in the set of density points of B provided the right hand (left hand) density of B at $a(b)$ equals 1.

LEMMA 4. Let the assumptions of Lemma 3 hold, let ρ be a point of density of $B(\alpha,\beta)$ and assume

(2.9) $y(t_0) = \rho$, for some $t_0 \in R$.

Then

(2.10) $y(t) \equiv \rho, \ t \in R$.

Observe that from Lemmas 2 and 4 one deduces

(2.11) $B_d(a,b) \subset \Gamma_c(x) \subset B(a,b)$

where $B_d(a,b)$ is the set of density points of $B(a,b)$.

Our next purpose is to define an interval $[\lambda,\mu] \subset [a,b]$ such that if

(2.12) $\Gamma_0(x) \overset{\text{def}}{=} [\lambda,\mu] \cap \Gamma_c(x)$,

then, under the assumptions (1.2)-(1.8) and (F_∞),

(2.13) $m(\Gamma_0(x)) = 0$,

and $\Gamma_0(x)$ has the property formulated below in Lemma 5.

Suppose at first that $m(B(a,b)) = 0$. In this case let

(2.14) $\lambda = \inf_{y\in\Gamma_c(x)} y \quad , \qquad \mu = \sup_{y\in\Gamma_c(x)} y \quad .$

Next suppose $m(B(a,b)) > 0$ and assume that there exists $\eta \in (a,b)$ such that

(2.15) $g(\eta) \neq 0, \quad [a,\eta) \cap B_d(a,b) \neq \emptyset, \; (\eta,b] \cap B_d(a,b) = \emptyset.$

In this case define

(2.16) $\lambda = \sup\{y|\; y < \eta,\; y \in B_d\},\; \mu = \inf\{y|\; y > \eta,\; y \in B_d\}.$

Finally let $m(B(a,b)) > 0$ and assume that there does not exist $\eta \in (a,b)$ such that (2.15) holds. Now take any $\eta \in [a,b]$ satisfying $g(\eta) \neq 0$. If $\eta > y$, for all $y \in B_d$ then define

(2.17) $\lambda = \sup_{y\in B_d} y, \quad \mu = \sup\{y|\; \lambda \leq y \leq b,\; y \in \Gamma_c(x)\}.$

If $\eta < y$ for all $y \in B_d$, then let

(2.18) $\lambda = \inf\{y \mid a \le y \le \inf_{z\in B_d} z,\ y \in \Gamma_c(x)\}$, $\mu = \inf_{y\in B_d} y$.

Any of the definitions (2.14), (2.17), (2.18) may yield $\lambda = \mu$. This however is an easy case and therefore, without loss of generality, assume in each of the above cases $\lambda < \mu$ and define $\Gamma_0(x)$ by (2.12). Observe that the hypothesis $\lambda < \mu$ contains the assumption that i) in the Theorem does not hold. The property (2.13) now follows by (2.11) and the above construction of $[\lambda,\mu]$. Consequently we may continue to

LEMMA 5. *Let* λ,μ *be as above and assume* (1.2)-(1.8) *and* (F_∞) *are valid. Suppose there exist* t_n, T_n *both* $\to\infty$ *as* $n\to\infty$ *and such that either*

(2.19) $x(t) \le \lambda,\ t_n \le t \le t_n + T_n;\quad x(t_n) = x(t_n+T_n) = \lambda$

or

(2.20) $x(t) \ge \mu,\ t_n \le t \le t_n + T_n;\quad x(t_n) = x(t_n+T_n) = \mu,$

holds. Then there exist $[p_n,q_n],\ [r_n,s_n] \subset [t_n,t_n+T_n]$ *satisfying*

(2.21) $(q_n-p_n)\to\infty,\ (s_n-r_n)\to\infty,\ \int_{t_n}^{q_n} + \int_{r_n}^{t_n+T_n} |f(s)|ds \to 0$

as $n\to\infty$, *and such that*

(2.22) $x(t) \to \begin{cases} \lambda & \textit{if } (2.19) \textit{ holds} \\ \mu & \textit{if } (2.20) \textit{ holds} \end{cases}$

uniformly on $[p_n,q_n] \cup [r_n,s_n]$.

From Lemma 5 follows that one may for our purpose almost ignore the solution $x(t)$ for all t such that $x(t) \notin [\lambda,\mu]$. The subsequent analysis therefore essentially looks only at such t-values for which $x(t) \in [\lambda,\mu]$.

Having stated the necessary Lemmas we turn to the second part of the proof. The first step is here to construct sets $\Gamma_k(x)$; $k = 1,2\ldots$; which approximate $\Gamma_0(x)$, see (2.13), (2.25), (2.26) and each of which has a finite number of components.

Let the open intervals (c_i,d_i) be such that

$$
(2.23)\quad \begin{cases} \Gamma_0(x) = [\lambda,\mu] \setminus \bigcup_{i=1}^{N} (c_i,d_i) \\ \\ c_i,d_i \in \Gamma_0(x);\ (c_i,d_i) \cap (c_j,d_j) = \emptyset,\ i \neq j, \end{cases}
$$

where $1 \leq N \leq \infty$. As $\Gamma_0(x)$ is closed such intervals exist. By (2.13) and by (2.23) there exist integers $N_k < \infty$; $k = 1,2,\ldots$; $2 \leq N_k \leq N_{k+1} \leq N+1$ such that if

$$
(2.24)\quad \Gamma_k(x) \overset{\text{def}}{=} [\lambda,\mu] \setminus \bigcup_{i=1}^{N_k - 1} (c_i,d_i);\quad k = 1,2,\ldots
$$

then

$$
(2.25)\quad m(\Gamma_k(x)) \leq 2^{-k}.
$$

From (2.23), (2.24) follows

$$
(2.26)\quad \Gamma_0(x) \subset \Gamma_{k+1}(x) \subset \Gamma_k(x) \subset [\lambda,\mu],\quad \forall k,
$$

and also that each Γ_k consists of N_k pairwise disjoint components Γ_{jk}; $j = 1,2,\ldots,N_k$; each Γ_{jk} being either a point or a closed interval. Define, for $k = 1,2,\ldots$; ρ_k by

$$(2.27)\quad \rho_k = \frac{2^{-k}}{N_k},$$

and choose α_k so that

$$(2.28)\quad \int_{\alpha_k}^{\infty} \{ \int_{v}^{\infty} |da(\tau)| \} dv \le \rho_k .$$

Take $\tilde{z} \in (\lambda,\mu)$ satisfying

$$(2.29)\quad g(\tilde{z}) \neq 0 \quad (\text{thus } \tilde{z} \notin \Gamma_0)$$

and pick $t_n \to \infty$ satisfying

$$(2.30)\quad x(t_n) = \tilde{z}.$$

After the construction (2.23)-(2.30) we proceed to single out the pieces of the original solution $x(t)$ that will be put together to form the new solution $\tilde{x}(t)$.

By (F_∞) and Lemma 1 applied to $x(t)$ and by Lemma 5 there exist (for a subsequence n_ℓ of n which without loss of generality we take equal to n) $p \in \Gamma_0(x)$, a sequence $s_n \to \infty$ and for each k a positive integer n_k such that if $n \ge n_k$ then

$$(2.31)\quad s_n + \alpha_k \le t_n < s_{n+1}$$

$$(2.32)\quad \text{dist}\,(x(t),p) \le \rho_k,\quad s_n - \alpha_k \le t \le s_n + \alpha_k,$$

$$\text{2.33)}\quad \int_{s_n-\alpha_k}^{s_n+\alpha_k} |g(x(s))|ds \le 2^{-1}\rho_k, \quad \int_{s_n-\alpha_k}^{t_n} |f(s)|ds \le 2^{-1}\rho_k.$$

ake any such p, s_n, n_k. For each k there exists a component of $_k$ such that p belongs to this component. Without loss of generality et this component be Γ_{1k}.

Fix an arbitrary k.

By (2.29)-(2.33); as $x(t)$ spends most of the time close to Γ_k, s there exist only a finite number of components of Γ_k and by (1.4) ne has roughly the following. There do exist intervals contained in $s_n, s_{n+1})$ on which $x(t)$ moves from one component of Γ_k to another, ventually completing a closed loop, and such that $x(t)$ on these ntervals can be singled out, translated and put together to form a ew solution y_k of (1.1) on $(0,T_k)$ satisfying $y_k(0) \approx y_k(T_k) \approx p$ T_k equals the measure of the union of these intervals) and to which orresponds a sufficiently small nonhomogeneous term, see (2.47). e above rough description is made precise in the following paragraphs.

From (F_∞) and Lemma 1 applied to $x(t)$ and by Lemma 5 one has at there exists an integer L_k and for each i; $i = 1,2,\ldots,L_k$; equences s_{in}, t_{in} satisfying the following relations (2.34)-(2.40) r $n \ge n_k$. (Note that $s_{1n} = s_n$.)

$$\text{.34)}\quad 1 \le L_k \le N_k,$$

$$\text{.35)}\quad s_{1n} < t_n < t_{1n} \le s_{2n} < t_{2n} \le s_{3n} <\ldots< s_{L_k n} < t_{L_k n} \le s_{1,n+1}.$$

$$\text{.36)}\quad \text{dist}\,(x(t),\Gamma_{ik}) \le \rho_k,\ s_{in}-\alpha_k \le t \le s_{in}+\alpha_k;\ i = 1,\ldots,L_k,$$

$$(2.37)\quad \operatorname{dist}(x(t),\Gamma_{i+1,k}) \le \rho_k,\ t_{in}-\alpha_k \le t \le t_{in}+\alpha_k,\ i = 1,\dots,L_k-1,$$

$$(2.38)\quad \operatorname{dist}(x(t),\Gamma_{1k}) \le \rho_k,\ t_{L_kn}-\alpha_k \le t \le t_{L_kn}+\alpha_k,$$

$$(2.39)\quad \int_{s_{in}-\alpha_k}^{s_{in}+\alpha_k} + \int_{t_{in}-\alpha_k}^{t_{in}+\alpha_k} |g(s)|ds \le \rho_k,\quad i = 1,\dots,L_k,$$

$$(2.40)\quad \int_{s_{in}-\alpha_k}^{t_{in}+\alpha_k} |f(s)|ds \le \rho_k,\quad i = 1,\dots,L_k.$$

Above one may have to take a subsequence of n, increase n_k, and possibly relabel the components of Γ_k. Also observe that as we take $s_{1n} = s_n$ then (2.36) with $i = 1$ follows from (2.32). If $L_k = 1$ then (2.37) should be dropped.

Let φ_i; $i = 1,2$; be any functions satisfying

$$(2.41)\quad \begin{cases} \varphi_i \in C^\infty[0,1];\quad \varphi_1(0) = \varphi_2(1) = 1;\ \varphi_1(1) = \varphi_2(0) = 0; \\ \varphi_i'(0) = \varphi_i'(1) = 0,\ \varphi_i(x) \ge 0,\ i = 1,2;\ (\varphi_1+\varphi_2)(x) \equiv 1, \\ \varphi_1'(x) \le 0,\ \varphi_2'(x) \ge 0,\ 0 \le x \le 1. \end{cases}$$

Choose $n = n_k$ and define y_k as follows

$$(2.42)\quad y_k(s) = x(s+\tau_{jk});\ T_{j-1,k} + 1 \le s \le T_{jk}\quad \text{for } j = 1,\dots,L_k,$$

and

$$(2.43)\quad y_k(s) = \varphi_1(s-T_{jk})x(s+\tau_{jk}) + \varphi_2(s-T_{jk})x(s+\tau_{j+1,k});$$

$$T_{jk} \le s \le T_{jk}+1,$$

for $\tau = 1,\ldots,L_k-1$, where

$$(2.44)\quad T_{jk} \overset{\text{def}}{=} \sum_{i=1}^{j} (t_{in_k} - s_{in_k}),\ \tau_{jk} \overset{\text{def}}{=} \sum_{i=1}^{j} (s_{in_k} - t_{i-1,n_k}),\ j = 1,\ldots,L_k,$$

with $t_{on_k} \overset{\text{def}}{=} 0$, $T_{ok} \overset{\text{def}}{=} -1$. Clearly $y_k \in L^\infty \cap AC[0,T_k]$ where

$$(2.45)\quad T_k \overset{\text{def}}{=} T_{L_k,k}\ .$$

Making use of y_k one now defines f_k by

$$(2.46)\quad f_k(t) = y_k'(t) + \int_0^t g(y_k(t-\tau))da(\tau),\ 0 \le t \le T_k\ .$$

It now takes some lengthy but routine calculations which require the use of (1.4), (1.8), (2.25), (2.27)-(2.46) to show that the first part of (2.47) holds. The second part is obvious.

$$(2.47)\quad \int_0^{T_k} |f_k(t)|dt \le C\cdot 2^{-k},\ \sup_{0\le s\le T_k} |g(y_k(s))| \ge |g(\tilde{z})| > 0.$$

Above C is an a priori constant independent of k.

The above construction of y_k is now carried out for every k.

The final step of the construction of $\tilde{x},\tilde{f}$ is then to translate and put together the different y_k's in the manner described below. Recall that each y_k has the properties

$$(2.48)\quad \begin{cases} \int_0^{\alpha_k} |g(y_k(s))|ds \le 2^{-k},\ \int_{T_k-\alpha_k}^{T_k} |g(y_k(s))|ds \le 2^{-k}, \\ |y_k(0) - p| \le 2\cdot 2^{-k},\ |y_k(T_k) - p| \le 2\cdot 2^{-k}. \end{cases}$$

Let

$$(2.49)\quad \tau_k \overset{\text{def}}{=} \sum_{i=1}^{k} T_i - k;\quad k = 1,2,\ldots\ ;\ \tau_0 \overset{\text{def}}{=} 0,$$

and define $\tilde{x},\tilde{f}$ by

$$(2.50)\quad \begin{cases} \tilde{x}(s) = y_1(s), & 0 \le s \le \tau_1, \\ \tilde{x}(s) = y_k(s-\tau_{k-1}), & \tau_{k-1}+1 \le s \le \tau_k,\ k = 2,3,\ldots\ , \\ \tilde{x}(s) = \varphi_1(s-\tau_k)y_k(s-\tau_{k-1}) + \varphi_2(s-\tau_k)y_{k+1}(s-\tau_k); & \\ \qquad \tau_k \le s \le \tau_k+1,\quad k = 1,2,\ldots & \end{cases}$$

$$(2.51)\quad \tilde{f}(s) = \tilde{x}'(s) + \int_0^s g(\tilde{x}(s-\tau))da(\tau),\quad s \in R^+.$$

Clearly $\tilde{x},\tilde{f}$ satisfy (1.11) and (1.14). From (1.4), (2.41), (2.46)-(2.51) one gets after some calculations the assertion (1.10). A combination of (1.10), (1.11) and (1.14) with Lemma 1 will yield (1.12). The relation (1.13), which excludes i), is contained in the assumption $\lambda < \mu$.

As the final point observe that if i) does not hold and $\lambda = \mu$ then the above treatment can be significantly simplified and will again yield (1.10)-(1.13).

The proof of the Theorem is now complete.

3. PROOF OF LEMMA 1

By (1.2) and (2.2) it is evident that

$$\text{(3.1)}\qquad \sup_{t\in R} |g(y(t))| < \infty; \quad \sup_{t\in R} |G(y(t))| < \infty,$$

where $G(u) \overset{\text{def}}{=} \int_0^u g(z)dz$. Multiply (2.1) by $g(y(t))$ and integrate over $[0,t]$. This gives

$$\text{(3.2)}\qquad G(y(t)) - G(y(0)) + \int_0^t g(y(\tau)) \int_0^\tau g(y(\tau-s))da(s)d\tau =$$

$$= -\int_0^t g(y(\tau)) \int_\tau^\infty g(y(\tau-s))da(s)d\tau.$$

Define, for $t > 0$,

$$\text{(3.3)}\qquad z_t(\tau) = \begin{cases} g(y(\tau)) & 0 \le \tau \le t, \\ 0 & \tau \notin [0,t]. \end{cases}$$

Clearly $z_t \in L_1 \cap L_2(R)$. Then, by (1.3)-(1.5), (3.1)-(3.3),

$$\text{(3.4)}\qquad c_1 \overset{\text{def}}{=} \sup_{t>0} \int_{-\infty}^\infty |\hat{z}_t(\omega)|^2[\operatorname{Re} \hat{a}(\omega)]^2 d\omega < \infty.$$

Define

$$\text{(3.5)}\qquad u_t(\tau) = \begin{cases} y'(\tau), & 0 \le \tau \le t, \\ 0, & \tau \notin [0,t], \end{cases}$$

$$(3.6)\qquad f_t(\tau) = \begin{cases} 0 & \tau < 0, \\ -\int_{\tau}^{\infty} g(y(\tau-s))da(s), & 0 \le \tau \le t, \\ \int_{\tau-t}^{\tau} g(y(\tau-s))da(s), & t < \tau < \infty . \end{cases}$$

If (2.1), (3.3), (3.5), (3.6) are used one easily verifies that

$$(3.7)\qquad u_t(\tau) + \int_{-\infty}^{\infty} z_t(\tau-s)da(s) = f_t(\tau) \quad \text{for} \quad \tau \in R,$$

and after invoking also (1.4) and (3.1) one finds that $u_t, f_t \in (L_1 \cap L_2)(R)$ with

$$(3.8)\qquad c_2 \overset{\text{def}}{=} \sup_{t>0} 4\pi \int_{-\infty}^{\infty} |f_t(\tau)|^2 d\tau < \infty.$$

Consequently it follows that

$$(3.9)\qquad \hat{u}_t(\omega) + \hat{z}_t(\omega)\hat{a}(\omega) = \hat{f}_t(\omega), \; \omega \in R.$$

For $\alpha > 0$ let $k(\tau) = 2^{-1}\alpha e^{-\alpha|\tau|}$, $\tau \in R$. Then

$$(3.10)\qquad \hat{k}(\omega) = \alpha^2[\alpha^2+\omega^2]^{-1},$$

and by (3.8)-(3.10)

$$(3.11)\qquad \int_{-\infty}^{\infty} |\hat{u}_t\hat{k}|^2 d\omega = \int_{-\infty}^{\infty} |\hat{f}_t\hat{k} - \hat{z}_t\hat{a}\hat{k}|^2 d\omega \le c_2 + 2\int_{-\infty}^{\infty} |\hat{z}_t\hat{a}\hat{k}|^2 d\omega.$$

Observe that from (1.3), (2.1) and the fact that $g(y(t)) \in BUC(R)$ it follows that

$$(3.12)\qquad y' \in BUC(R).$$

Suppose (2.3) does not hold, use (3.5), (3.12), take α sufficiently large and recall the definition of k. This yields, that for some $\beta > 0$ and some $t_n \to \infty$ one has

$$(3.13) \quad \beta t_n \leq 2\pi \int_{-\infty}^{\infty} |(u_{t_n} * k)(\tau)|^2 d\tau = \int_{-\infty}^{\infty} |\hat{u}_{t_n} \hat{k}|^2 d\omega .$$

Take any such β, t_n and choose $\gamma, \delta > 0$ satisfying

$$(3.14) \quad |\hat{a}\hat{k}|^2 \leq \beta[16\pi h]^{-1} \quad \text{for} \quad |\omega| \geq \gamma$$

$$(3.15) \quad |\mathrm{Im}\ \hat{a}(\omega_2) - \mathrm{Im}\ \hat{a}(\omega_1)|^2 \leq \beta[16\pi h]^{-1} \quad \text{for} \quad |\omega_1 - \omega_2| < \delta,$$

where $h \stackrel{\text{def}}{=} \sup_{\tau \in R} |g(y(\tau))|^2$. By (3.3), (3.14) and by Parseval's relation it is obvious that

$$(3.16) \quad \int_{\gamma}^{\infty} + \int_{-\infty}^{-\gamma} |\hat{z}_t \hat{a}\hat{k}|^2 d\omega \leq 8^{-1}\beta t.$$

Also, from (3.4), (3.10)

$$(3.17) \quad \int_{-\gamma}^{\gamma} |\hat{z}_t \hat{a}\hat{k}|^2 d\omega \leq c_1 + \int_{S_1} + \int_{S_2} |\hat{z}_t \mathrm{Im}\ \hat{a}|^2 d\omega,$$

where

$$(3.18) \quad S_1 \stackrel{\text{def}}{=} \{\omega |\ |\omega| \leq \gamma,\ \mathrm{dist}(\omega, S) < \delta\},\ S_2 \stackrel{\text{def}}{=} [-\gamma, \gamma] \setminus S_1,$$

where S is as in (1.6). Clearly by (1.6), (3.15), (3.18),

$$|\mathrm{Im}\ \hat{a}(\omega)|^2 < \beta[16\pi h]^{-1},\ \omega \in S_1,$$

and so

$$(3.19) \quad \int_{S_1} |\hat{z}_t \mathrm{Im}\ \hat{a}|^2 d\omega \leq 8^{-1}\beta t.$$

Let $\varepsilon \overset{\text{def}}{=} \inf_{\omega \in S_2} |\text{Re } \hat{a}|^2$ and note that $\varepsilon > 0$. Then by (3.4) and as $c_3 \overset{\text{def}}{=} \sup_{\omega \in R} |\text{Im } \hat{a}|^2 < \infty$,

$$\int_{S_2} |\hat{z}_t \text{ Im } \hat{a}|^2 d\omega \leq c_1 c_3 \varepsilon^{-1}. \tag{3.20}$$

But from (3.11), (3.13), (3.16), (3.17), (3.19), (3.20) we obtain

$$\beta t_n \leq 2[c_2 + 2c_1 + 2c_1 c_3 \varepsilon^{-1}]$$

which cannot possibly hold for all n. Consequently (2.3) is true.

4. PROOF OF LEMMA 2

Note at first that (1.1), (1.7) and (F_∞) give us

$$g(y) = 0, \quad y \in \Gamma_c(x). \tag{4.1}$$

Let $G_m = \inf_{y \in \Gamma_c(x)} G(y)$ and suppose $G_m < G_M \overset{\text{def}}{=} \sup_{a \leq y \leq b} G(y)$. Take $p \in \Gamma_c(x)$ and $r_n \to \infty$ such that $G(p) = G_m$, $x(t+r_n) \to p$ uniformly on compact sets. Let q be any number such that

$$G_m < q < G_M, \; q \notin \{G(y) |\; a \leq y \leq b, \; g(y) = 0\}. \tag{4.2}$$

Such numbers do exist, see e.g. [3, Lemma 2.3]. Define s_n by $s_n = \sup\{s |\; G(x(t)) < q, \; r_n \leq t < s\}$. Then

$$G(x(s_n)) = q, \tag{4.3}$$

(4.4) $\quad G(x(t)) \leq q, \quad r_n - T_n \leq t \leq s_n,$

for some $T_n \to \infty$. By essentially redoing the proof of Lemma 1 one can show that $\lim_{n\to\infty} (\beta_n - \alpha_n)^{-1} \int_{\alpha_n}^{\beta_n} |x'(\tau)| d\tau = 0$ for any $[\alpha_n, \beta_n] \subset R^+$ satisfying $\lim_{n\to\infty} (\beta_n - \alpha_n) = \infty$. From this fact and from (F_∞) follows the existence of $[\lambda_n, \mu_n] \subset [r_n - T_n, s_n]$ and of y_0 such that

(4.5) $\quad x(t) \to y_0 \in \Gamma_c(x)$ uniformly on $[\lambda_n, \mu_n]$, $\lim_{n\to\infty} (\mu_n - \lambda_n) = \infty$,

(4.6) $$\lim_{n\to\infty} \int_{\lambda_n}^{s_n} |f(s)| ds = 0,$$

and recalling also (1.2), (1.4), (4.1) one may assume

(4.7) $$\lim_{n\to\infty} \int_{\lambda_n}^{s_n} g(x(\tau)) \int_{\tau-\lambda_n}^{\tau} g(x(\tau - s)) da(s) d\tau = 0.$$

The relations (4.1), (4.4) and (4.5) combined with the second part of (4.2) imply

(4.8) $\quad G(y_0) < q.$

Multiply (1.1) by $g(x(t))$ and integrate over $[\lambda_n, s_n]$. This gives

(4.9) $$G(x(s_n)) - G(x(\lambda_n)) + \int_{\lambda}^{s_n} g(x(\tau)) \int_0^{\tau} g(x(\tau - s)) da(s) d\tau = \int_{\lambda_n}^{s_n} g(x(s)) f(s) ds.$$

From (4.3), (4.5)-(4.9) follows

$$\int_{\lambda_n}^{s_n} g(x(\tau)) \int_0^{\tau-\lambda_n} g(x(\tau - s)) da(s) d\tau < 0,$$

for all sufficiently large n. But by (1.5) this cannot be true and so (2.6) is proved.

5. PROOF OF LEMMA 3

Suppose (2.8) does not hold. Then, for example,

(5.1) $y'(\hat{t}) = \varepsilon > 0,$

(5.2) $G(y(\hat{t})) = G_M \overset{\text{def}}{=} \sup_{\alpha \le y \le \beta} G(y).$

Without loss of generality let $\hat{t} = 0$. By (1.7) and Lemma 1 there exist $[\tau_n, t_n] \subset R^-$ such that

(5.3) $\int_{\tau_n}^{t_n} |g(y(s))| ds \to 0, \ (t_n - \tau_n) \to \infty, \text{ as } n \to \infty.$

Take any such intervals. Then define w by

(5.4) $w(\tau, z) = \begin{cases} g(z) & -\infty < \tau \le 0, \\ z - y(0), & 0 < \tau < \infty \end{cases}$

and consider the equation, with $z(t)$ as the unknown,

(5.5) $z'(t) + \int_0^\infty w(t-\tau, z(t-\tau)) da(\tau) = 0, \ t \ge 0.$

Select any C^1-solution of (5.5) which exists on $(-\infty, \hat{\delta}]$ for some $\hat{\delta} > 0$ and is such that

(5.6) $z(t) = y(t), \quad -\infty < t \le 0.$

By (2.1) and (5.4) such a solution $z(t)$ can be found. From (5.1), (5.6) follows that there exists δ satisfying $0 < \delta \le \hat{\delta}$ and such that

$$(5.7)\qquad z'(t) \geq \frac{\varepsilon}{2}, \quad 0 \leq t \leq \delta.$$

Choose any such δ, multiply (5.5) by $w(t,z(t))$ and integrate over $[\tau_n,\delta]$. This gives

$$\int_0^\delta z'(\tau)w(\tau,z(\tau))d\tau + G(y(0)) - G(y(\tau_n)) +$$

$$(5.8)\qquad + \int_{\tau_n}^{\delta} w(\tau,z(\tau)) \int_0^{\tau-\tau_n} w(\tau-s,z(\tau-s))da(s)d\tau$$

$$= -\int_{\tau_n}^{\delta} w(\tau,z(\tau)) \int_{\tau-\tau_n}^{\infty} w(\tau-s,z(\tau-s))da(s)d\tau.$$

By (5.2) and as $\alpha \leq y(\tau_n) \leq \beta$ one has

$$(5.9)\qquad G(y(0)) - G(y(\tau_n)) \geq 0.$$

From (5.4), (5.7) and as $z(0) = y(0)$ one gets, for some positive constant λ,

$$(5.10)\qquad \int_0^\delta z'(\tau)w(\tau,z(\tau))d\tau \geq \lambda.$$

By (1.5) the last term on the left side of (5.8) is nonnegative. From (1.4), (5.3), (5.6) follows that the right side of (5.8) tends to zero with increasing n. But these facts, when combined with (5.9) and (5.10) and used in (5.8) immediately produce a contradiction. Consequently (5.1) is false and Lemma 3 holds.

6. PROOF OF LEMMA 4

Suppose for example that there exist t_1, ρ_1 such that

(6.1) $\quad y(t_1) = \rho_1 < \rho, \quad t_1 < t_0.$

As ρ is a point of density of $B(\alpha,\beta)$ we can take $\rho-\rho_1$ small enough so that

(6.2) $\quad m(B \cap [\rho_1,\rho]) \geq \frac{1}{2} [\rho-\rho_1].$

But then, by Lemma 3 and by (6.2),

(6.3) $\quad \rho - \rho_1 = y(t_0) - y(t_1) = \int_S y'(\tau)d\tau = \int_M y'(\tau)d\tau \leq \frac{1}{2} [\rho-\rho_1]$

where

(6.4) $\quad S = \{\tau |\ t_1 < \tau < t_0,\ y'(\tau) > 0,\ y(s) < y(\tau);\ t_1 \leq s < \tau\}$

(6.5) $\quad M = \{\tau |\ t_1 < \tau < t_0;\ \tau \in S,\ y(\tau) \notin B(\alpha,\beta)\}.$

Clearly (6.3) cannot hold and so (6.1) is false. The remaining possibilities are handled in the same way and so Lemma 4 follows.

7. PROOF OF LEMMA 5

Consider at first the case when λ,μ are defined by (2.14). By essentially redoing the proof of Lemma 1 one can show that $\lim_{n\to\infty} (\beta_n-\alpha_n)^{-1} \int_{\alpha_n}^{\beta_n} |x'(\tau)|d\tau = 0$ for any $[\alpha_n,\beta_n] \subset R^+$ satisfying $\lim_{n\to\infty} (\beta_n-\alpha_n) = \infty$. Consequently there exist subintervals $[\gamma_n,\delta_n] \subset [\alpha_n,\beta_n]$ such that $\lim_{n\to\infty} (\delta_n-\gamma_n) = \infty$ and such that $x(t) \to$ some constant $y \in \Gamma_c(x)$ uniformly on $[\gamma_n,\delta_n]$. But if $x(t) \leq \lambda$ $(x(t) \geq \mu)$ on $[\alpha_n,\beta_n]$ then by (2.14) this constant must necessarily equal $\lambda(\mu)$. It is also evident that the subintervals can be chosen to satisfy the last part of (2.21).

Next suppose λ,μ are defined by (2.16) and let (2.19) hold. (The case when (2.20) is valid can be treated in a similar way.) Assume that for any intervals $[p_n,q_n]$, $[r_n,s_n] \subset [t_n,t_n+T_n]$ such that all of (2.21) holds one has $x(t) \not\to \lambda$ uniformly on $[p_n,q_n] \cup [r_n,s_n]$ as $n \to \infty$. Then there exist τ_n such that (for example)

$$(7.1) \qquad t_n < \tau_n; \ \sup_n [\tau_n - t_n] < \infty, \ x(\tau_n) \leq \lambda - \varepsilon,$$

for some $\varepsilon > 0$. Pick τ_n such that $\lim_{n\to\infty} (\tau_n - t_n) = T$ exists. By (2.16) one can without loss of generality take $x(\tau_n) \in B_d(a,b)$. Choose $y(t)$ such that $x(t+\tau_n) \to y(t)$ uniformly on compact intervals. Let $\alpha = \inf_{t\in R} y(t) \leq \sup_{t\in R} y(t) = \beta$. Then $x(\tau_n) \to y(0) \in B_d(a,b) \cap [\alpha,\beta] \subset B_d(\alpha,\beta)$ and by the second part of (2.19), and by (7.1) $\lambda = x(t_n) = y(-T)$ and $y(0) \leq \lambda-\varepsilon$. This however violates Lemma 4.

Finally suppose λ,μ are defined by (2.17) or (2.18). Let at first (2.17) and (2.20) or (2.18) and (2.19) hold. But then the arguments in the first paragraph of the present proof can be applied. In case (2.18) and (2.20) or (2.17), (2.19) hold one argues as in the second paragraph above.

The proof of Lemma 5 is thus complete.

REFERENCES

[1] J.J. LEVIN, On some geometric structures for integrodifferential equations, Adv. in Math., 22 (1976), 146-186.

[2] O.J. STAFFANS, On the asymptotic spectra of the bounded solutions of a nonlinear Volterra equation, J. Differential Eqs. 24 (1977), 365-382.

[3] O.J. STAFFANS, On a nonlinear Volterra integrodifferential equation with a nonintegrable perturbation. To be published.

Numerical Approximations for Volterra Integral Equations*

R. C. MacCamy and Philip Weiss
Department of Mathematics
Carnegie-Mellon University
Pittsburgh, PA 15213/USA

1. Introduction

This paper continues work begun in [8] on numerical solutions of a class of Volterra equations. Suppose $a \in C[0,\infty)$ and M_a represents the linear operator,

$$M_a[v](t) = \int_0^t a(t-\tau)\, v(\tau)\, d\tau. \tag{1.1}$$

For functions $u(x,t)$ on $(0,1)\times(0,\infty)$ we denote by g the operator,

$$g(u) = \frac{\partial}{\partial x}\sigma(u_x) + \lambda u. \tag{1.2}$$

We consider the problem,

$$u = M_a[g(u)] + f; \quad u(0,t) \equiv u(1,t) \equiv 0, \qquad 0 < x < 1,\ t > 0. \tag{P'}$$

(P') serves as a model for a general class of Volterra equations on a Hilbert space as we describe in section five. This class of equations has been much studied (see [1],[2] and [4]) from the point of view of existence and uniqueness of solutions and of qualitative theory, that is stability and asymptotic stability. In [8] we discussed numerical procedures for this class when $H = R^N$. The procedures were designed to preserve the qualitative theory. Here we present a comparable theory for infinite dimensional H's. In the interest of clarity, however, we will work with (P') and simply outline the general theory in section five.

We transform (P') by a device that has been used before. Suppose $a(0) \neq 0$, $a \in C^{(2)}[0,\infty)$ and f_t exists. Then there is a $k \in C^{(1)}[0,\infty)$ such that if we differentiate the equation in (P') and invert we obtain,

*This work was supported by the National Science Foundation under Grant MCS77-01449.

$$\mathcal{L}[u(x,\cdot)](t) - g(u(x,t)) = \mathcal{L}[f(x,\cdot)](t) \equiv F(x,t) \qquad \text{(P)}$$

$$u(0,t) \equiv u(1,t) \equiv 0, \; u(x,0) = f(x,0),$$

where,

$$\mathcal{L}[w](t) = \frac{d}{dt}\{a(0)^{-1}w(t) + M_k[w](t)\} \qquad (1.3)$$

When $a(t) \equiv 1$ and $\sigma' > 0$ then (P) is a nonlinear parabolic problem. Our numerical procedure is a direct extension to (P) of that in [3] and [9] for this special case.

Remark 1.1. The term λu is included in $g(u)$ in order to illustrate the role of the memory effect as expressed by M_a. In the case of the parabolic problem one can expect asymptotic stability for all σ with $\sigma' > 0$ only if $\lambda < 0$. We show that if a is properly chosen we can still have asymptotic stability for positive λ's.

Throughout the paper we make the following assumptions:

$$a \in C^{(2)}[0,\infty) \text{ and } a(0) > 0; \qquad (a_1)$$

$$k(0) > 0, \; \dot{k} \in L_1(0,\infty); \qquad (a_2)$$

there exists a $\gamma > 0$ such that for any $v \in C(0,\infty)$ and any $T > 0$

$$\int_0^T v(t) \frac{d}{dt} M_k[v](t)\,dt \geq \gamma \int_0^T v^2(t)\,dt; \qquad (a_3)$$

$$\sigma \in C^{(1)}(-\infty,\infty) \text{ and } \sigma'(\zeta) > 0; \qquad (\sigma_1)$$

there is a $p \geq 2$ such that,

$$|\sigma(\zeta)| \leq \alpha(1+|\zeta|^{p-1}); \; \sigma(\zeta)\zeta \geq \beta|\zeta|^p, \; \beta > 0; \qquad (\sigma_2)$$

$$f, f_t \in L_q((0,1)\times(0,\infty)), \; p^{-1} + q^{-1} = 1. \qquad (f)$$

Remark 1.2. In the appendix we give conditions on a which guarantee (a_2) and (a_3). A prototype a is $e^{-\gamma t}$ for which $k \equiv \gamma$.

Under the above conditions one has a qualitative theory for (P). We introduce some notation. For $\varphi(x)$ on $[0,1]$ and $\psi(x,t)$

on $[0,1]\times[0,\infty]$ set,

$$\|\varphi\|_r = \left(\int_0^1 |\varphi(x)|^r dx\right)^{1/r}; \quad \|\psi\|_{r,T} = \left(\int_0^T \int_0^1 |\psi(x,t)|^r dxdt\right)^{1/r} \tag{1.5}$$

Define $[u]_{p,T}$ and $|||f|||_{q,T}$ by the formulas,

$$[u]_{p,T} = \sup_{t\le T} \|u(\cdot,t)\|_2^2 + \|u\|_{2,T}^2 + \|u_x\|_{p,T}^p, \tag{1.6}$$

$$|||f|||_{q,T} = \|f\|_{q,T} + \|f_t\|_{q,T}. \tag{1.7}$$

Observe that since $k \in L_1(0,\infty)$ equation (1.3) implies that there is a constant C independent of T such that,

$$\|F\|_{q,T} \le C\,|||f|||_{q,T}. \tag{1.8}$$

<u>Theorem</u> 1.1. <u>Suppose</u> $\gamma > \lambda$. <u>Then any solution of</u> (P) <u>satisfies the following</u>.

(i) <u>There is a</u> C, <u>independent of</u> $T > 0$ <u>such that</u>,

$$[u]_{p,T} \le C\,\{|||f|||_{q,T}^q + \|f(\cdot,0)\|_2^2\}^*: \tag{1.9}$$

(ii) <u>If</u> $\|u(\cdot,t)\|_2$ $(\|u_x(\cdot,t)\|_p)$ <u>is uniformly continuous on</u> $[0,\infty)$ <u>then</u> $\|u(\cdot,t)\|_2 \to 0$ $(u(x,t) \to 0)$ as $t \to \infty$.

<u>Proof</u>. It follows from (1.9) and (f) that $\|u(\cdot,t)\|_2 \in L_2(0,\infty)$ and $\|u_x(\cdot,t)\|_p \in L_p(0,\infty)$. Hence the hypotheses of (ii) imply $\|u(\cdot,t)\|_2 \to 0$ $(\|u_x(\cdot,t)\|_p \to 0)$. But the boundary conditions imply $|u(x,t)| \le \|u_x(\cdot,t)\|_p$. Hence (ii) follows from (i) and (f).

Conclusion (i) follows from a simple energy argument. Multiply the equation in (P) by u and integrate over $Q_{T'} = (0,1)\times(0,T')$, $T' \le T$. We integrate the term $\int_0^{T'}\int_0^1 \frac{\partial}{\partial x}\sigma(u_x)u\,dxdt$ by parts, using the boundary condition. Then by (a_3), $(\sigma_2)_2$, Young's inequality and (1.8), we obtain,

$$\frac{1}{2}a(0)^{-1}\{\|u(\cdot,T')\|_2^2 - \|f(\cdot,0)\|_2^2\} + (\gamma-\lambda)\|u\|_{2,T'}^2 + \beta\|u_x\|_{p,T'}^p$$

$$\le \int_0^{T'}\int_0^1 F\,u\,dxdt \le \frac{\beta}{2}\|u\|_{p,T'}^p + C\|F\|_{q,T'}^q \le \frac{\beta}{2}\|u_x\|_{p,T'}^p + C|||f|||_{q,T'}^q.$$

The result 1.9 follows.

---

*Throughout we will use C to denote a generic constant.

In section two we introduce a Galerkin procedure which reduces (P) to the finite dimensional case studied in [8]. We obtain existence, uniqueness of solutions and a qualitative theory for the approximations. In section three we show that the Galerkin procedure is optimal and obtain convergence rates for piecewise linear finite elements. In section four we give a brief discussion of implementation, reviewing some of the work in [8].

2. Galerkin Approximations

We let V denote the space of continuous and piecewise differentiable functions of x which vanish at 0 and 1. For two functions $\varphi(x)$, $\Phi(x)$ we write $(\varphi,\Phi) = \int_0^1 \varphi(x)\Phi(x)\,dx$. Suppose that u is a solution of (P) and $v\in V$. Then we have,

$$(\mathcal{L}[u](t),v) + (\sigma(u_x(\cdot,t),v') - \lambda(u(\cdot,t),v) = (F(\cdot,t),v), \quad (2.1)$$

$$(u(\cdot,0),v) = (f(\cdot,0),v). \quad (2.2)$$

Equations (2.1) and (2.2) form the basis for Galerkin's method. Let V^h denote a family of finite dimensional subspaces depending on a parameter h. We set

$$W^h = C^{(1)}([0,\infty):V^h) \quad (2.3)$$

<u>Definition</u>. A <u>V^h approximation for</u> (P) is a function $u^h \in W^h$ such

$$(\mathcal{L}[u^h](t),v_h) + \sigma(u_x^h(\cdot,t),v_h') - \lambda(u^h(\cdot,t),v_h) = (F(\cdot,t)v_h), \quad (2.1)'$$

$$(u^h(\cdot,0),v_h) = (f(\cdot,0),v_h), \quad (2.2)'$$

for all $v_h\in V^h$.

We show that (2.1)' and (2.2)' are equivalent to an equation on R^N. Suppose $\varphi_1^h,\ldots,\varphi_{N^h}^h$ is a basis for V^h. Then clearly (2.1)' and (2.2)' hold if and only if they are true for $v = \varphi_j^h$, $j = 1,\ldots,N_h$. We put $\underset{\sim}{U}^h = (u_1^h,\ldots,u_{N^h}^h)$, where,

$$u^h(x,t) = \sum_{i=1}^{N^h} u_i^h(t)\varphi_i^h(x) \quad (2.4)$$

Then $U^h \in C^{(1)}([0,\infty):R^{N^h})$. For ease of writing we think of h as fixed and suppress the superscripts h on the φ_i's, u_i's,

$\underset{\sim}{U}$ and N. Now define linear transformations G and $K(t)$ of R^N, into R^N, a nonlinear map G of R^N into R^N, a function $\mathfrak{F}$ from R^+ into R^N and $\underset{\sim}{U}_0$ in R^N by the formulas,

$$(G\underset{\sim}{V})_j = a(0)^{-1}\,\Sigma(\varphi_i,\varphi_j)v_i\,,$$

$$(K(t)\underset{\sim}{V})_j = k(t)\;\Sigma(\varphi_i,\varphi_j)v_i\,,$$

$$(G(\underset{\sim}{V}))_j = (\sigma(\Sigma\, v_i\varphi_i'),\varphi_j') - \lambda\Sigma(\varphi_i,\varphi_j)v_i\,, \tag{2.5}$$

$$(\underset{\sim}{\mathfrak{F}}(t))_j = (F(\cdot,t),\varphi_j)\,,$$

$$(\underset{\sim}{U}_0)_j = (f(\cdot,0),\varphi_j).$$

Then (2.1)' and (2.2)' are equivalent to the system;

$$G\dot{\underset{\sim}{U}}(t) + \frac{d}{dt}\int_0^t K(t-\tau)\underset{\sim}{U}(\tau)\,d\tau + G(\underset{\sim}{U}(t)) = \underset{\sim}{\mathfrak{F}}(t)\,;\quad \underset{\sim}{U}(0) = \underset{\sim}{V}_0 \tag{E}$$

Equation (E) is on R^N and is of the type considered in [8]. From results in that paper we can establish an analog of Theorem 1.1 for (E). We set,

$$[\underset{\sim}{U}]_{p,T} = \sup_{0\le t\le T}\|\underset{\sim}{U}(t)\|^2 + \int_0^T \|\underset{\sim}{U}(t)\|^2 dt + \int_0^T \|U(t)\|^p dt.$$

<u>Theorem</u> 2.1. <u>Suppose</u> $\gamma > \lambda$. <u>Then we have the following conclusions</u>:

(i) <u>There exists a constant</u> C_h, <u>independent of</u> T <u>such that if</u> $\underset{\sim}{U}$ <u>satisfies</u> (E) <u>then</u>,

$$[U]_{p,T} \le C_h\{|||f|||^q_{q,T} + \|f(\cdot,0)\|_2^2\} \tag{2.6}$$

(ii) <u>If</u> $\underset{\sim}{U}$ <u>is a solution of</u> (E) $\underset{\sim}{U}(t) \to 0$ <u>as</u> $t \to \infty$

(iii) <u>There exists a unique solution of</u> (E).

<u>Theorem</u> 2.2. <u>Suppose</u> $\gamma > \lambda$. <u>Then we have the following conclusions</u>:

(i) <u>There exists a constant independent of</u> h <u>and</u> T <u>such that if</u> u^h <u>is a</u> V^h-<u>approximation then</u> u^h <u>satisfies</u> (1.9).

(ii) <u>If</u> u^h <u>is a</u> V^h-<u>approximation</u> $u^h(x,t) \to 0$ <u>as</u> $t \to \infty$.

(iii) <u>There exists a unique</u> V^h <u>approximation</u>.

Remark 2.1. Note that the constant depends on h in Theorem 2.1 but not in Theorem 2.2.

Remark 2.2. Note that the asymptotic stability result in Theorem 2.2 contains no provisional hypotheses as did Theorem 1.1 (ii). Note also that Theorem 2.2 (iii) has no counterpart in Theorem 1.1 (The ideas of this section can, however, be used in establishing the existence of a unique generalized solution, see [5] and [11]).

In order to apply the theory of [8] to (E) we need some estimates for the quantities in (2.5). The results are these. There exist positive constants, $\underline{a},\gamma',\beta',\lambda',C$, independent of h and T such that $\gamma' > \lambda'$ and we have,

$$\begin{aligned}
&(G\underset{\sim}{V},\underset{\sim}{V}) \geq \underline{a}\|\underset{\sim}{V}\|^2, \\
&\int_0^T \Big(\frac{d}{dt}\int_0^t K(t-\tau)\underset{\sim}{V}(\tau)\,d\tau,\underset{\sim}{V}(t)\Big)\,dt \geq \gamma' \int_0^T \|\underset{\sim}{V}(t)\|^2 dt, \\
&(G(\underset{\sim}{V}),\underset{\sim}{V}) \geq \beta'\|\underset{\sim}{V}\|^p - \lambda'\|\underset{\sim}{V}\|^2, \\
&\Big(\int_0^T \|\mathcal{F}(t)\|^q dt\Big)^{1/q} \leq C\,|||f|||_{q,T}.
\end{aligned} \tag{2.7}$$

The proofs of these estimates becomes quite transparent once we make a preliminary observation. We define ϑ_N: $R^N \to V$ by the formula.

$$\vartheta_N(\underset{\sim}{V}) = \vartheta_N((v_1,\dots,v_N)) = \sum_{i=1}^{N} v_i\varphi_i \tag{2.8}$$

Then there exist positive constants $\underline{\mu}_N$, $\overline{\mu}_N$ and $\underline{\tau}_N$, $\overline{\tau}_N$ such that

$$\underline{\mu}_N\|\underset{\sim}{V}\| \leq \|\vartheta_N(\underset{\sim}{V})\|_2 \leq \overline{\mu}_N\|\underset{\sim}{V}\|,\quad \underline{\tau}_N\|\underset{\sim}{V}\| \leq \|(\vartheta_N(\underset{\sim}{V}))'\|_p \leq \overline{\tau}_N\|\underset{\sim}{V}\| \tag{2.9}$$

Let us verify $(2.7)_3$. We have by $(2.5)_3$, (σ_2) and (2.9),

$$\begin{aligned}
(G(\underset{\sim}{V}),\underset{\sim}{V}) &= (\sigma(\Sigma\, v_i\varphi_i'),\Sigma\, v_j\varphi_j') - \lambda\, \Sigma\Sigma(\varphi_i,\varphi_j)v_iv_j \\
&= (\sigma((\vartheta_N(\underset{\sim}{V}))'),(\vartheta_N(\underset{\sim}{V}))') - \lambda(\vartheta_N(\underset{\sim}{V}),\vartheta_N(\underset{\sim}{V})) \\
&\geq \beta\|(\vartheta_N(\underset{\sim}{V}))'\|_p^p - \lambda\|\vartheta_N(\underset{\sim}{V})\|_2^2 \geq \underline{\tau}_N^p\ \beta\|\underset{\sim}{V}\|^p - \lambda\overline{\mu}_N^2\|\underset{\sim}{V}\|^2
\end{aligned}$$

This is $(2.7)_3$ with $\beta' = \underline{\tau}_N^p\ \beta$ and $\lambda^1 = \lambda\overline{\mu}_N^2$. Estimates $(2.7)_1$ and $(2.7)_2$ follow in the same way if we use (a_3). $(2.7)_4$ follows immediately from $(2.5)_4$ and (1.8).

According to [8] Theorem 2.1 follows from the estimates (2.7). We sketch the proof. First the estimates (2.7) imply the a priori estimate (2.6) of Theorem 2.1. A minor variation of a standard Schauder-Leray argument then yields the existence of at least one solution of (E). That the solution is unique is implied by the fact that the operator on the left is monotone. For the linear parts this follows from (2.7). For the nonlinear term one has by (σ_1), $(G(\underset{\sim}{V}) - G(\underset{\sim}{W}), \underset{\sim}{V}-\underset{\sim}{W}) = (\sigma((\mathcal{J}_N(\underset{\sim}{V}))') - \sigma((\mathcal{J}_N(\underset{\sim}{W}))'), (\mathcal{J}_N(\underset{\sim}{V}) - \mathcal{J}_N(\underset{\sim}{W}))') \geq 0$
This is the basic proof of Theorem 2.1 (i) and (iii).

To establish (ii) we observe once again that (2.6) and (f) imply $\|\underset{\sim}{U}(t)\| \in L_2(0,\infty)$. Hence (ii) follows if $\|\underset{\sim}{U}(t)\|$ is uniformly continuous on $[0,\infty)$. But now we can draw this conclusion from (E). Indeed we have,

$$\dot{\underset{\sim}{U}}(t) = -a(0)\{K(0)\underset{\sim}{U}(t) + \int_0^t \dot{K}(t-\tau)\underset{\sim}{U}(\tau)\,d\tau + G(\underset{\sim}{U}(t)) - \underset{\sim}{\mathfrak{F}}(t)\} \qquad (2.10)$$

By (2.6) the first and second terms are bounded and the last term is in $L_q((0,\infty): R^N)$. Let us estimate the third term. By $(2.5)_3$ (σ_2), Holder's inequality (2.9) and (2.6)

$$|G(\underset{\sim}{U}(t))_j| \leq \alpha(\|\varphi_j'\|_1 + \|(\mathcal{J}_N(U))'\|_p^{p-1}\|\varphi_j'\|_p) + C'\|\underset{\sim}{U}(t)\|$$

$$\leq C'' + C'''\|\underset{\sim}{U}(t)\|^{p-1} + C'\|\underset{\sim}{U}(t)\| \leq C$$

Thus $\|\dot{\underset{\sim}{U}}(t)\|$ is bounded by a constant plus a function in $L_q(0,\infty)$ and hence $\underset{\sim}{U}$ is uniformly continuous. This concludes the proof of Theorem 2.1.

Theorem 2.2 (iii) follows from Theorem 2.1 (iii). Theorem 2.2 (i) comes from equations (2.1)' and (2.2)'. In those formulas one can choose $v_h = u^h$ and perform exactly the same calculation we did in deriving (1.9). In order to prove Theorem 2.2 (ii) we recall that there is a constant $\overline{\tau}_N$ such that $\|(\mathcal{J}_N(V))'\|_p \leq \overline{\tau}_N\|\underset{\sim}{V}\|$. Hence by Theorem 2.1 (ii) we have $\|(\mathcal{J}_N(U^h(t))'\|_p \to 0$. But $\mathcal{J}_N(U^h) = u^h$ and $|u^h(x,t)| \leq \|u^h_x(\cdot,t)\|_p$ so (ii) of Theorem 2.2 follows.

3. Error Estimates

Our object here is to estimate the difference $z = u - u^h$ between the solution of (P) (assuming it exists) and the V^h approximation. Specifically we want to estimate the quantity

$[z]_{2,T}$ of (1.6), which we recall is,

$$[z]_{2,T} = \sup_{t\leq T} \|z(\cdot,t)\|_2^2 + \|z\|_{2,T}^2 + \|z_x\|_{2,T}^2 \tag{3.1}$$

The first step is to establish what is called <u>optimality</u> of the approximation. We let $e = u - w^h$ where w^h is an arbitrary element of $W^h = C^{(1)}([0,\infty):V^h)$ and we introduce a quantity $E_T(e)$. Let $Q_T = [0,1]x[0,\infty)$. Then,

$$E_T(e) = \sup_{t\leq T} \|e(\cdot,t)\|_2^2 + \|e\|_{L_2(Q_T)}^2 + \|e_x\|_{L_1(Q_T)} + \|e_x\|_{L_p(Q_T)} + \|e_t\|_{L_2(Q_T)}^2 \tag{3.2}$$

<u>Remark</u> 3.1. The form (3.2) should be studied carefully. Note the non-homogeneity of the terms and the presence of a term involving e_t.

To establish our optimality result we need a strengthening of condition (σ_1):

$$\sigma\epsilon C^{(1)}(-\infty,\infty), \ \sigma'(\zeta) \geq \epsilon > 0 \tag{σ_1*}$$

($\sigma(\zeta) = \epsilon\zeta + |\zeta|^{p-2}\zeta$ satisfies (σ_1*) for $\epsilon > 0$ but only (σ_1) if $\epsilon = 0$.

<u>Theorem</u> 3.1. <u>Suppose</u> (σ_1*) <u>holds and</u> $\gamma > \lambda$. <u>Then there exists a constant</u> C; <u>independent of</u> h <u>and</u> T, <u>such that</u>

$$[z]_{2,T} \leq C\, E_T(e) \tag{3.3}$$

This theorem reduces the estimate of $[z]_{p,T}$ to the question of how small $E_T(e)$ can be made by appropriate choices of w^h. This is a question in approximation theory and depends on the choice of basis φ_i^h. We discuss this question for the particular choice of piecewise linear finite elements, that is,

$$\varphi_1^h(x) = x/h \ \text{ for } \ 0 \leq x \leq h; \ 2-x/h \ \text{ for } \ h \leq x \leq 2h; \ 0 \ \text{ for } \ x\leq 0; x\geq 2h$$
$$\varphi_j^h(x) = \varphi_1^h(x-(j-1)h) \ \text{ for } \ j \leq N^n = h^{-1} - 1 \tag{3.4}$$

For this system we have the following result.

<u>Theorem</u> 3.2. <u>There exists a constant</u> C <u>independent of</u> h <u>such that for any</u> $u\epsilon C^{(3)}(0,1]x[0,\infty)$ <u>and any</u> r <u>there exists a</u> $U^h\epsilon W^h$ <u>such that</u>,

$$\|u(\cdot,t) - U^h(\cdot,t)\|_r \le Ch^2\|u_{xx}(\cdot,t)\|_r$$
$$\|u_x(\cdot,t) - U^h_x(\cdot,t)\|_r \le Ch\;\|u_{xx}(\cdot,t)\|_r \qquad (3.5)$$
$$\|u_t(\cdot,t) - U^h_t(\cdot,t)\|_r \le Ch^2\|u_{txx}(\cdot,t)\|_r$$

If we combine Theorem 3.1 and Theorem 3.2 we obtain an estimate for $[z]_{p,T}$ for the specific choice (3.4) of the φ_i^h's. The result is:

Theorem 3.3. Suppose $\gamma > \lambda$ and $(\sigma_1{*})$ holds and φ_i^h is as in (3.4). Suppose in addition that u satisfies the following re-requirements: If $Q = (0,1)\times(0,\infty)$, then,

$$u_{xx}\in L_1(Q)\cap L_2(Q)\cap L_p(Q)\cap L_\infty(Q)$$
$$u_{xxt}\subset L_2(Q) \qquad (3.6)$$

Then there exists a constant C, independent of h and T such that

$$[z]_{2,T} \le Ch \qquad (3.7)$$

Remark 3.2. Conditions (3.6) represent hypotheses on the solution. They cannot be verified from the problem. Thus (3.7) is provisional in the same sense as Theorem 1.1 (ii).

Remark 3.3. When $p < 4$ it is possible to refine the argument to obtain h^2 in (3.7). This calculation is carried out in [11].

Remark 3.4. If one uses smoother basis functions one can increase the order in (3.5) and hence in (3.7) provided one assumes more smoothness for the solution; see [10].

Proof of Theorem 3.1. Consider formulas (2.1) and (2.1)'. Both formulas hold for $v\in V^h$. If we subtract the results and integrate from 0 to T' we have,

$$\int_0^{T'} (\mathcal{L}[z](t),\ v^h(\cdot,t))\,dt + \int_0^{T'} (\sigma(u_x(\cdot,t)) - \sigma(u_x^h(\cdot,t)),\ v^h(\cdot,t))\,dt$$

$$-\lambda\int_0^{T'} (z(\cdot,t),\ v^h(\cdot,t))\,dt = 0 \quad \text{for all} \quad v^h \epsilon W^h \tag{3.8$_2$}$$

In this formula we choose $v^h = z - e = w^h - u^h$ and write the result as,

$$(I) = \int_0^{T'} (\mathcal{L}[z](t),\ z(\cdot,t))\,dt + \int_0^{T'} (\sigma(z_x(\cdot,t) + u_x^h(\cdot,t))$$

$$- \sigma(u_x^h(\cdot,t)),\ z_x(\cdot,t))\,dt - \lambda \int_0^{T'} (z(\cdot,t),(z(\cdot,t))\,dt$$

$$= \int_0^{T'} (\mathcal{L}[z](t),\ e(\cdot,t))\,dt + \int_0^{T'} (\sigma(u_x(\cdot,t)) - \sigma(u_x^h(\cdot,t)),\ e_x(\cdot,t))\,dt$$

$$- \lambda \int_0^{T'} (z(\cdot,t),\ e(\cdot,t))\,dt = (II) \tag{3.9$_2$}$$

Our task is to estimate (I) from below and (II) from above. The lower estimate on the linear terms in (I) is obtained just as in the proof of (1.9) and is,

$$\int_0^{T'} (\mathcal{L}[z](t),\ z(\cdot,t))\,dt - \lambda \int_0^{T'} (z(\cdot,t),\ z(\cdot,t))\,dt$$

$$\geq \frac{1}{2}\, a(0)^{-1}\{\|z(\cdot,T')\|_2^2 - \|z(\cdot,0)\|_2^2\} + (\gamma-\lambda)\,\|z\|_{2,T'}^2 \tag{3.10$_1$}$$

For the nonlinear term we use (σ_1^*) and obtain,

$$\int_0^{T'} (\sigma(z_x(\cdot,t) + u_x^h(\cdot,t)) - \sigma(u^h(\cdot,t)),z_x(\cdot,t))\,dt \geq \epsilon\|z_x\|_{2,T'}^2 \tag{3.10$_2$}$$

We estimate each term of (II) from above (we use C here to denote various constants). From $(3.9)_1$, if we use (a_2), (σ_2) and Young's inequality we find:

$$|a(0)^{-1} \int_0^{T'} (z_t(\cdot,t),e(\cdot,t))\,dt| \leq a(0)^{-1}(|(z(\cdot,T'),e(\cdot,T'))| +$$

$$+ |(z(\cdot,0),e(\cdot,0))|) + a(0)^{-1} \int_0^{T'} (z(\cdot,t),e_t(\cdot,t))\,dt \leq \frac{a(0)^{-1}}{2}\|z(\cdot,T')\|_2^2$$

$$+ \frac{\gamma-\lambda}{4}\|z\|_{2,T'}^2 + C\{\|e(\cdot,T')\|_2^2 + \|e(\cdot,0\|_2 + \|e_t\|_{2,T'}^2\}; \tag{3.11$_1$}$$

$$\left|\int_0^{T'} \left(\frac{d}{dt}\int_0^t k(t-\tau)\,z(\cdot,\tau)\,d\tau - \lambda z(\cdot\tau),\ e(\cdot,t)\right)dt\right|$$

$$\leq (k(0)+|\lambda|)\int_0^{T'} |z(\cdot,t),\ e(\cdot t))|\,dt +$$

$$\int_0^{T'} \left(\int_0^t |\dot{k}(t-\tau)|\ \ |z(\cdot,\tau),e(\cdot,t))|\,d\tau dt\right.$$

$$\leq \frac{\gamma-\lambda}{4}\ \|z\|^2_{2,T'} + C\ \|e\|^2_{2,T'}: \qquad (3.11)_2$$

$$\left|\int_0^{T'} (\sigma(u_x(\cdot,t)) - \sigma(u_x^h(\cdot,t)),\ e_x(\cdot,t))\,dt\right| \leq \alpha\{\int_0^{T'}\int_0^1 |e_x(x,t)|\,dxdt$$

$$+ \int_0^{T'}\int_0^1 (|u_x(x,t)|^{p-1} + |u_x^h(x,t)^{p-1})\,|e_x(\cdot,t)|\,dxdt$$

$$\leq \alpha\{\|e_x\|_{1,T'} + (\|u_x\|^{\frac{p-1}{p}}_{p,T'} + \|u_x^h\|^{\frac{p-1}{p}}_{p,T'})\,\|e_x\|_{p,T'}\} \qquad (3.11)_3$$

We note that, by (1.9) and Theorem 2.2 (i) the coefficient of $\|e_x\|_{p,T}$ in $(3.11)_3$ is bounded independently of T'. We also need an estimate for $\|z(\cdot,0)\|_2$ in $(3.11)_1$. To obtain this we subtract (2.2)' from (2.2) and obtain $(z(\cdot,0),v^h) = 0$ for all $v^h \in V^h$. In particular take $v^h(x) = z(x,0) - e(x,0) = w^h(x,0) - u^h(x,0)$. Then we obtain $(z(\cdot,0),z(\cdot,0)) = (z(\cdot,0),\ e(\cdot,0))$ or $\|z(\cdot,0)\|_2 \leq \|e(\cdot,0)\|_2$. Thus in $(3.11)_1$ we have $\|e(\cdot,0)\|_2\|z(\cdot,0)\|_2 \leq \|e(\cdot,0)\|^2_2$. Now the estimate (3.3) follows from the inequalities (3.10) and (3.11) together with the fact that $\|z\|_{2,T'} \leq \|z_x\|_{2,T'}$.

<u>Proof of Theorem</u> 3.2. We choose U^h as the interpolant of u that is,

$$U^h(x,t) = \sum_{j=1}^{N_h} u(jh,t)\,\varphi_j(x) \qquad (3.12)$$

Thus, by (3.4), $e(jh,t) = u(jh,t) - U^h(jh,t) \equiv 0$. Consider one interval $[(k-1)h,kh]$. Since $e((k-1)h,t) \equiv e(kh,t) \equiv 0$ there is an $x'(t)$ such that $e_x(x',t) = 0$ and consequently we have, for $r' = r/r-1$,

$$|e_x(x,t)|^r \leq \left(\int_{x'}^x |e_{xx}(x,t)|\,dx\right)^r \leq \left(\int_{(k-1)h}^{kh} |e_{xx}|^r dx\right) h^{r-1} \qquad (3.13)$$

We integrate (3.13) over $((k-1)h,kh)$ and sum over k to obtain,

$$\int_0^1 |e_x|^r dx \le h^r \int_0^1 |e_{xx}|^r dx$$

This is estimate $(3.5)_2$. We obtain $(3.5)_1$ and $(3.5)_3$ by noting that on $[(k-1)h,kh]$ we have $|e(x,t)|^r \le (\int_{(k-1)h}^{x} |e_x| dx)$ and using (3.13).

4. Implementation

The work of the preceding sections reduces the problem (P) to the solution of the finite dimensional integro-differential equation (E). To this system one must, of course, apply a second numerical approximation scheme. This step was discussed in detail in [8]. Here we will make just a few remarks on the implementation question, emphasizing the fact that the choice (3.4) of basis makes the formulas fairly simple.

We want to study the structure of the operators A, K and G in (2.5). Let $\underset{\sim}{\Phi}$ denote the matrix $\Phi_{ji} = (\varphi_i, \varphi_j)$. Then (3.4) yields

$$\underset{\sim}{\Phi} = \frac{h}{6} \begin{pmatrix} 4 & 1 & 0 & . & \cdots & & \cdots & 0 \\ 1 & 4 & 1 & 0 & & & \cdots & 0 \\ 0 & 1 & 4 & 1 & 0 & & \cdots & 0 \\ \vdots & & & & & & & \\ 0 & 0 & \cdots & & & 0 & 1 & 4 \end{pmatrix} \tag{4.1}$$

Then if $\underset{\sim}{\mathfrak{a}}$ and $\underset{\sim}{K}(t)$ are the matrices of $\mathfrak{a}$ and $K(t)$ we have,

$$\underset{\sim}{\mathfrak{a}} = a(0)^{-1} \underset{\sim}{\Phi};\ K(t) = k(t)\Phi, \tag{4.2}$$

so both are tridiagonal.

The function G is also very simple for the choice (3.4). The functions $\sigma(\Sigma v_i \varphi_i')\varphi_j'$ are piecewise constant on (0,1). Hence the integrals in G can be computed explicitly. One obtains then,

$$G(\underset{\sim}{V}) = \begin{pmatrix} \sigma(\frac{v_1}{h}) - \sigma(\frac{v_2}{h} - \frac{v_1}{h}) & - \frac{h\lambda}{6}(4v_1+v_2) \\ \sigma(\frac{v_2}{h} - \frac{v_1}{h}) - \sigma(\frac{v_3}{h} - \frac{v_2}{h}) & - \frac{h\lambda}{6}(v_1+4v_2+v_3) \\ \sigma(\frac{v_3}{h} - \frac{v_2}{h}) - \sigma(\frac{v_4}{h} - \frac{v_3}{h}) & - \frac{h\lambda}{6}(v_2+4v_3+v_4) \\ \vdots & \\ \sigma(\frac{v_N}{h} - \frac{v_{N-1}}{h}) - \sigma(-\frac{v_N}{h}) & - \frac{h\lambda}{6}(v_{n-1}+4v_N) \end{pmatrix}$$

In [8] we consider two different numerical schemes for (E), an implicit difference procedure and a time-Galerkin method. The second has considerable theoretical advantages over the first but is harder to implement. Here we confine ourselves to a brief description of the difference scheme.

Let $\tau > 0$ be given. Then we seek an approximation to $\underset{\sim}{U}(m\tau)$, $m = 1,2,\ldots,$. Call these approximations $\underset{\sim}{U}^m$. The implicit scheme has the form,

$$\mathfrak{a}\underset{\sim}{U}^m + \tau G(\underset{\sim}{U}^m) + \tau K(0)\underset{\sim}{U}^m = R^m(\underset{\sim}{U}^0, \underset{\sim}{U}^1, \ldots, \underset{\sim}{U}^{m-1}) + \tau \mathfrak{F}^m_m \tag{4.4}$$

$$m = 1,\ldots \quad \underset{\sim}{U}^o = \underset{\sim}{U}(0), \quad \mathfrak{F}^m = \underset{\sim}{\mathfrak{F}}(m\tau)$$

Here the quantities R^m are given by,

$$R^m(\underset{\sim}{U}^0, \ldots, \underset{\sim}{U}^{m-1}) = \mathfrak{a}\underset{\sim}{U}^{m-1} - \tau \sum_{i=0}^{m-1} \overline{K}_{m-i-1} \underset{\sim}{V}^i, \tag{4.5}$$

$$\overline{K}_j = K((j+1)\tau) - K(j\tau)$$

The idea behind this formula is simple. We replace the derivative in (E) by a difference and evaluate the integral term by a quadrature formula. The choice of quadrature formula has the following consequence. If k in (P) satisfies $k(t) > 0$, $\dot{k}(t) < 0$ then there will be a qualitative theory for (4.4) completely analogous to the ones obtained earlier in the paper. Remark (A.1) of the appendix indicates that such k's form a proper sub-class of all k's for which (a_3) holds. In this respect the time-Galerkin procedure in [8] is superior since it yields a qualitative theory in the general case.

It is shown in [8] that the scheme (4.4) is convergent to solutions of (E), but only order τ. The time-Galerkin schemes have higher order convergence rates.

The scheme (4.4) is relatively simple to implement. It is a recursive scheme in which one must invert the operator $A\underset{\sim}{U} + G(\underset{\sim}{U})$ at each step, that is one must solve a nonlinear equation on R^N. (4.1)-(4.3) show, however, that these equations are not very complicated. Moreover our conditions guarantee that they have unique solutions. Numerical details appear in [11].

5. The Abstract Problem

The problem (P') is a very special case of the following problem. Let H be a Hilbert space and $S(H)$ denote the bounded symmetric* linear operators on H. For $A \in C([0,\infty):S(H))$ define M_A by

$$M_A[v](t) = \int_0^t A(t-\tau)\, v(\tau)\, d\tau \tag{5.1}$$

Let $g: D_g \subset H \to H$ be a map which can be nonlinear and unbounded. Then we consider the equation,

$$u(t) = - M_A[g(u)](t) + f(t) \quad t > 0 \tag{I'}$$

In order to fit (P') into this general form take,

$$H = L_2(0,1)\, ; \; A(t) = a(t)I$$

$$D_g = \{v: v \in C^{(2)}[0,1]\, ; \; v(0) = v(1) = 0\} \tag{5.2}$$

$$g(v) = - \frac{d}{dx}\sigma(v') - \lambda v$$

If $A \in C^{(2)}([0,\infty):S(H))$, $A(0)$ is invertible and $\dot{f}$ exists we can perform the analog of the inversion that led to (P) and obtain,

$$\mathcal{L}[u](t) + g(u(t)) = \mathcal{L}[f](t) \equiv F(t)\, , \; u(0) = u_0 \tag{I}$$

where

$$\mathcal{L}[w](t) = \frac{d}{dt}\{A(0)^{-1}w(t) + M_K[w](t)\}\, , \tag{5.3}$$

---

*The reason for restriction to symmetric operators is indicated in the appendix.

where $K \in C^{(1)}([0,\infty):S(H))$.

It is possible to give to give an analysis of (I) which exactly parallels that for the special problem (P). The conditions for A are direct abstractions of (a_i):

$$A \in C^{(2)}[0,\infty) \text{ and } (A(0)u,u) \geq a_0\|u\|^2,\ a_0 > 0, \qquad (A_1)$$

$$(K(0),u,u) \geq k_0\|u\|^2,\ k_0 > 0;\ \dot{K} \in L_1((0,\infty):S(H)), \qquad (A_2)$$

there is a $\gamma > 0$ such that for any $v \in C([0,\infty):H)$ and $T > 0$

$$\int_0^T \left(\frac{d}{dt} M_k[V](t),\ V(t)\right) dt \geq \gamma \int_0^T \|V(t)\|^2 dt \qquad (A_3)$$

The abstraction of the conditions on σ requires greater attention. A careful analysis of what we did for (P) shows that there were two different norms entering. The abstraction of this is the following. We suppose that there is a separable Banach space V which is dense in H with $\|u\| \leq \alpha\|u\|_V$ for $u \in V$ and with $D_g \subset V$. One can imbed H in V', the dual of V by the formula $\langle h,v\rangle = (h,v)$ for $h \in H$ and $v \in V$.

Now we assume that there is a map $G\colon V \to V'$ such that,

$$\langle Gu,v\rangle = (g(u),v) \text{ for all } u \in D_g \text{ and } v \in V. \qquad (5.4)$$

We assume $A(0)^{-1}$ maps V onto V. Then we can define a map $A\colon V' \to V'$ by the formula,

$$\langle A[v'],v\rangle = \langle v',Av\rangle \qquad (5.5)$$

With this structure we can interpret (I) as an equation on V', namely,

$$A\dot{u}(t) + \frac{d}{dt}\int_0^t K(t-\tau)u(\tau)\,d\tau + G(u(t)) = F(t),\ u(0) = f(0). \qquad (5.6)$$

A "solution" of (5.6) will be a function of t with values in V. It will have a (generalized) derivative $\dot{u}(t)$ with values in V'. The second term on the left and $F(t)$, having values in H, can also be interpreted as elements of V' so the equation makes sense.

Now we can state the abstractions of the properties for σ. (Remember that in the example g also contains the term $-\lambda u$).

$$\langle G(u) - G(v), u - v\rangle \geq -\lambda\|u - v\|^2 \qquad (G_1)$$

$$\|G(u)\|' \leq \alpha(1+\|u\|_V^{p-1}) \quad p > 2 \qquad (G_2)_1$$

$$\langle G(u), u\rangle \geq \beta\|u\|_V^p - \lambda\|u\|^2 \qquad (G_2)_2$$

The condition on f is,

$$f, \dot{f} \in L_q((0,\infty) : H), \quad q = p/p-1, \qquad (f)$$

With the conditions above the entire theory can be carried through. One obtains the following analog of estimate (1.9) for solutions of (5.6):

$$\|u\|^2_{L_\infty((0,T):H)} + \|u\|^2_{L_2((0,T):H)} + \|u\|^p_{L_p((0,T):V)}$$
$$\leq C(\|f\|^q_{L_q((0,T):H)} + \|f_t\|^q_{L_q((0,T):H)}). \qquad (5.7)$$

From this one gets a provisional asymptotic stability theorem like Theorem 1.2.

There is a version of the Galerkin procedure. One chooses finite dimensional subspaces V^h of V and then defines V^h approximates u^h as solutions of the equation,

$$\langle A\dot{u}^h(t), v\rangle + (\frac{d}{dt}\int_0^t K(t-\tau)u^h(\tau)\,d\tau, v) + \langle G(u^h(t)), v\rangle = (F(t), v), \quad (5.6)$$

$$(u^h(0), v) = (f(0), v), \qquad (5.7)$$

for all $v \in V^h$. It can be shown that these finite dimensional problems are uniquely solvable and there is a qualitative theory for the solutions which parallels that for the solution. Finally one can establish an optimality theorem analogous to (3.1) with the understanding that the norms involving derivatives in [z] and E_T are to be replaced by $\|z\|^2_{L_2((0,T):V)}$, $\|e\|_{L_1((0,T):V)}$ and $\|e\|_{L_p((0,T):V)}$ respectively. In order to obtain this result, however, one must sharpen (G_1) to an analog of (σ_1^*). This is,

$$\langle G(u) - G(v)\rangle \geq \epsilon\|u-v\|_V^2 - \lambda\|u-v\|^2 \qquad (G_1^*)$$

The details of this abstract analysis can be found in [11]. The existence and uniqueness portion is a fairly straightforward extension of the theory in [5] for differential equations in the same setting.

Appendix Frequency Domain Methods

In this appendix we review very briefly the ideas of [6] and [7]. These give conditions on a which imply (a_2) and (a_3) and use what are called frequency domain methods, that is Laplace transforms. Let us proceed formally. It is not difficult to verify that the transforms $a^\wedge$ and $k^\wedge$ of the functions a and k are related by the formula,

$$k^\wedge(s) = (s\, a^\wedge(s))^{-1} - a(0)^{-1} \tag{A.1}$$

The idea is to translate (a_2) and (a_3) into conditions on $k^\wedge$ and then, by (A.1), into conditions on a. The first observation is this. Suppose $k^\wedge(s)$ is analytic in $\Re e\, s > -m$, $m > 0$ save for a simple pole $k_\infty s^{-1}$ at $s = 0$, and $k^\wedge$ satisfies,

$$k^\wedge(s) \sim \sum_{n=0}^{\infty} \frac{k_n}{s^{n+1}} \quad \text{as} \quad s \to \infty \quad \text{in} \quad \Re e\, s > -m. \tag{A.2}$$

Then, by using the complex inversion formula, one shows that $k^\wedge$ is the transform of a function $k(t)$ having derivatives of all orders and of the form

$$k(t) = k_\infty + K(t),\ K^{(j)} = O(e^{-\mu t})\quad \mu < m, \text{ as } \quad t \to \infty$$

$$k_n = k^{(n)}(0) \tag{A.3}$$

Clearly then (a_2) will be satisfied if $k_o > 0$. In the above situation one can give an argument involving the inversion formula and Parseval's theorem to show that sufficient conditions for (a_3) are $k_o > 0$, $k_\infty > 0$ and

$$\eta\, \Im m\, k^\wedge(i\eta) > 0 \qquad \eta \neq 0 \tag{A.4}$$

One can insure that the conditions in the preceding paragraph hold as follows. Assume a has derivatives of all orders and that these derivatives are all $O(e^{-m't})$ for some $m' > 0$. Then $a^\wedge(s)$ exists in $\Re e\, s > -m'$ and,

$$a^{\wedge}(s) \sim \sum_{n=0}^{\infty} \frac{a^{(n)}(0)}{s^{n+1}}, \quad \text{as} \quad s \to \infty \quad \text{in} \quad \Re e\, s > m' \tag{A.5}$$

It follows from (A.5) that the right side of (A.1) exists in $\Re e\, s > -m', |s|$ sufficiently large, and that (A.2) holds with $k_o = -\dot{a}(0)/a(0)^2$. The next observation is that if a satisfies,

$$\Re e\, a^{\wedge}(i\eta) > 0 \quad \text{for all} \quad \eta \tag{A.6}$$

then there must be some $m < m'$ such that $a^{\wedge}(s) \neq 0$ in $\Re e\, s > -m$. We note, however, that $(sa^{\wedge}(s))^{-1}$ has a pole, $k_{\infty} s^{-1}$, $k_{\infty} = a^{\wedge}(0)$ at $s = 0$. If (A.6) holds we will have $k_{\infty} > 0$. Finally we observe that (A.6) implies (A.4). Thus if we define $k^{\wedge}$ by (A.1) it will satisfy all the conditions of the preceding paragraph if $a(0) > 0$ and $\dot{a}(0) < 0$. This yields the following result.

<u>Theorem</u> A.1. <u>Sufficient conditions for</u> (a_2) <u>and</u> (a_3) <u>are</u>:

(i) $a \in C^{\infty}[0,\infty)$ $a^{(j)}(t) = O(e^{-m't})$ $m' > 0$

(ii) $a(0) > 0$, $\dot{a}(0) < 0$

(iii) a <u>satisfies</u> (A.6).

It is shown in [6] that a sufficient condition for (A.6) is

$$(-1)^j a^{(j)}(t) > 0 \qquad j = 0,1,2 \tag{A.7}$$

However (A.7) is not necessary as the example $a(t) = e^{-m't}\cos \beta t$ shows. It also follows from the work of [6] that a sufficient condition for (A.4)

$$(-1)^j k^{(j)}(t) > 0 \qquad j = 0,1 \tag{A.8}$$

This condition is not necessary for (A.4) either as the example $k(t) = 1 + e^{-m't}\sin \beta t$ shows.

<u>Remark</u> A.1. The mapping from a into k preserves conditions (A.4) and (A.6); that is k satisfies (A.4) if and only if a satisfies (A.6). It is almost certainly not true, however, that k satisfies (A.8) if and only if a satisfies (A.7).

All of the above theory can be extended to general Hilbert spaces H provided all operators are in $S(H)$. Essentially one needs only replace $>$ by the requirement that the operator be positive definite. In these arguments, however, the symmetry is essential, see [6].

References

[1] V. Barbu, Nonlinear Volterra equations in a Hilbert space, SIAM Journ. Math Anal. 6 (1975), 728-741.

[2] M. G. Crandell, S.-O. Londen and J. A. Nohel, An abstract nonlinear Volterra integro-differential equation, Journ. Math. Anal. and Appl. (to appear).

[3] J. Douglas and T. Dupont, Galerkin methods for parabolic equations, SIAM Journ. Num. Anal. 7 (1970), 575-626.

[4] J. Levin. On a nonlinear Volterra equation, Journ. Math. Anal. and Appl. 39 (1972), 458-476.

[5] J. L. Lions, Quelques Methodes de Résolution des Problemes aux Limites Non Linéaires, Gauthier-Villars, Paris, 1969.

[6] R. C. MacCamy, Remarks on frequency domain methods for Volterra integral equations, Journ. Math. Anal. Appl. 55 (1976), 555-575.

[7] R. C. MacCamy, An integro-differential equation with applications in heat flow, Quart. Appl. Math. 35 (1977), 1-19.

[8] R. C. MacCamy and Philip Weiss, Numerical solutions of Volterra integral equations, to appear.

[9] B. Neta, Finite element approximation of a nonlinear diffusion problem, Thesis, Department of Mathematics, Carnegie-Mellon University (1977).

[10] G. Strang and G. J. Fix, An Analysis of the Finite Element Method, Prentice-Hall (1971).

[11] Philip Weiss, Numerical solutions of Volterra integral equations, Thesis, Department of Mathematics, Carnegie-Mellon University (1978).

WELL POSEDNESS OF ABSTRACT VOLTERRA PROBLEMS

R. K. Miller
Iowa State University
Ames, Iowa 50011 / USA

1. INTRODUCTION

Let A be an n×n matrix, B(t) a matrix valued function in $L^1(0,\infty)$, $\gamma\geq 0$, R^n = real n-space and $(y_0,\phi)\varepsilon R^n\times L^1(-\infty,0)$. It is well known [1,2] that the solution y(t) of

$$y'(t) = \gamma Ay(t) + \int_{-\infty}^{t} B(t-s)y(s)ds \quad (t\geq 0)$$
$$y(0) = y_0,\ y(t) = \phi(t) \ \text{a.e. in} \ -\infty<t<0$$

can be used to define a C_0-semigroup on $R^n\times L^1(-\infty,0)$. Indeed if y_t is defined by $y_t(s) = y(t+s)$ for $-\infty<s<0$, then

$$T(t)(y_0,\phi) = (y(t),y_t) \quad (0\leq t<\infty).$$

The inhomogenous problem

$$(1.1)\qquad z'(t) = \gamma Az(t) + \int_0^t B(t-\tau)z(\tau)d\tau+f(t)$$

for $t\varepsilon R^+ = [0,\infty)$ and $z(0) = z_0$ determines the adjoint semigroup (see [2,12]). We put B U = all bounded uniformly continuous functions, let $f\varepsilon BU$ and define

$$T^*(t)(z_0,f) = (z(t),f_t + \int_0^t B_s z(t-s)ds)$$

for all $t\varepsilon R^+$. The infinitesimal generator C of the semigroup $T^*(t)$ is defined by

$$(1.2)\qquad C(z_0,f) = (\gamma Az_0 + f(0),f' + Bz_0).$$

The formal calculations discussed here work equally well when R^n is replaced by any Banach space X and $A : D(A)\to X$ is any closed linear map with dense domain. This point of view is exploited in [13] where properties of the operator (1.2) are used in order to prove existence, uniqueness and continuity theorems for (1.1) in the Banach space case. In [13] it is required that $\gamma>0$, A generate a C_0-semigroup, $B(t)x\varepsilon C^1$ for all $x\varepsilon D(A)$, and $B'(t) : D(A)\to X$ be locally L^1-subordinate to A (e.g., $B(t) = b(t)A$ where b is a C^1 scalar function). The case $\gamma = 0$ can not be analyzed using the techniques in [13]. Our purpose here is to study the case $\gamma = 0$.

We study here (and in [15]) an abstract Volterra integrodifferential equation of the form

$$(E)\quad x'(t) = cx(t) + \int_0^t K(t-s)x(s)ds + F(t),\ x(0) = x_0.$$

Here ' denotes the strong derivative w.r.t. the variable $t\varepsilon R^+$, $F = R^+\to X$ and c is a real scalar. Note that the transformation $z = \exp(-ct)x$ reduces (E) to (1.1) with $\gamma = 0$, $B(t) = K(t)\exp(-ct)$ and $f = \exp(-ct)F$. Also note that this equation can be considered as the integrated form of the abstract second order problem

$$(E_1) \qquad x''(t) = cx'(t) + K(0)x(t) + \int_0^t K'(t-s)x(s)ds + f(t).$$

Equations of the form (E_1) arise in a model for heat flow proposed by Coleman and Gurtin [3] (see also [10]) and have been studied by many authors (see the bibliography in [14]). For related work on well posedness and stability see for example Dafermos [5] and Slemrod [16] for results using Lyapunov techniques, Gripenberg [9], London [11], Grimmer and Miller [8], and Crandall and Nohel [4]. The paper [4] is a particularly interesting and smoothly presented study of well posedness for nonlinear equations.

2. CONTINUOUS DEPENDENCE OF SOLUTIONS OF (E)

Define $L = K(0)$, $K_0(t) = K'(t)$. Assume throughout this section that the forcing function $F(t)$ belongs to $C^1(R^+;X)$, and that

(A1) The operator L is a closed linear operator with domain $D(L)$ dense in X while for each $t\varepsilon R^+$, the domain of the linear operator $K(t)$ contains $D(L)$. Moreover, for each $x\varepsilon D(L)$, $K(t)x\varepsilon C^1(R^+;X)$, and there exist continuous positive functions κ and κ_0 with $\kappa\varepsilon L^\infty(R^+)$ and $\kappa_0\varepsilon L^1(R^+)$ so that

$$(2.1) \qquad \|K(t)x\| \le \kappa(t)\|x\|_L, \quad \|K_0(t)x\| \le \kappa_0(t)\|x\|_L$$

for all $t\varepsilon R^+$ and all $x\varepsilon D(L)$. Here $\|x\|_L = \|x\| + \|Lx\|$.

Let BU denote the Banach space $BU = \{f\varepsilon C(R^+;X) : f$ is bounded and uniformly continuous$\}$ with norm $\|f\| = \sup\{\|f(t)\| : t\varepsilon R^+\}$. Define the Banach space $Z = X \times X \times BU \times BU$ with the norm $\|(x,u,f,F)\| = \|x\| + \|u\| + \|f\| + \|F\|$. Define the linear map C on the domain $D(C) = \{(x,u,f,F)\varepsilon Z : x$ and $F(0)\varepsilon D(L), F\varepsilon C^2,$ $F'\varepsilon BU$, $F'' + K(\cdot)[cx + F(0)] + K_0(\cdot)x\varepsilon BU\}$ by

$$(2.2)\ C(x,u,f,F) = \Big[[L + c^2 I]x + cF(0) + F'(0),\ LF(0),\ F',\ F'' + K(\cdot)[cx + F(0)] + K_0(\cdot)x\Big].$$

<u>LEMMA 2.1</u> <u>The operator C is closed.</u>

<u>Proof</u>. Let $z_n = (x_n,u_n,f_n,F_n)\varepsilon D(C)$, $z_n \to z_0 = (x_0,u_0,f_0,F_0)\varepsilon Z$ and $Cz_n \to (y,w,g,G)$ in Z. Since L is closed and $LF_n(0) \to w$, we see that $F_0(0)\varepsilon D(L)$ and $LF_0(0) = w$. Also $x_0\varepsilon D(L)$ and $Lx_0 = y - c^2x_0 - cF_0(0) - g(0)$. Next we observe that

$$(2.3) \quad F_n''(t) \to G(t) - K(t)[cx_0 + F_0(0)] - K_0(t)x_0 \ , \ n \to \infty,$$

uniformly for t on compact intervals in R^+. We apply the fundamental theorem and use (2.3) to take limits under the integral to obtain

$$g(t) = g(0) + \int_0^t \{G(\tau) - K(\tau)[cx_0 + F_0(0)] - K_0(\tau)x_0\}d\tau.$$

A second use of the fundamental theorem and passage to the limit yields

$$F_0(t) = F_0(0) + tg(0) + \int_0^t \int_0^s \{G(\tau) - K(\tau)[cx_0 + F_0(0)] - K_0(\tau)x_0\} d\tau\, ds.$$

We can differentiate this last line to see that $F_0 \varepsilon C^2$, $F_0' = g \varepsilon BU$ and $F_0'' + K(\cdot)[cx_0 + F_0(0)] + K_0(\cdot)x_0 = G \varepsilon BU$. Thus, $z_0 \varepsilon D(C)$ and $Cz_0 = (y,w,g,G)$; hence, C is closed.

LEMMA 2.2 D(C) is dense in Z.

Proof. Let $z = (x,u,f,F) \varepsilon Z$ and $\varepsilon > 0$. We must find $z_0 = (x_0,u,f,F_0) \varepsilon D(C)$ so that $\|z - z_0\| < \varepsilon$. Since $\overline{D(L)} = X$, pick $x_0 \varepsilon D(L)$ so that $\| x - x_0\| < \varepsilon/2$. To find an F_0, we begin by using the denseness of $D(L)$ to pick $\alpha \varepsilon D(L)$ satisfying $\| F(0) - \alpha\| < \varepsilon/6$. Next, let $\xi(t)$ be a nonnegative function so that $\xi \varepsilon C^\infty(-\infty,\infty)$, $\xi(0) = 1$, $\xi(t) = 0$ for $|t| \geq 1$ and $\xi'(-1) = \xi'(0) = 0$. Set $\phi(t) = [cx_0 + \alpha]te^t + \xi(t)x_0$, and define

$$g(t) = \int_t^\infty K(\tau)\phi(t-\tau)d\tau.$$

It follows using the first inequality in (2.1) and the definition of ϕ, that $g \varepsilon BU$. Moreover,

$$g'(t) = -K(t)x_0 + \int_t^\infty K(\tau)\phi'(t-\tau)d\tau,$$

$$g''(t) = -K_0(t)x_0 - K(t)[cx_0 + \alpha] + \int_t^\infty K(\tau)\phi''(t-\tau)d\tau.$$

Hence, g' and $g'' + K(\cdot)[cx_0 + \alpha] + K_0(\cdot)x_0$ lie in BU. Let ψ be a C^2-function with $\psi^{(j)} \varepsilon BU$ for $j = 0,1,2$, and $\| F-g-\psi\| < \varepsilon/6$. Define $F_0(t) = \alpha + \psi(t) - \psi(0) + g(t) - g(0)$. Then $F_0 \varepsilon BU$, $F_0(0) = \alpha \varepsilon D(L)$, $F_0' \varepsilon BU$ and $F_0'' + K(\cdot)[cx_0 + F_0(0)] + K_0(\cdot)x_0 \varepsilon BU$. Also,

$$\begin{aligned} \| F - F_0\| &\leq \| F - \psi - g\| + \|\alpha - \psi(0) - g(0)\| \\ &< \varepsilon/6 + \| \alpha - F(0)\| + \|F(0) - \psi(0) - g(0)\| < \varepsilon/2. \end{aligned}$$

Thus, $z_0 = (x_0,u,f,F_0) \varepsilon D(C)$ and $\|z - z_0\| < \varepsilon$. Q.E.D.

Define $\rho(\lambda) = [L + K_0^*(\lambda) + cK^*(\lambda) + c^2 I - \lambda^2 I]^{-1}$ at all points λ with $\mathrm{Re}\lambda > 0$ for which the inverse exists as a bounded linear map on X. Here * denotes the Laplace transform.

LEMMA 2.3 Suppose $\rho(\lambda)$ exists for some λ with $\mathrm{Re}\lambda > 0$. Then, given $z \varepsilon Z$, there exists $z_1 \varepsilon D(C)$ so that

$$(C - \lambda^2 I)z_1 = z. \tag{2.4}$$

Moreover, if $\mathrm{Re}\lambda$ is sufficiently large, z_1 can be chosen so that

$$\| z_1\| \leq M(\lambda)\| z\| \tag{2.5}$$

where $M(\lambda)$ does not depend on z.

Proof. Suppose that λ, $\mathrm{Re}\lambda > 0$, is so that $\rho(\lambda)$ exists. Given

$z = (x,u,f,F) \varepsilon Z$, we must find $z_1 = (x_1,u_1,f_1,F_1)$ in $D(C)$ so that (2.4) holds. Set $F_1(0) = 0$, $x_1 = \rho(\lambda)[F^*(\lambda) + x]$, and $F_1'(0) = \lambda^2 x_1 - Lx_1 - c^2 x_1 + x$. Also, define $u_1 = -u/\lambda^2$ and $f_1 = (F_1' - f)/\lambda^2$ where

$$2\lambda F_1(t) = e^{\lambda t}\left\{F_1'(0) + \int_0^t e^{-\lambda\tau}G(\tau)d\tau\right\} - e^{-\lambda t}\left\{F_1'(0) + \int_0^t e^{\lambda\tau}G(\tau)dt\tau\right\},$$

with $G(t) = F(t) - K(t)cx_1 - K_0(t)x_1$. Our choice of x_1 and $F_1'(0)$ yields $F_1'(0) + G^*(\lambda) = 0$; hence, we can rewrite $F_1(t)$ as

$$(2.6)\quad 2\lambda F_1(t) = -e^{\lambda t}\int_t^\infty e^{-\lambda\tau}G(\tau)d\tau - e^{-\lambda t}\left\{F_1'(0) + \int_0^t e^{\lambda\tau}G(\tau)d\tau\right\}.$$

It is easy to verify that this $z_1 \varepsilon D(C)$ and that equation (2.4) holds.

It remains to show that if $\sigma = \mathrm{Re}\lambda$ is large enough, then there exists a constant $M(\lambda)$. We begin by finding a constant $M_1(\lambda)$ which is independent of z so that

$$(2.7)\qquad \| F_1'(0)\| \le M_1(\lambda)\| z\| .$$

To obtain (2.7), recall that $F_1'(0) = -G^*(\lambda)$. Then, using the definition of $G(t)$, the bounds (2.1), and the definition $F_1'(0) = \lambda^2 x_1 - Lx_1 - c^2 x_1 + x$, we obtain

$$\| F_1'(0)\| \le \|F\|\sigma^{-1} + [|c|\kappa^*(\sigma) + \kappa_0^*(\sigma)]\| x_1\|_L \le \|F\|\sigma^{-1} + [|c|\kappa^*(\sigma) + \kappa_0^*(\sigma)][(1 + |\lambda|^2 + |c|^2)\| x_1\| + \| x\| + \| F_1'(0)\|].$$

For σ sufficiently large, $|c|\kappa^*(\sigma) + \kappa_0^*(\sigma) < 1/2$; hence, using the definition of x_1, (2.7) follows with

$$M_1(\lambda) = 2\sigma^{-1} + (1+|\lambda|^2 + |c|^2)\|\rho(\lambda)\|(1+\sigma^{-1}) + 1.$$

Next, starting with equation (2.6), and using (2.1) and (2.7), we deduce that

$$\begin{aligned}\| F_1(t)\| &\le \Big\{e^{\sigma t}\int_t^\infty e^{-\sigma\tau}[\| F\| + (\kappa(\tau)|c| + \kappa_0(\tau))\| x_1\|_L]\, d\tau \\ &\quad + e^{-\sigma t}[\| F_1'(0)\| + \int_0^t e^{\sigma\tau}[\| F\| + (\kappa(\tau)|c| + \kappa_0(\tau))\|x_1\|_L]d\tau]\Big\}/2|\lambda| \\ &\le |\lambda|^{-1}\{\sigma^{-1}\| F\| + [\| \kappa\|_\infty |c|\sigma^{-1} + \| \kappa_0\|_1]\| x_1\|_L + \| F_1'(0)\|/2\} \\ &\le |\lambda|^{-1}\Big\{\sigma^{-1} + [\|\kappa\|_\infty|c|\sigma^{-1} + \|\kappa_0\|_1]\Big[[1 + |\lambda|^2 + |c|^2]\|\rho(\lambda)\|(1+\sigma^{-1}) \\ &\quad + 1 + M_1(\lambda)\Big] + M_1(\lambda)/2\Big\}\| z\| = M_2(\lambda)\| z\| .\end{aligned}$$

Also, if we differentiate both sides of equation (2.6), we find that

$$F_1'(t) = -\lambda F_1(t) - e^{-\lambda t}\int_t^\infty e^{-\lambda\tau}G(\tau)d\tau.$$

The second term can be bounded using the methods of the previous paragraph, and we obtain

$$\|F_1'(t)\| \le \Big\{|\lambda| M_2(\lambda) + \sigma^{-1} + [\|\kappa\|_\infty |c|\sigma^{-1} + \|\kappa_0\|_1] \Big[(1 + |\lambda|^2 + |c|^2)\|\rho(\lambda)\| (1 + \sigma^{-1}) + M_1(\lambda) + 1\Big]\Big\} \|z\|$$

$$= M_3(\lambda)\|z\| \qquad \text{for} \qquad t \ge 0.$$

Finally, using the definitions of x_1, u_1, f_1 and F_1 as well as the last two estimates, we have

$$\|z_1\| = \|x_1\| + \|u_1\| + \|f_1\| + \|F_1\|$$

$$\le \|\rho(\lambda)\| (1 + \sigma^{-1})\|z\| + (\|u\| + \|F_1'\| + \|f\|)|\lambda|^{-2} + \|F_1\|$$

$$\le \Big\{\|\rho(\lambda)\| (1 + \sigma^{-1}) + (1 + M_3(\lambda)) |\lambda|^{-2} + M_2(\lambda)\Big\} \|z\| = M(\lambda) \|z\|,$$

where $M(\lambda)$ does not depend on z. Q.E.D.

By a <u>solution</u> of equation (E) we mean a function $x: R^+ \to D(L)$ so that $x(0) = x_0$, $x(t) \in C^1(R^+:X)$, $Lx(t) \in C(R^+:X)$ and (E) is true for all $t \ge 0$. By a <u>generalized solution</u> we mean the limit uniformly on compact subsets of $0 \le t < \infty$ of a sequence of solutions. We make the additional assumption.

(A2) For any point $(x_0,u,f,F) \in D(C)$, there exists a unique solution $x(t,x_0,F)$ of equation (E) on R^+.

Let π_1 denote the projection onto the first coordinate of Z. For $(x_0,u,f,F) \in D(C)$ define

$$U(t)(x_0,u,f,F) = [\, x(t), u, f, F_t + \int_0^t K_{t-\tau} x(\tau)\, d\tau\,] \tag{2.8}$$

where $x(t) = x(t,x_0,F)$, and $F_t(s) = F(t+s)$ denotes translation.
Thus, for each $t \ge 0$

$$[F_t + \int_0^t K_{t-\tau} x(\tau) d\tau\,](s) = F(t+s) + \int_0^t K(t+s-\tau) x(\tau) d\tau \quad (s \ge 0).$$

<u>LEMMA 2.4</u> <u>Let</u> $z \in D(C^2)$ <u>and</u> $x(t) = \pi_1 U(t) z$. <u>Then</u> $x''(t) = \pi_1 U(t)\, Cz$ <u>for</u> $t \ge 0$.

<u>Proof</u>. Fix $z = (x_0,u,f,F) \in D(C^2)$ and let $x(t) = \pi_1 U(t) z$. Let $X(t)$ be the solution of (E) with $X(0) = [L + c^2 I]x_0 + cF(0) + F'(0)$ and forcing function $\phi = F'' + K(\cdot)[cx_0 + F(0)] + K_0(\cdot)x_0$; thus, $X(t) = \pi_1 U(t) Cz$.

Let $*$ denote convolution. An application of the fundamental theorem yields

$$X(t) = [L + c^2 I]x_0 + cF(0) + F'(0) + 1 * cX(t) + 1 * K * X(t) + 1 * \phi(t).$$

Define $y' = cx_0 + F(0) + 1 * X$, $y(0) = x_0$. Then $y''(t) = X(t)$. Also,

$$y'(t) = cx_0 + F(0) + t\{[L + c^2 I]x_0 + cF(0) + F'(0)\} + 1 * 1 * cX(t) + 1 * 1 * K * X(t) + 1 * 1 * \phi(t)$$

$$= cx_0 + F(0) + t\{[L + c^2 I]x_0 + cF(0) + F'(0)\}$$

$$+K * y(t) - 1 * Kx_0 - 1 * 1 * Kcx_0 - 1 * 1 * KF(0) + 1 * 1 * cX(t)$$
$$+ 1 * 1 * \{F''(t) + K(t)cx_0 + K(t)F(0) + K_0(t)x_0\}$$
$$= cx_0 + tc^2x_0 + tcF(0) + 1 * 1 * cX(t) + K * y(t) + F(t)$$
$$= cy(t) + K * y(t) + F(t).$$

By hypothesis (A2), (E) has a unique solution with initial value x_0 and forcing function F; hence $y(t) = x(t)$, and $x''(t) = X(t)$. Q.E.D.

Define the normed linear space

$$W = \{(x,u,f,F) \in Z: x \in D(L), F \in C^1, F' \in BU\}$$

with norm $\|(x,u,f,F\|_W = \|(x,u,f,F)\| + \|[L+c^2I]x + cF(0) + F'(0)\| + \|F'\|$. For fixed $T > 0$ define $C(W) = \{(y(t),u,f,F(t,\cdot)): [0,T] \to W: \text{continuous}\}$ with norm $\|z\|_{C(W)} = \sup\{\|z(t)\|_W: 0 \leq t \leq T\}$.

LEMMA 2.5 C(W) is a Banach space.

The proof of the completeness of W is a routine exercise which uses the definition of the W-norm and the fact that the operator L is closed. We are now ready to state and prove our result on continuous dependence.

THEOREM 2.1. Suppose (A1) and (A2) hold, and let $\rho(\lambda)$ exist for some λ with $\mathrm{Re}\lambda > 0$ so large that inequality (2.5) of Lemma 2.3 is valid. Then the solution $x(t,x_0,F)$ of (E) depends continuously on (x_0,F) in the sense that for any $T > 0$, there is an $M = M(T) > 0$ such that for $0 \leq t \leq T$,

$$(2.9) \qquad \|x(t,x_0,F)\| \leq M(\|x_0\| + \sup\{\|F(\tau)\| : 0 \leq \tau < \infty\})$$

for all $(x_0,u,f,F) \in D(C)$.

Remark. In the next section we exhibit a class of kernels K satisfying (A1) for which the existence-uniqueness hypothesis (A2) is valid, and for which $\rho(\lambda)$ exists for all sufficiently large positive λ.

Proof. Fix $T > 0$. It is not difficult to check that $U: D(C) \to C(W)$ where, for $0 \leq t \leq T$, $U(t)(x_0,u,f,F)$ is defined by (2.8) when $(x_0,u,f,F) \in D(C)$. Moreover, U is a closed operator for D(C) with its graph norm $\|z\|_C = \|z\| + \|Cz\|$ into C(W). To verify that U is closed, first observe that since C is closed, D(C) with its graph norm is complete. Thus, let the sequence $\{z_n = (x_n,u_n,f_n,F_n)\}$ and $z_0 = (x_0,u_0,f_0,F_0)$ be so that $\|z_n - z_0\|_C \to 0$. Also, suppose that

$$Uz_n = (x_n(t),u_n,f_n,F_n(t,\cdot)) \to (y(t),v,g,G(t,\cdot))$$

in C(W). Here, $F_n(t,\cdot)$ is defined by

$$F_n(t,s) = F_n(t+s) + \int_0^t K(t+s-\tau)x_n(\tau)d\tau, \quad s \geq 0.$$

Then, $\|x_n(t) - y(t)\| \to 0$,

$$\| F_n(t,\cdot) - G(t,\cdot)\| \to 0, \qquad \| \tfrac{\partial}{\partial s} F_n(t,\cdot) - \tfrac{\partial}{\partial s} G(t,\cdot)\| \to 0,$$

$$\| [L + c^2 I][x_n(t) - y(t)] + cF_n(t,0) - cG(t,0) + \tfrac{\partial}{\partial s} [F_n(t,0) - G(t,0)]\| \to 0,$$

where each convergence is uniform for t in $[0,T]$. It follows that $\| Lx_n(t) - Ly(t)\| \to 0$ uniformly for $0 \leq t \leq T$. Thus, we may integrate equation (E) for $x_n'(t)$ from 0 to t, apply the fundamental theorem, use the first bound in inequality (2.1) and let $n \to \infty$ to obtain

$$y(t) = x_0 + c\int_0^t y(s)ds + \int_0^t \int_0^s K(s-\tau)y(\tau)d\tau\, ds + \int_0^t F_0(s)ds.$$

Hence, y satisfies (E) with initial value x_0 and forcing function F_0. By the uniqueness hypothesis in (A2) $y(t) = x_0(t)$ for $0 \leq t \leq T$. Finally, to see that $G(t,\cdot) = F_0(t,\cdot)$, observe that for $0 \leq t \leq T$ and $s \in R^+$,

$$\begin{aligned} \| G(t,s) - F_0(t,s)\| &\leq \| F_n(t,s) - F_0(t,s)\| + \| F_n(t,s) - G(t,s)\| \\ &= \| F_n(t,s) - F_0(t,s)\| + o(1) \end{aligned}$$

as $n \to \infty$. Then, using the definition of F_n and F_0, and the first bound of (2.1), let $n \to \infty$ to obtain $G(t,s) = F_0(t,s)$. This completes the proof that U is closed.

To continue with the proof of Theorem 2.1, let λ, $\mathrm{Re}\lambda > 0$, be such that $\rho(\lambda)$ exists and inequality (2.5) holds. Fix $z = (x_0,u,f,F) \in D(C)$. Choose $z_1 = (x_1,u_1,f_1,F_1) \in D(C^2)$ satisfying (2.4) and (2.5), and write $x_1(t) = \pi_1 U(t)z_1$. Using (2.4) and Lemma 2.4, we find that for $0 \leq t \leq T$,

$$\begin{aligned} \| x(t)\| &= \|\pi_1 U(t)\ z\| = \| \pi_1 U(t)(C - \lambda^2 I)\ z_1\| \\ &\leq \| x_1''(t)\| + \| \pi_1 U(t)\ \lambda^2 z_1\| \qquad (2.10)\\ &\leq (1 + |\lambda|^2)\{\| x_1''(t)\| + \| \pi_1 U(t)\ z_1\| \}. \end{aligned}$$

From the definition of the norm $\|\cdot\|_W$, we have $\|\pi_1 U(t)\ z_1\| \leq \| U(t)z_1\|_W$, $0 \leq t \leq T$. Also, differentiating equation (E) for $x_1(t)$ and then substituting equation (E) for $x_1'(t)$ into the resulting expression yields

$$\begin{aligned} x_1''(t) &= [L + c^2 I]x_1(t) + c(F_1(t) + \int_0^t K(t-\tau)x_1(\tau)d\tau) \\ &+ \int_0^t K_0(t-\tau)x_1(\tau)d\tau + F_1'(t). \end{aligned}$$

Hence, using the definition of $\|\cdot\|_W$, we have $\|U(t)\ z_1\|_W \geq \|x_1''(t)\|$, $0 \leq t \leq T$. Thus, after substituting the last two inequalities into inequality (2.10), and using the definition of $\|\cdot\|_{C(W)}$, we find that

$$\| x(t)\| \leq 2(1 + |\lambda|^2)\| Uz_1\|_{C(W)}, \qquad 0 \leq t \leq T.$$

Since U is a closed linear map from $D(C)$ with its graph norm to $C(W)$, U is bounded [18, p. 181]. Combining this with (2.4) and (2.5), we obtain

$$\|x(t)\| \le 2(1+|\lambda|^2)\|U\|\,[\|z_1\| + \|Cz_1\|]$$
$$\le 2(1+|\lambda|^2)\|U\|\,[\|z\| + (1+|\lambda|^2)\|z_1\|]$$
$$\le 2(1+|\lambda|^2)\|U\|\,[1+M(\lambda)(1+|\lambda|^2)]\|z\| = M\|z\|$$

where $M = M(T)$. This inequality is true for all choices of u and f in $z = (x_0,u,f,F)$; in particular, it is true with $z = (x_0,0,0,F)$. But, for this choice of z, the last inequality is simply (2.9). Q.E.D.

3. WELL-POSED PROBLEMS.

We say that (E) is uniformly well-posed if for each pair (x_0,F) with $(x_0,0,0,F) \in D(C)$, there exists a unique solution $x(t,x_0,F)$ of equation (E) on R^+, and if for any $T>0$, $\|x(t,x_0,F)\| \to 0$ uniformly on $[0,T]$ as $\|(x_0,0,0,F)\| \to 0$. In this section, we give an example of a class of problems which are uniformly well-posed. We begin by obtaining conditions that guarantee that the existence and uniqueness hypothesis (A2) holds.

Consider the initial value problem

(3.1) $$x'(t) = cx(t) + \int_0^t [B(t-\tau)Ax(\tau) + G(t-\tau)x(\tau)]d\tau + F(t),\quad x(0) = x_0,$$

where c is a real scalar, $B(t)$ and $G(t)$ are real-valued functions in $C^2(R^+)$ with $B(0)>0$, and $F \in C^1(R^+;X)$. The closed, densely defined linear operator A will be assumed to satisfy

(A3) A is the infinitesimal generator of a strongly continuous cosine family $C(t)$, $t \in R = (-\infty,\infty)$.

The fundamental work on cosine families is that of H. O. Fattorini [6,7] and M. Sova [17]. For more recent results concerning cosine families and abstract nonlinear second order differential equations, we refer the reader to the paper by C. C. Travis and G. F. Webb [19]. Since $x(t)$ satisfies (3.1) if and only if $y(t) = \exp(-ct)\,x(t)$ satisfies

$$y'(t) = \int_0^t \exp(-c(t-\tau))[B(t-\tau)Ay(\tau) + G(t-\tau)y(\tau)]d\tau + \exp(-ct)F(t),$$

we may assume that $c = 0$ in (3.1) Moreover, since $B(0)>0$, we can assume that $B(0) = 1$. Finally, it is clear that solving equation (3.1) with $c=0$ and $B(0)=1$ is equivalent to finding $x(t) \in C^2(R^+;x)$ with $x(t) \in D(A)$ and $Ax(t) \in C(R^+;X)$ so that

(3.2) $$x''(t) = [A+\gamma I]x(t) + \int_0^t [b(t-\tau)Ax(\tau) + g(t-\tau)x(\tau)]d\tau + f(t),$$

$x(0) = x_0$, $x'(0) = v_0$, where $b = B'$, $g = G'$, $f = F'$, $\gamma = g(0)$ and $v_0 = F(0)$.

Let A be the infinitesimal generator of a strongly continuous cosine family $C(t)$. The operator $S(t)$, defined by

$S(t)x = \int_0^t C(\tau)x\,d\tau$, $x \in X$, $t \in R$, is called the corresponding sine family. Also, define the sets $D(A)$ and E by

$$D(A) = \{x \in X: C(t)x \in C^2(R;X)\}, \qquad E = \{x \in X: C(t)x \in C^1(R;X)\}.$$

With this notation, Travis and Webb [19] prove

LEMMA 3.1. *Assume that* (A3) *holds and that* $f \in C^1(R^+;X)$. *Then for each* $x_0 \in D(A)$ *and* $v_0 \in E$, *there is a unique function* $u(t) \in C^2(R^+;X)$ *with* $u(t) \in D(A)$ *which solves the initial value problem*

$$(3.3) \qquad u''(t) = Au(t) + f(t), \qquad u(0) = x_0, \qquad u'(0) = v_0,$$

for $t \in R^+$. *Furthermore, this* $u(t)$ *can be written as*

$$(3.4) \qquad u(t) = C(t)x_0 + S(t)v_0 + \int_0^t S(t-\tau)f(\tau)d\tau, \qquad t \geq 0.$$

Using this lemma we prove

THEOREM 3.1. *Let* (A3) *hold*, $f \in C^1(R^+;X)$, *and both* b *and* g *belong to* $C^1(R^+)$. *Then for each* $x_0 \in D(A)$ *and* $v_0 \in E$, *the initial value problem* (3.2) *has a unique solution* $x(t)$ *on* R^+.

Proof. Our proof is similar to the proof of Lemma 7.2 in [13].

Fix $T > 0$. Let $0 < t_0 \leq T$ and define $d(t_0) = \{w: w \text{ maps } [0,t_0] \text{ into } D(A) \text{ with } w(t), w'(t) \text{ and } Aw(t) \text{ continuous}\}$. Since A is closed, it is easy to see that $d(t_0)$ with norm $|||w||| = \sup\{\|w(t)\|_A + \|w'(t)\| : 0 \leq t \leq t_0\}$ is a Banach space. For $w \in d(t_0)$ define

$$\Psi_0 w(t) = \gamma w(t) + \int_0^t [b(t-\tau)Aw(\tau) + g(t-\tau)w(\tau)]d\tau$$

on $0 \leq t \leq t_0$. Then $\Psi_0 w$ is strongly continuously differentiable with

$$(\Psi_0 w)'(t) = \gamma w'(t) + b(0)Aw(t) + g(0)w(t)$$
$$+ \int_0^t [b'(t-\tau)Aw(\tau) + g'(t-\tau)w(\tau)]d\tau.$$

From Lemma 3.1 it follows that for any $w \in d(t_0)$,

$$u''(t) = Au(t) + \{\Psi_0 w(t) + f(t)\}, \quad u(0) = x_0, \quad u'(0) = v_0,$$

has a unique solution $u = \Psi w$ which is again in $d(t_0)$.

Since $\Psi_0 w$ is linear, we see, that we can decide whether Ψw is a contraction map by computing the norm when $x_0 = v_0 = 0$ and $f(t) \equiv 0$. If $w \in d(t_0)$ and $|||w||| \leq 1$, then

$$\|\Psi_0 w(t)\| \leq |\gamma| + \int_0^t |b(\tau)| + |g(\tau)|d\tau.$$

Thus, using formula (3.4) with $x_0 = v_0 = 0$ and forcing function $\Psi_0 w(t)$, we obtain

$$\|\Psi w(t)\| \le \int_0^t \|S(t-\tau)\| \, \|\Psi_0 w(\tau)\| \, d\tau$$

$$\le M \int_0^t \{|\gamma| + \int_0^s (|b(\tau)| + |g(\tau)|) d\tau\} ds$$

$$= M\{|\gamma| t + \int_0^t (t-s)(|b(s)| + |g(s)|) ds\},$$

where M is a bound for $\|C(t)\|$ and $\|S(t)\|$ on $[0,T]$ and $0 \le t \le t_0$ If we differentiate the expression for $\Psi w(t)$, we obtain

$$(\Psi w)'(t) = \int_0^t C(t-\tau)\Psi_0 w(\tau) d\tau.$$

Then, as above, we find that

$$\|(\Psi w)'(t)\| \le M\{|\gamma| t + \int_0^t (t-s)(|b(s)| + |g(s)|) ds\}$$

for $0 \le t \le t_0$. Finally, if we integrate the expression for $\Psi w(t)$ by parts, and observe that $\Psi_0 w(0) = \gamma w(0)$, we obtain

$$\Psi w(t) = \int_0^t (t-\tau) C(\tau) \gamma w(0) d\tau$$

$$+ \int_0^t \int_0^{t-s} (t-s-u) C(u) (\Psi_0 w)'(s) du \, ds .$$

Since A is closed, $w(0) \varepsilon D(A)$, and A commutes with $C(t)$ on $D(A)$

$$A \int_0^t (t-\tau) C(\tau) \gamma w(0) \, d\tau = \int_0^t (t-\tau) C(\tau) \gamma A w(0) d\tau.$$

Also, it is well-known [17, p. 14] that for any $x \varepsilon X$,

$$\int_0^t (t-\tau) C(\tau) x \, d\tau \, \varepsilon \, D(A) \quad \text{and} \quad A \int_0^t (t-\tau) C(\tau) x \, d\tau = C(t)x - x.$$

Thus, again observing that A is closed, we may apply A to the above expression for $\Psi w(t)$ and get

$$A\Psi w(t) = \int_0^t (t-\tau) C(\tau) \gamma A w(0) d\tau$$

$$+ \int_0^t [C(t-\tau) - I] (\Psi_0 w)'(\tau) d\tau.$$

Then, using the expression for $(\Psi_0 w)'$, and recalling the definition of $|||w|||$ and the fact that $|||w||| \le 1$, we obtain

$$\| A\Psi w(t) \le \int_0^t (t-\tau) \|C(\tau)\| \, |\gamma| \, \|Aw(0)\| d\tau$$

$$+ \int_0^t \|C(t-\tau) - I\| \, \|(\Psi_0 w)'(\tau)\| \, d\tau$$

$$\le M|\gamma| t + (M+1)\{(|\gamma| + |b(0)| + |g(0)|) t$$

$$+ \int_0^t (t-s)(|b'(s)| + |g'(s)|) ds\}.$$

It follows that

$$|||\Psi w||| \leq (M+1)\{(4|\gamma| + |b(0)| + |g(0)|)t_0 + \int_0^{t_0}(t_0 - s)(2|b(s)| + 2|g(s)| + |b'(s)| + |g'(s)|)ds\}<1$$

for t_0 sufficiently small. The contraction mapping theorem implies the existence and uniqueness of a solution of (3.2) on $[0,t_0]$.

Translate (3.2) by t_0 to see that $y(t) = x(t+t_0)$ must satisfy

$$y''(t) = [A+\gamma I]y(t) + \int_0^t (b(t-\tau)Ay(\tau) + g(t-\tau)y(\tau))d\tau + \{f(t+t_0) + \int_0^{t_0}(b(t+t_0-\tau)Ax(\tau)+g(t+t_0-\tau)x(\tau))d\tau\},$$

$y(0) = x(t_0)$, $y'(0) = x'(t_0)$. Clearly the term in brackets, that is, the new forcing function, is strongly continuously differentiable, and $y(0) = x(t_0) \in D(A)$. To see that $y'(0) = x'(t_0)$ belongs to E, we note that $\Psi x = x$, and hence, using the representation (3.4),

$$x(t) = C(t)x_0 + S(t)v_0 + \int_0^t S(t-\tau)\{\Psi_0 x(\tau) + f(\tau)\}d\tau$$

for $0 \leq t \leq t_0$. From the theory of cosine functions [19], we can differentiate the last expression and write

$$x'(t_0) = S(t_0)Ax_0 + C(t_0)v_0 + \int_0^{t_0} S(t_0-\tau)\{(\Psi_0 x)'(\tau) + f'(\tau)\}d\tau.$$

It is well-known [19] that $S(t_0)Ax_0 \in E$. Since $C(r)C(t_0) = 2^{-1}(C(r+t_0) + C(r-t_0))$ and $v_0 \in E$, we have that

$$\frac{d}{dr}C(r)C(t_0)v_0 = 2^{-1}A(S(r+t_0)v_0 + S(r-t_0)v_0),$$

and, hence, $C(t_0)v_0 \in E$. Finally, if $g(t)$ is strongly continuous

$$C(r)\int_0^{t_0} S(t_0-\tau)g(\tau)d\tau$$

$$= \int_0^{t_0}\int_0^{t_0-\tau} C(r)C(s)g(\tau)ds\,d\tau$$

$$= 2^{-1}\int_0^{t_0}\int_0^{t_0-\tau}(C(r+s) + C(r-s))g(\tau)ds\,d\tau .$$

Using this last expression, it is easy to show that

$$h^{-1}(C(r+h) - C(r))\int_0^{t_0} S(t_0-\tau)g(\tau)d\tau$$

$$= \int_0^{t_0}(2h)^{-1}\int_0^h (C(r+t_0-\tau+u) - C(r-t_0+\tau+u))g(\tau)du\,d\tau$$

It follows using the strong continuity of $g(\tau)$, the continuity of $C(\tau)x$ for each $x \in X$, and the dominated convergence theorem, that

$$\frac{d}{dr} C(r) \int_0^{t_0} S(t_0 - \tau)g(\tau)d\tau$$
$$= 2^{-1} \int_0^{t_0} (C(r + t_0 - \tau) - C(r - t_0 + \tau))g(\tau)d\tau.$$

Thus, since $(\Psi_0 x)' + f'$ is continuous, the last term in the expression for $x'(t_0)$ belongs to E. Hence, $x'(t_0)$ belongs to E.

Therefore, the same argument can be repeated to obtain a solution to (3.2) on $[t_0,2t_0]$, $[2t_0,3t_0]$, ... until $[Nt_0,T]$ where $(N+1)t_0 > T$. Since T is an arbitrary positive number, this proves existence and uniqueness on R^+. Q.E.D.

In order to give our example of a class of problems for which (E) is uniformly well-posed, we make the assumption

(A4) The scalar function B and G are of the form

$$B(t) = \beta + \int_0^t b(\tau)d\tau, \quad G(t) = \gamma + \int_0^t g(\tau)d\tau$$

with b and g in $C^1(R^+) \cap L^1(R^+)$ and $\beta > 0$ and γ real constants.

THEOREM 3.2. *Assume that (A3) and (A4) hold. Then the initial value problem (3.1) is uniformly well-posed.*

Proof. Since $\beta > 0$, we may assume that $\beta = 1$. The existence of a unique solution $x(t,x_0,F)$ to (3.1) for each $(x_0,0,0,F) \in D(C)$ follows from Theorem 3.1 and the remarks preceding equation (3.2). Define $K(t)$ by $K(t) = B(t)A + G(t)I$. Since (A3) and (A4) hold, $L = K(0) = A + \gamma I$ is closed and densely defined, $K(t)x$ is strongly continuously differentiable for each $x \in D(L)$, and there exist positive functions $\kappa(t) \in L^\infty(R^+)$ and $\kappa_0(t) \in L^1(R^+)$ such that the inequalities (2.1) hold. Thus, $K(t)$ satisfies the assumption (A1).

Finally, $\rho(\lambda)$ of Theorem 2.1 becomes

$$\rho(\lambda) = \{[A + \gamma I] + b^*(\lambda)A + g^*(\lambda) + c(B^*(\lambda)A + G^*(\lambda)) + c^2 I - \lambda^2 I\}^{-1}.$$

Since $B^*(\lambda)$, $G^*(\lambda)$, $b^*(\lambda)$ and $g^*(\lambda)$ all tend to zero as $\mathrm{Re}\lambda \to \infty$, $\rho(\lambda)$ exists if and only if

$$(\lambda^2 - \gamma - g^*(\lambda) - cG^*(\lambda) - c^2)/(1 + b^*(\lambda) + cB^*(\lambda))$$

is in the resolvent set of A when $\mathrm{Re}\lambda$ is sufficiently large. Since A is the infinitesimal generator of a strongly continuous cosine family, it is known [6, p.90] that the resolvent set of A contains λ^2 whenever $\mathrm{Re}\lambda$ is sufficiently large. Thus, since the last expression is asymptotic to λ^2 as $\mathrm{Re}\lambda \to \infty$, $\rho(\lambda)$ exists for all sufficiently large $\mathrm{Re}\lambda$.

Therefore the hypotheses of Theorem 2.1 are satisfied and problem (3.1) is well-posed. Q.E.D.

REFERENCES

1. V. Barbu and S. I. Grossman, asymptotic behavior of linear integrodifferential systems, Trans. Amer. Math. Soc. 171(1972), 277-288.
2. J. A. Burns and T. L. Herdman, Adjoint semigroup theory for a class of functional differential equations, SIAM J. Math. Anal. 7(1976), 729-745.
3. B. D. Coleman and M. E. Gurtin, Equipresence and constitutive equations for rigid heat conductors, Z. Angew, Math. Phys. 18 (1967), 199-208.
4. M. G. Crandall and J. A. Nohel, An abstract functional differential equation and a related nonlinear Volterra equation, Math. Res. Center, Univ. of Wisconsin, Tech. Summary Report #1765, 1977.
5. C. M. Dafermos, Asymptotic stability in viscoelasticity, Arch. Rational Mech. Anal. 37(1970), 297-308.
6. H. O. Fattorini, Ordinary differential equations in linear topological spaces, I. J. Differential Equations 5(1968), 72-105.
7. ____________, Ordinary differential equations in linear topological spaces, II, J. Differential Equations 6(1969), 50-70.
8. R. C. Grimmer and R. K. Miller, Existence, uniqueness, and continuity for integral equations in a Banach space, J. Math. Anal. Appl. 57(1977), 429-447.
9. G. Gripenberg, An existence result for a nonlinear Volterra integral equation in Hilbert space, SIAM J. Math. Anal. (to appear).
10. M. E. Gurtin and A. C. Pipkin, a general theory of heat conduction with finite wave speeds, Arch. Rational Mech. Anal. 31(1968), 113-126.
11. S.-O., Londen, On an integral equation in Hilbert space, SIAM J. Math. Anal. (to appear).
12. R. K. Miller, Linear Volterra integrodifferential equations as semigroups, Funk. Ekvac. 17(1974), 39-55.
13. ____________, Volterra integral equations in a Banach space, Funckcial. Ekvac. 18(1975), 163-193.
14. ____________, An integrodifferential equation for rigid heat conductors with memory, J. Math. Anal. Appl.,(to appear).
15. R. K. Miller and R. L. Wheeler, Well-posedness and stability of linear Volterra integrodifferential equations in abstract space, Funckcial Ekvac. (to appear).
16. M. Slemrod, A hereditary partial differential equation with applications in the theory of simple fluids, Arch. Rational Mech. Anal. 62(1976b), 303-322.
17. M. Sova, Cosine operator functions, Rozprawy Matematiyczne 49 (1966), 1-47.
18. A. E. Taylor, "Introduction to Functional Analysis", Wiley, New York, 1958.

9. C. C. Travis and G. F. Webb, Cosine families and abstract non linear second order differential equations, submitted.

NONLINEAR VECTOR-VALUED HEREDITARY EQUATIONS ON THE LINE

R.C. MacCamy[1] and Victor J. Mizel[2]
Department of Mathematics
Carnegie-Mellon University
Pittsburgh, PA 15213/USA

This investigation furnishes an existence theorem for the equation

$$(\underset{\sim}{E}) \qquad u(t) + \int_{-\infty}^{t} \underset{\sim}{a}(t-s)g(s,u(s))ds = f(t), \quad t \in R.$$

Here the solution u takes values in a reflexive Banach space V, the forcing term f and g take values in the dual space V' (assumed to contain V), and the values of $\underset{\sim}{a}$ lie in the set of bounded linear operators on V'. In particular, V can be a Sobolev space and V' its dual, so that $(\underset{\sim}{E})$ is sufficiently general to include cases in which each $g(t,\cdot)$ is a nonlinear partial differential operator. The hypothesis on the forcing term f is that each of the functions f^t, $t \in R$, defined by

$$f^t(s) = f(t-s), \quad s \in R^+,$$

belongs to a certain prescribed vector space $\mathcal{F}$ of strongly measurable V'-valued functions, but no assumption is made concerning the asymptotic behavior of f at $t = +\infty$.

Equations of the form $(\underset{\sim}{E})$ constitute important examples of hereditary laws, in that the present value of f (the output) depends on the present and all past values of u (the input). Indeed we have in mind as a prototype an equation governing one-dimensional heat flow with memory [16]. In such an application V is a Sobolev space on a real interval Ω, g is a partial differential operator on V of the form

$$g(t,u) = -\frac{\partial}{\partial x}\sigma(t,u_x), \qquad \sigma : R^2 \to R,$$

[1] Research partially supported by the U.S. National Science Foundation under Grant MCS77-01449.

[2] Research partially supported by the U.S. National Science Foundation under Grant MCS77-03643 A01.

and $\underset{\sim}{a} = aI$ is a scalar decay function times the identity operator on V'.

Furthermore, as has recently been emphasized [18]-[21], [9] one very important byproduct of an existence theory for $(\underset{\sim}{E})$ is that information concerning solutions of $(\underset{\sim}{E})$ can be directly applied to the study of asymptotic stability for non-linear Volterra equations such as

$$(\underset{\sim}{V}) \qquad v(t) + \int_0^t \underset{\sim}{a}(t-s)h(s,v(s))ds = e(t), \quad t \in R^+.$$

In this article we limit ourselves to a few remarks concerning these connections with asymptotic stability results for $(\underset{\sim}{V})$, referring the reader to the references cited above for more details.

The present work can be viewed as an extension to the Banach space context of results recently obtained by Leitman and Mizel [7], [8] for the scalar version of $(\underset{\sim}{E})$. The overall approach is that of [8], in that $(\underset{\sim}{E})$ is approximated by a family of equations of type $(\underset{\sim}{V})$. Here, however, the delicate a priori bounds obtained by Levin [10], [11] for the scalar version of $(\underset{\sim}{V})$ are no longer available, and instead the requisite a priori estimates for solutions of $(\underset{\sim}{V})$ are obtained via a frequency domain technique which was originally developed for this equation by MacCamy [14], [15], following ideas of Popov [3]. Another new feature in the present work is that an appeal to the Ascoli theorem, which was heavily utilized in [8], is no longer possible. Hence the approximation scheme has to be carried out by different methods.

The balance of the paper is organized as follows. Section 2 provides some notational conventions and mathematical preliminaries, including an existence-uniqueness theorem for $(\underset{\sim}{V})$. Section 3 contains the statement and proof of our existence theorem for $(\underset{\sim}{E})$.

2. Notations and an Existence Theorem for $(\underset{\sim}{V})$

Hereafter V denotes a fixed reflexive (real) Banach space and $H \supset V$ denotes a Hilbert space such that V is continuously imbedded and dense in H. Thus H, when identified with its own dual, acts as a "pivot" space for the pair V,V', $V \subset H \subset V'$. Generally, we follow the notation in [1], so that norms in H,V,V' will be denoted by $|\cdot|, \|\cdot\|, \|\cdot\|_*$, respectively. The inner product on H will be denoted $(\cdot,\cdot)$ while the evaluation map for V,V' will be denoted $\langle\cdot,\cdot\rangle$. Thus $\langle v,v'\rangle = (v,v')$ whenever $v' \in H$, $v \in V$. Finally, the space of bounded linear operators on V' will be denoted by $[V']$ and the space of bounded self-adjoint operators on H will be denoted $BS[H]$.

We proceed to give an existence theorem for $(\underset{\sim}{V})$ which will be needed in our study of $(\underset{\sim}{E})$. See [1], [4], [5], [13], for related results stated for the autonomous case.

Suppose we have a family $\{A(t)\}$ of non-linear operators from V to V',

$$A(t) : v \to h(t,v), \qquad t \in R^+, \; v \in V,$$

satisfying for some p, $2 \leq p < \infty$:

(A1) $A(t)$ is monotone and hemicontinuous, a.e. $t \in R^+$,

(A2) $t \to A(t)v(t)$ is measurable whenever $v(\cdot) \in L^p_{loc}(R^+;V)$,

(A3) There are constants c, $\omega > 0$ and $b \in L^{p'}_{loc}(R^+;R^+)$ such that

$$\|A(t)v\|_* \leq c\|v\|^{p-1} + b(t)$$

$$\langle A(t)v,v\rangle \geq \omega\|v\|^p, \qquad v \in V, \quad \text{a.e. } t \in R^+.$$

Furthermore suppose we have a family $\{\underset{\sim}{a}(t)\}$ of linear operators on V' satisfying:

(a1) $\underset{\sim}{a} \in W^{2,1}(R^+;[V'])$, $\underset{\sim}{a}|_H \in W^{2,1}(R^+;BS[H])$.

(a2) $\underset{\sim}{a}(0)$ is such that: $\underset{\sim}{a}(0)^{-1} \in [V']$, $\underset{\sim}{a}(0)^{-1}(V) = V$, and there are constants $\alpha, \beta > 0$ for which

$$\alpha|x|^2 \leq (\underset{\sim}{a}(0)^{-1}x,x) \leq \beta|x|^2, \qquad \forall x \in H,$$

$\dot{\underset{\sim}{a}}(0)|_H$ is negative definite.

It then follows [15], [16] that there exists a kernel $\underset{\sim}{k} \in W^{1,2}_{loc}(R^+;[V'])$ with $\underset{\sim}{k}|_H \in W^{1,2}_{loc}(R^+;BS[H])$, such that

$$(2.1) \qquad \underset{\sim}{a}(0)v(t) + \dot{\underset{\sim}{a}}*v(t) = u(t) \Leftrightarrow v(t) = \underset{\sim}{a}(0)^{-1}u(t) + \underset{\sim}{k}*u(t),$$

$$u \in L^{p'}_{loc}(R^+;V').$$

Further, suppose (see [16] for regularity hypotheses on $\underset{\sim}{a}$ ensuring this) that

(a3) $\underset{\sim}{k} \in W^{1,1}_{loc}(R^+;BS[H])$, $\dot{\underset{\sim}{k}} \in L^1(R^+;BS[H])$, and there exists $m > 0$ such that for each $T > 0$

$$(2.2) \qquad \int_0^T \langle \frac{d}{dt}\int_0^t \underset{\sim}{k}(t-s)x(s)ds, x(t)\rangle dt \geq m\int_0^T |x(t)|^2 dt,$$

$$x \in L^2(0,T;H).$$

Then the following result holds.

Theorem 2.1. Let V,H,V' and the mappings $h,\underset{\sim}{a}$ be as above. Then for each $f \in W^{1,p'}_{loc}(R^+;V')$ with $f(0) \in H$ there exists a unique function $v \in L^p_{loc}(R^+;V) \cap W^{1,p'}_{loc}(R^+;V') \cap C(R^+;H)$ satisfying

$$(\underset{\sim}{V}) \qquad v(t) + \int_0^t \underset{\sim}{a}(t-s)h(s,v(s))ds = f(t), \qquad \forall t \in R^+.$$

Sketch of Proof: It suffices to consider $(\underset{\sim}{V})$ on a fixed but arbitrary interval $[0,T] \subset R^+$. Denote

$$\mathcal{V} = L^p[0,T;V], \ \mathcal{H} = L^2[0,T;H], \ \mathcal{V}' = L^{p'}[0,T;V'] \ (\frac{1}{p'} + \frac{1}{p} = 1),$$

$$\langle\langle v,v'\rangle\rangle = \int_0^T \langle v(t),v'(t)\rangle dt.$$

Clearly $\mathcal{V},\mathcal{V}'$ are a dual pair of reflexive Banach spaces with $\mathcal{V}$ dense and continuously imbedded in $\mathcal{H}$.

It follows from (a1), (A2), (A3) that the integral term in $(\underset{\sim}{V})$ is a locally absolutely continuous V'-valued function. (In fact, this would follow from the weaker condition $\underset{\sim}{a} \in BV[0,T;[V']]$, [7], [8].) Thus the reflexivity of V' ensures [6] that all terms in $(\underset{\sim}{V})$ are primitives. That is, $(\underset{\sim}{V})$ is equivalent to:

$$(2.3) \qquad \dot{v}(t) + \frac{d}{dt}\int_0^t \underset{\sim}{a}(t-s)h(s,r(s))ds = \dot{f}(t), \qquad \text{a.e. } t \in [0,T],$$

$$v(0) = f(0).$$

Applying the relation (2.1) to this equation and simplifying we get

$$(2.4) \quad \underset{\sim}{a}(0)^{-1}\dot{\underset{\sim}{v}}(t) + A(t)v(t) + \frac{d}{dt}\int_0^t \underset{\sim}{k}(t-s)v(s)ds = F(t), \quad v(0) = f(0),$$

where $F(t) = a(0)^{-1}\dot{f}(t) + \frac{d}{dt}\int_0^t \underset{\sim}{k}(t-s)f(s)ds$. Now our hypotheses on f and $\underset{\sim}{a}$ ensure that the right side of (2.4) is an element in $\mathcal{V}'$. Hence the desired result will follow once we show that (2.4) possesses a unique solution $v \in W^{p'}(0,T;V') \cap L^p(0,T;V) \cap C(0,T;H)$

whatever $F \in \mathcal{V}'$ be prescribed.

We will prove that (2.4) has a unique solution by making use of a perturbation theorem for maximal monotone operators (compare [2, pp. 167-8]).

Proposition 2.2. The operator $B : \mathcal{V} \to \mathcal{V}'$ defined by

$$Bv = \underset{\sim}{a}(0)^{-1}\dot{v}, \text{ for } v \in D_B = \{u \in \mathcal{V} \cap W^{1,p'}(0,T;V') : u(0) = f(0)\}$$

is maximal monotone in $\mathcal{V} \times \mathcal{V}'$. Moreover,

$$v \in D_B \Rightarrow v \in C(0,T;H).$$

Proof: The monotonicity follows by a simple computation. Now let $v_o \in \mathcal{V}$, $w_o \in \mathcal{V}'$ be a purported pair extending B. That is,

$$\text{(2.5)} \qquad \langle\langle \varphi - v_o, B\varphi - w_o \rangle\rangle \geq 0 \qquad \forall \varphi \in D_B.$$

By restricting attention to the subclass of D_B consisting of functions having the form, for some fixed $h \in D_B$

$$\varphi = h + \chi c, \qquad c \in V, \qquad \chi \in \overset{\circ}{W}{}^{1,P}(0,T;R)$$

one obtains from (2.5) a relation equivalent to the system

$$\text{(a)} \quad \int_0^T [\langle c, w_o\rangle \chi + \langle v_o, a(0)^{-1}c\rangle \dot{\chi}]\,dt = 0 \qquad \forall \chi \in \overset{\circ}{W}{}^{1,P}(0,T;R)$$

(2.6)

$$\text{(b)} \quad \langle\langle v_o - f(0), w_o - Bh \rangle\rangle \geq 0.$$

A standard result in the calculus of variations now leads from (2.6a) to the conclusion that $\langle v_o, \underset{\sim}{a}(0)^{-1}c\rangle$ is in $W^{1,p'}(0,T;R)$ and

$$\text{(2.7)} \quad \langle c, -\int_0^t w_o(s)\,ds\rangle + \langle v_o(t), a(0)^{-1}c\rangle = K_c = \text{const.}$$

$$\text{a.e. } t \in [0,T], \ \forall c \in V.$$

Using essential separability of the range of v_o, w_o and the symmetry of $a(0)^{-1}$ one infers that there exists a function $\bar{v}_o$ equivalent to v_o and an element $b \in V'$ satisfying:

$$(2.8)\qquad \langle c, a(0)^{-1}\bar{v}_o(t) - b - \int_0^t w_o(s)ds\rangle = 0, \quad \forall t \in [0,T],\ c\in V.$$

It follows that actually $b\in V$ and

$$(2.9)\qquad a(0)^{-1}\bar{v}_0(t) = b + \int_0^t w_0(s)ds, \quad t \in [0,T].$$

Insertion of (2.9) into (2.5) readily leads to the relation $\bar{v}_0(0) = f(0)$. That is, $v_0 \in D_B$, $w_0 = Bv_0$, as claimed. For the remaining assertion concerning elements of D_B see [2, Lemma 4.1, pp. 167-8].

The remainder of the proof of Theorem 2.1 consists in verifying that the nonlinear operator $C : \mathcal{V} \to \mathcal{V}'$ given by

$$(2.10)\qquad Cv(t) = A(t)v(t) + \frac{d}{dt}\int_0^t \underset{\sim}{k}(t-s)v(s)ds, \quad t \in [0,T],$$

is monotone, hemicontinuous, bounded and coercive. Then a perturbation argument [2, Cor. 1.1, p. 39 and Theorem 1.6, p. 45] implies that $B + C$ is maximal monotone and onto $\mathcal{V}'$. Thus (2.4) always possesses a unique solution of the kind claimed, which completes the argument.

3. The Existence Theorem for $(\underset{\sim}{E})$

In this section we carry out the analysis of the equation

$$(\underset{\sim}{E})\qquad u(t) + \int_{-\infty}^t \underset{\sim}{a}(t-s)g(s,u(s))ds = f(t), \quad t\in R.$$

Here $\underset{\sim}{a}$ satisfies (a1)-(a3) as before, while $G(t) : v \to g(t,v)$ is a nonlinear mapping from V to V' for a.e. $t\in R$, where V,H,V' are spaces related as in Section 2.

Observe that finding a solution of this equation differs considerably from the task of finding a solution to the Volterra equation $(\underset{\sim}{V})$, since no initial data on u are provided by $(\underset{\sim}{E})$. On the other hand, the existence of a solution for $(\underset{\sim}{E})$ on the entire axis will be seen to follow from the demonstration that a solution exists on each half-line $(-\infty,T]$, $T\in R$. [In fact, once a solution of $(\underset{\sim}{E})$ is known to exist up to any <u>one</u> time T, its value for times $t > T$ can be determined through a Volterra equation which, after translation of origin, is of the form $(\underset{\sim}{V})$.] This feature of $(\underset{\sim}{E})$ makes the following notations quite convenient (see [7], [8]).

<u>Definition</u> 3.1. Let x be a function on R with values in a Banach space X, and let $\underset{\sim}{\mathfrak{X}}$ be a given function space consisting of X-valued functions (or equivalence classes of such) defined on R^+. For each $t \in R$, x induces an X-valued function x^t on R^+ according to

$$x^t(s) = x(t-s), \qquad s \in R^+.$$

The function $x : R \to X$ will be called <u>$\underset{\sim}{\mathfrak{X}}$-admissible</u> provided that x^t is in $\underset{\sim}{\mathfrak{X}}$ for every $t \in R$. The vector space of all $\underset{\sim}{\mathfrak{X}}$-admissible functions is denoted $ad(\underset{\sim}{\mathfrak{X}})$, so that

$$(3.1) \qquad x \in ad(\underset{\sim}{\mathfrak{X}}) \Leftrightarrow x^t \in \underset{\sim}{\mathfrak{X}}, \qquad \forall t \in R.$$

We proceed with the formulation of the existence theorem for $(\underset{\sim}{E})$. We shall suppose that the family of nonlinear operators $G(t) : v \to g(t,v) \in V'$, $t \in R$, satisfies for some p, $2 \leq p < \infty$,

(G1) $G(t)$ is monotone and demicontinuous, and $G(t)(0) = 0$, a.e. $t \in R$.

(G2) For each $T \in R$, $v \in L^p(-\infty,T;V) \Rightarrow$

$$t \to G(t)v(t) \in V' \text{ is measurable on } (-\infty,T],$$

(G3) For each $T \in R$ there are constants $c_T, \omega_T > 0$ and a function $b_T \in L^{p'}(-\infty,T;R^+)$, $\frac{1}{p'} + \frac{1}{p} = 1$, such that

$$\text{(i)} \qquad \|G(t)v\|_* \leq c_T\|v\|^{p-1} + b_T(t)$$

$$\text{(ii)} \qquad \langle G(t)v,v\rangle \geq \omega_T\|v\|^p, \qquad v \in V, \qquad \text{a.e. } t \leq T.$$

Note that (G2) and (G3i) imply that G induces a mapping

$$G : ad(L^p(R^+;V)) \to ad(L^{p'}(R^+;V')).$$

In addition, we suppose that the kernel $\underset{\sim}{a}$ satisfies (a1)-(a3) and that the forcing term f satisfies $f \in ad(W^{1,p'}(R^+;V'))$ with $f(t) \in H$, $\liminf_{t \to -\infty} |f(t)| = 0$; that is, f is H-valued, $f \in W^{1,p'}(-\infty,T;V')$ for each $T \in R$, and $\liminf_{t \to -\infty} |f(t)| = 0$.

<u>Theorem</u> 3.1. Let V,H,V' and the mappings G and $\underset{\sim}{a}$ be as above. Given $f \in ad(W^{1,p'}(R^+;V'))$ such that f is H-valued and $\liminf_{t \to -\infty} |f(t)| = 0$, there exists a unique function

$v \in ad(L^p(R^+;V) \cap W^{1,p'}(R^+;V'))$ satisfying

$$(\underset{\sim}{E}) \qquad v(t) + \int_{-\infty}^{t} \underset{\sim}{a}(t-s)g(s,v(s))ds = f(t), \quad t \in R.$$

Moreover, the function v satisfies $v \in C(R;H)$.

<u>Proof</u>: Fix $T \in R$ and select any number $\nu \in R^+$ such that $-\nu < T$. Consider the equation

$$(3.2) \qquad z(t) + \int_{-\nu}^{t} \underset{\sim}{a}(t-s)g(s,z(s))ds = f(t), \quad t \in [-\nu,T].$$

It is easy to see that (3.2) is equivalent to the following equation for functions on $[0,\nu + T]$

$$(\underset{\sim}{V}^\nu) \qquad w^\nu(t) + \int_{0}^{t} \underset{\sim}{a}(t-s)g^\nu(s,w^\nu(s))ds = f^\nu(t), \quad t \in [0,\nu + T],$$

where $w^\nu(t) = z(t-\nu)$, $g^\nu(s,v) = g(s-\nu,v)$, $f^\nu(t) = f(t-\nu)$.

Now the hypotheses on g and f ensure that $(\underset{\sim}{V}^\nu)$ fulfills the hypotheses of Theorem 2.1. Hence (3.2) has a unique solution $z^\nu \in L^p(-\nu,T;V) \cap W^{1,p'}(-\nu,T;V') \cap C(-\nu,T,H)$. Moreover, by the proof of Theorem 2.1, z^ν satisfies the differentiated equation

$$(3.3) \qquad \dot{z}^\nu(t) + \underset{\sim}{a}(0)g(t,z^\nu(t)) + \int_{-\nu}^{t} \dot{\underset{\sim}{a}}(t-s)g(s,z^\nu(s))ds = \dot{f}(t), \quad \text{a.e. } t \in [-\nu,T]$$

$$z^\nu(-\nu) = f(-\nu),$$

as well as the inverted version

$$(3.4) \qquad \underset{\sim}{a}(0)^{-1}\dot{z}^\nu(t) + \frac{d}{dt}\int_{-\nu}^{t} \underset{\sim}{k}(t-s)z^\nu(s)ds + g(t,z^\nu(t)) = F^\nu(t), \quad \text{a.e. } t \in [-\nu,T]$$

$$z^\nu(-\nu) = f(-\nu),$$

where $F^\nu(t) = \underset{\sim}{a}(0)^{-1}\dot{f}(t) + \frac{d}{dt}\int_{-\nu}^{t} \underset{\sim}{k}(t-s)f(s)ds$ defines a function in $L^{p'}(-\nu,T;V')$.

By taking the 'product' of (3.4) with $z^\nu(t)$ and integrating from $-\nu$ to t_1 we obtain (reversing order in $\langle\cdot,\cdot\rangle$, for convenience):

$$(3.5) \qquad \tfrac{1}{2}\langle\underset{\sim}{a}(0)^{-1}z^\nu(t_1),z^\nu(t_1)\rangle + \int_{-\nu}^{t_1} \langle\frac{d}{dt}\int_{-\nu}^{t} \underset{\sim}{k}(t-s)z^\nu(s)ds,z^\nu(t)\rangle dt + \int_{-\nu}^{t_1} \langle g(t,z^\nu(t)),z^\nu(t)\rangle dt =$$

$$= \frac{1}{2}\langle \underset{\sim}{a}(0)^{-1} f(-\nu), f(-\nu)\rangle + \int_{-\nu}^{t_1} \langle F^\nu(t), z^\nu(t)\rangle dt.$$

Hence (a2), (a3), (G3) and Young's inequality yield the relation

$$(3.6) \qquad \frac{1}{2}\alpha|z^\nu(t_1)|^2 + m\int_{-\nu}^{t_1} |z^\nu(t)|^2 dt + \omega_T \int_{-\nu}^{t_1} \|z^\nu(t)\|^p dt$$
$$\leq \frac{1}{2}\beta|f(-\nu)|^2 + \frac{\omega_T}{2}\int_{-\nu}^{t_1} \|z^\nu(t)\|^p dt + K(\omega_T)\int_{-\nu}^{t_1} \|F^\nu(t)\|_*^{p'} dt.$$

If we now extend z^ν and F^ν to the half-line $(-\infty,T)$ by putting

$$\bar{z}^\nu(t) = \begin{cases} z^\nu(t), & t \in [-\nu,T] \\ 0, & t < -\nu, \end{cases} \qquad \bar{F}^\nu(t) = \begin{cases} F^\nu(t), & t \in [-\nu,T] \\ 0, & t < -\nu, \end{cases}$$

then (3.6) ensures

$$(3.7) \qquad \|\bar{z}^\nu\|_{L^p(-\infty,T;V)}, \|\bar{z}^\nu\|_{L^2(-\infty,T;H)}, \|\bar{z}^\nu\|_{L^\infty(-\infty,T;H)} \leq K_T,$$

where $K_T = K(\omega_T, f^T)$ is independent of ν. Use of (G3) and the subsequent application of (a1), (a2) to (3.3) yield

$$(3.8) \qquad \|g(\cdot,\bar{z}^\nu(\cdot))\|_{L^{p'}(-\infty,T;V')} \leq K_T', \quad \|\dot{\bar{z}}^\nu\|_{L^{p'}(-\infty,T;V')} \leq K_T'',$$

where K_T' and K_T'' are also independent of ν. Hence (3.2) implies

$$(3.9) \qquad \|\bar{z}^\nu(\cdot)\|_{L^{p'}(-\infty,T;V')} \leq K_T''',$$

with K_T''' independent of ν.

According to (3.7)-(3.9) and our hypotheses on f, we can select a sequence $\nu_j \to \infty$ so that there exist $u \in L^p(-\infty,T;V) \cap L^\infty(-\infty,T;H) \cap L^{p'}(-\infty,T;V')$, $\dot{u} \in L^{p'}(-\infty,T;V')$ and $\gamma \in L^{p'}(-\infty,T;V')$ satisfying:

$$(3.10) \qquad \begin{aligned} &\bar{z}^{\nu_j} \rightharpoonup u \text{ weakly in } L^p(-\infty,T;V) \text{ and in } L^{p'}(-\infty,T,V'), \\ &\qquad\qquad \text{weak* in } L^\infty(-\infty,T;H) \\ &\dot{\bar{z}}^{\nu_j} \rightharpoonup \dot{u} \text{ weakly in } L^{p'}(-\infty,T;V') \\ &G(\cdot)\bar{z}^{\nu_j}(\cdot) \rightharpoonup \gamma \text{ weakly in } L^{p'}(-\infty,T;V') \\ &f(-\nu_j) \to 0 \text{ in } H. \end{aligned}$$

[The conclusion that the weak limit of the $\{\dot{\bar{z}}^{\nu_j}\}$ is the distribution derivative of u, and hence that $\dot{u} \in L^{p'}(-\infty,T;V')$, follows from the validity for large j of

$$\int_{-\infty}^{T} \langle \dot{\bar{z}}^{\nu_j},\varphi\rangle = -\int_{-\infty}^{T} \langle \bar{z}^{\nu_j},\dot{\varphi}\rangle, \quad \varphi \in C_0^\infty(-\infty,T;V)].$$

We proceed to show that the function u selected in this manner is actually a solution of $(\underset{\sim}{E})$ on $(-\infty,T]$. Note that $u \in L^p(I;V) \cap W^{1,p'}(I;V')$ for each bounded subinterval $I \subset (-\infty,T]$, so that by a previously cited result [2, p. 160] $u \in C(-\infty,T;H)$, as well.

The first step is to show that the weak limit of the $\{g(\cdot,\bar{z}^{\nu_j}(\cdot))\}$ is $g(\cdot,u(\cdot))$, i.e.,

$$\gamma(t) = g(t,u(t)) \quad \text{a.e. } t \in (-\infty,T]. \tag{3.11}$$

Let us hereafter abbreviate $\mathfrak{w}_T = L^p(-\infty,T;V) \cap W^{1,p'}(-\infty,T;V')$ and denote

$$\begin{aligned} L^\nu w(t) &= \underset{\sim}{a}(0)^{-1}\dot{w}(t) + \frac{d}{dt}\int_{-\nu}^{t} \underset{\sim}{k}(t-s)w(s)ds \\ Lw(t) &= \underset{\sim}{a}(0)^{-1}\dot{w}(t) + \frac{d}{dt}\int_{-\infty}^{t} \underset{\sim}{k}(t-s)w(s)ds, \qquad w \in \mathfrak{w}_T. \end{aligned} \tag{3.12}$$

By (a3) we have

$$\begin{aligned} \int_{-\nu}^{t_1} \langle L^\nu w,w\rangle dt &= \tfrac{1}{2}\langle \underset{\sim}{a}(0)^{-1}w(t_1),w(t_1)\rangle - \tfrac{1}{2}\langle \underset{\sim}{a}(0)^{-1}w(-\nu),w(-\nu)\rangle + \\ &\quad + \int_{-\nu}^{t_1} \langle \frac{d}{dt}\int_{-\nu}^{t} \underset{\sim}{k}(t-s)w(s)ds,w(t)\rangle dt \\ &\geq -\tfrac{1}{2}\langle \underset{\sim}{a}(0)^{-1}w(-\nu),w(-\nu)\rangle. \end{aligned} \tag{3.13}$$

Hence

$$\begin{aligned} \int_{-\nu_j}^{T} [\langle L^{\nu_j}\bar{z}^{\nu_j} - L^{\nu_j}w,\ \bar{z}^{\nu_j}-w\rangle + \langle G(\cdot)\bar{z}^{\nu_j} - G(\cdot)w,\ \bar{z}^{\nu_j}-w\rangle]dt \\ \geq -\tfrac{1}{2}\langle \underset{\sim}{a}(0)^{-1}(f(-\nu_j) - w(-\nu_j)),\ f(-\nu_j)-w(-\nu_j)\rangle, \quad w \in \mathfrak{w}_T. \end{aligned} \tag{3.14}$$

By reference to (3.4) we can rewrite (3.14) in the form

$$\begin{aligned} \int_{-\nu_j}^{T} [\langle \bar{F}^{\nu_j},\ \bar{z}^{\nu_j}-w\rangle - \langle L^{\nu_j}w,\ \bar{z}^{\nu_j}-w\rangle - \langle G(\cdot)w,\ \bar{z}^{\nu_j}-w\rangle]dt \\ \geq -\tfrac{1}{2}\langle \underset{\sim}{a}(0)^{-1}(f(-\nu_j) - w(-\nu_j)),f(-\nu_j) - w(-\nu_j)\rangle. \end{aligned} \tag{3.15}$$

Now $\bar{F}^{\nu_j} \to F$, $L^{\nu_j}w \to Lw$ strongly in $L^{p'}(-\infty,T;V')$, where F is defined as in (3.4), but with $\nu = +\infty$. If we temporarily restrict attention to those functions $w \in \mathfrak{W}_T$ with compact support and proceed to the limit $j \to \infty$ in (3.15) then by (3.10) we obtain

$$(3.16) \qquad \int_{-\infty}^{T} [\langle F,u-w\rangle - \langle Lw,u-w\rangle - \langle G(\cdot)w,u-w\rangle]dt \geq 0,$$

$$w \in \mathfrak{W}_T \text{ of compact support.}$$

Since functions of compact support are dense in $\mathfrak{W}_T$, the demicontinuity condition (G1) ensures that (3.16) is actually valid for <u>all</u> $w \in \mathfrak{W}_T$. Moreover, passage to the limit $j \to \infty$ in (3.4) yields,

$$(3.4'') \qquad Lu + \gamma = F.$$

Substituting (3.4") into (3.16) yields

$$(3.17) \qquad \int_{-\infty}^{T} \langle Lu-Lw,u-w\rangle dt + \int_{-\infty}^{T} \langle \gamma-G(\cdot)w,u-w\rangle dt \geq 0, \quad \forall w \in \mathfrak{W}_T.$$

By taking $w = u-\lambda y$, $y \in \mathfrak{W}_T$, in (3.16) and proceeding to the limit $\lambda \downarrow 0+$, we obtain by use of (G1), (G3) and an appeal to the dominated convergence theorem

$$(3.18) \qquad \int_{-\infty}^{T} \langle \gamma-G(\cdot)u,y\rangle dt \geq 0, \qquad \forall y \in \mathfrak{W}_T.$$

Hence (3.11) holds and therefore u is a (weak) solution of (3.4)

$$(3.4) \quad \underset{\sim}{a}(0)^{-1}\dot{u}(t) + \frac{d}{dt}\int_{-\infty}^{t} \underset{\sim}{k}(t-s)u(s)ds + g(t,u(t))$$
$$= \underset{\sim}{a}(0)^{-1}f(t) + \frac{d}{dt}\int_{-\infty}^{t} \underset{\sim}{k}(t-s)f(s)ds.$$

In order to show that u is actually a (pointwise) solution of $(\underset{\sim}{E})$ on $(-\infty,T]$ we proceed as follows. Equation (3.3) implies that the functions $\bar{z}^{\nu_j}$ satisfy (3.2), which can be written

$$(3.19) \qquad \bar{z}^{\nu_j}(t) + \int_{-\infty}^{t} \underset{\sim}{a}(t-s)g(s,\bar{z}^{\nu_j}(s))ds = f(t)\chi_{[-\nu_j,T]}(t), \quad t \in (-\infty,$$

Now, by a theorem of Banach and Saks there is a subsequence $\{\bar{z}^{\nu_j'}\}$ of $\{\bar{z}^{\nu_j}\}$ such that the Cesaro means

$$\sigma_M(z) = \frac{1}{M}\sum_{j=1}^{M} z^{-\nu_j'}, \qquad \sigma_M(g) = \frac{1}{M}\sum_{j=1}^{M} g(\cdot, z^{-\nu_j'}(\cdot))$$

converge strongly to their respective limits $u \in \mathbb{W}_T$, $g(\cdot,u(\cdot)) \in L^{p'}(-\infty,T;V')$. In particular, it follows by the (local) Sobolev imbedding of $W^{1,p'}(-\infty,T;V')$ into $C(-\infty,T;V')$ that

$$\sigma_M(z)(t) \to u(t) \quad \text{in} \quad V', \qquad \forall t \in (-\infty,T].$$

Thus (3.19) leads to the relation

$$\sigma_M(z)(t) + \int_{-\infty}^{t} \underset{\sim}{a}(t-s)\sigma_M(g)(s)ds = \sigma_M(f\chi_{[-\nu_j,T]})(t)$$

and therefore, letting $M \to \infty$, to

$$(3.20) \qquad u(t) + \int_{-\infty}^{t} \underset{\sim}{a}(t-s)g(s,u(s))ds = f(t), \qquad t \in (-\infty,T].$$

At this point, $u = u_T$ could be continued as a solution of $(\underset{\sim}{E})$ to all of R by means of the Volterra equation $(\underset{\sim}{V})$ (with an appropriately modified f and shift of origin). However, we use the following procedure instead. Since $T \in R$ was chosen arbitrarily, there is for any sequence $T_n \to +\infty$ a family $\{u_{T_n}\}$ consisting of solutions to $(\underset{\sim}{E})$ defined on the respective half-lines $\{(-\infty,T_n]\}$. Moreover, our previous arguments reveal that, for each T, these solutions satisfy

$$(3.21) \qquad \|u_{T_n}\|_{L^p(-\infty,T;V)} \le K_T, \quad \|\dot{u}_{T_n}\|_{L^{p'}(-\infty,T;V)} \le K_T'',$$

$$\|g(\cdot,u(\cdot))\|_{L^{p'}(-\infty,T;V')} \le K_T', \quad \text{whenever} \quad T_n \ge T.$$

Hence a diagonalization argument patterned on that applied in the preceding paragraph shows that there exists a subsequence $\{u_{T_{n'}}\}$ converging weakly in $\mathbb{W}_T$ (for <u>each</u> $T \in R$) to a limit function v. It follows that v is everywhere defined on R and is a solution of $(\underset{\sim}{E})$ on the entire axis, as claimed.

Finally, the uniqueness follows, for each fixed T, by subtraction of equations, multiplication by a difference of solutions and integration over $(-\infty,T]$.

Acknowledgements. The second-named author wishes to express appreciation to John Nohel for rekindling his interest in the present vector-valued version of $(\underset{\sim}{E})$ and to Juan Schäffer for a helpful suggestion.

References

[1] Barbu, V. "Nonlinear Volterra equations in Hilbert space", Siam J. Math. Anal. 6 (1975), 728-741.

[2] Barbu, V., Nonlinear Semigroups and Differential Equations in Banach Spaces, Noordhoff International Publishing, Leyden, 1976.

[3] Corduneanu, C., Integral Equations and Stability of Feedback Systems, Academic Press, New York, 1973.

[4] Crandall, M. G. and Nohel, J. A., "An abstract functional differential equation and a related nonlinear Volterra equation", Math. Res. Center, Univ. of Wisconsin, Tech. Summary Report #1765 (1977).

[5] Gripenberg, G., "An existence result for a nonlinear Volterra integral equation in Hilbert space", Siam J. Math. Anal. (to appear).

[6] Komura, Y., "Nonlinear semigroups in Hilbert spaces", J. Math. Soc. Japan 19 (1967), 493-507.

[7] Leitman, M. J. and Mizel, V. J., "On fading memory spaces and hereditary integral equations", Arch. Rat. Mech. Anal. 55 (1974), 18-51.

[8] Leitman, M. J. and Mizel, V. J., "Hereditary laws and nonlinear integral equations on the line", Adv. in Math. 22 (1976), 220-266.

[9] Leitman, M. J. and Mizel, V. J., "Asymptotic stability and the periodic solutions of $x(t) + \int_{-\infty}^{t} a(t-s)g(s,x(s))ds = f(t)$, J. Math. Anal. Appl. (to appear), also in Technion Preprint Series #MT-339 (1977).

[10] Levin, J. J., "On a nonlinear Volterra equation", J. Math. Anal. Appl. 39 (1972), 458-476.

[11] Levin, J. J., "Some a'priori bounds for nonlinear Volterra equations", Siam J. Math. Anal. 7 (1976), 872-897.

[12] Levin, J. J. and Shea, D. F., "On the asymptotic behavior of some integral equations I, II, III", J. Math. Anal. Appl. 37 (1972), 42-82, 288-326, 537-575.

[13] Londen, S.-O., "On an integral equation in a Hilbert space", Siam J. Math. Anal. (to appear).

[14] MacCamy, R.C., "Nonlinear Volterra equations on a Hilbert space", J. Diff. Eqns. 16 (1974), 373-393.

[15] MacCamy, R. C., "Remarks on frequency domain methods for Volterra integral equations", J. Math. Anal. Appl. 55 (1976), 555-575.

[16] MacCamy, R. C., "An integro-differential equation with application in heat flow", Q. Appl. Math. 35 (1977), 1-19.

[17] MacCamy, R. C. and Weiss, P., "Qualitative numerical theory for Volterra equations", these proceedings.

[18] MacCamy, R. C. and Smith, R. L., "Limits of solutions of nonlinear Volterra equations", Applic. Anal. (1977), 19-27.

[19] Miller, R. K. and Sell, G. R., "The topological dynamics of Volterra integral equations", Proc. of Conf. for Qual. Th. Nonlin. Diff. and Integ. Eqns., Madison, Wisconsin, (1968).

[20] Miller, R. K. and Sell, G. R., Volterra Integral Equations and Topological Dynamics, Mem. Am. Math. Soc. 102 (1970).

[21] Miller, R. K., Nonlinear Volterra Integral Equations, Benjamin, Menlo Park, 1971.

A NONLINEAR HYPERBOLIC VOLTERRA EQUATION

John A. Nohel(*)

University of Wisconsin
Madison, WI 53706/USA

Abstract. A mathematical model for the motion of a nonlinear one dimensional viscoelastic rod is analysed by an energy method developed by C.M. Dafermos and the author. Global existence, uniqueness, boundedness, and the decay of smooth solutions as $t \to \infty$ are established for sufficiently smooth and "small" data.

1. Introduction. In this lecture which is based on joint work C.M. Dafermos [4], we use energy methods to discuss the global existence, uniqueness, boundedness, and decay as $t \to \infty$ of smooth solutions of the nonlinear Cauchy problem :

$$(\mathrm{VE}) \quad \begin{cases} u_{tt}(t,x) = \sigma(u_x(t,x))_x + \displaystyle\int_0^t a'(t-\tau)\,\sigma(u_x(\tau,x))_x\,d\tau + g(t,x) \\ \qquad\qquad\qquad\qquad\qquad\qquad\qquad\qquad (0 < t < \infty,\ x \in \mathbb{R}) \\ u(0,x) = u_0(x), \quad u_t(0,x) = u_1(x) \qquad (x \in \mathbb{R}), \end{cases}$$

for appropriately small, smooth data u_0, u_1, g; $a : [0,\infty) \to \mathbb{R}^+$, $\sigma : \mathbb{R} \to \mathbb{R}$ $(\sigma(0) = 0)$, $g : [0,\infty) \times \mathbb{R} \to \mathbb{R}$, $u_0, u_1 : \mathbb{R} \to \mathbb{R}$ are given functions satisfying assumptions motivated by physical considerations sketched below and partly by the method of analysis. In (VE) subscripts denote partial derivatives, and u is the unknown function. In addition to the Cauchy problem (VE), we will comment on several closely related initial-boundary value problems.

Problem (VE) arises in the following physical context. Consider one dimensional motion of an unbounded viscoelastic rod of unit density. According to the theory of materials of "fading memory" type (see Coleman and Gurtin [1]) the stress $S(t,x)$ at time t and position x is given by a functional

---

(*) sponsored by the United States Army under Grant No. DAAG 29-77-G-0004 and under Contract No. DAAG 29-75-C-0024.

of the history of the strain, $u_x(t-\tau, x)$ $(\tau \geq 0)$, where $x + u(t,x)$ denotes the position at time t of a section of the rod which is at position x in the unstretched configuration. In the nonlinear case the theory suggests assuming that the stress functional S has the form

$$(1.1) \qquad S(t,x) = \sigma(u_x(t,x)) - \int_0^\infty b(\tau)\varphi(u_x(t-\tau,x))d\tau \qquad (t > 0),$$

with the history of the displacement $u(t,x)$ prescribed for $t < 0$ and $x \in \mathbb{R}$. Relaxation experiments of materials indicate that $\sigma, \varphi : \mathbb{R} \to \mathbb{R}$ are smooth functions which satisfy the assumptions $\sigma(0) = \varphi(0) = 0$, $\sigma'(\xi) > 0$, $\varphi'(\xi) > 0$ $(\xi \in \mathbb{R})$, and that the "influence" (or memory) function $b : [0,\infty) \to \mathbb{R}^+$ satisfies $b(t) > 0$, $b'(t) < 0$ for $t \in \mathbb{R}^+$ and that $b \in L^1(0,\infty)$ (e.g. b is a linear combination of decaying exponentials with positive coefficients). We remark that a standard assumption of linear theory is that $\sigma(\xi) = c_1\xi$, $\varphi(\xi) = c_2\xi$ where $c_1, c_2 > 0$ are constants [2].

If the rod is also subjected to an external force $F(t,x)$, then the equation of motion for the rod is

$$(1.2) \qquad u_{tt}(t,x) = S_x(t,x) + F(t,x), \qquad (0 < t < \infty, \ x \in \mathbb{R}),$$

together with prescribed initial values $u(0,x)$, $u_t(0,x)$, where S is the stress functional defined by (1.1). Recalling that the history of displacement is prescribed for $t < 0$, and defining

$$(1.3) \qquad g(t,x) = F(t,x) - \int_t^\infty b(\tau)\varphi(u_x(t-\tau,x))_x d\tau$$

for $t > 0$, $x \in \mathbb{R}$ shows that the motion of the unbounded viscoelastic rod is described by the Cauchy problem

$$(1.4) \qquad \begin{cases} u_{tt} = \sigma(u_x)_x - b * \varphi(u_x)_x + g & (0 < t < \infty, \ x \in \mathbb{R}) \\ u(0,x) = u_0(x), \quad u_t(0,x) = u_1(x) & (x \in \mathbb{R}), \end{cases}$$

where $*$ denotes the convolution, defined by

$$(b * \varphi(u_x)_x)(t,x) = \int_0^t b(t-\tau)\,\varphi(u_x(\tau,x))_x\, d\tau .$$

Our method of analysis requires us to make the further assumption

$$(1.5) \qquad \varphi(\xi) = c\sigma(\xi) \qquad (\xi \in \mathbb{R}),$$

where $c > 0$ is a constant. Assumption (1.5) is satisfied in the linear case and is reasonable for certain nonlinear problems. We shall be primarily interested in the "genuinely nonlinear case" $\sigma''(\xi) \neq 0 \quad (\xi \in \mathbb{R})$.

Consider next the Cauchy problem (1.4) with $g \equiv 0$ (or $\lim_{t\to\infty} g(t,x) = 0$, uniformly in x), under assumption (1.5). The corresponding steady state problem is meaningful if it is assumed that

$$(1.6) \qquad 1 - c\int_0^\infty b(\tau)\,d\tau > 0 \;;$$

assumption (1.6) has the interpretation that the static modulus elasticity is positive (see Dafermos [2], [3] where the same assumption is made in the linear case). With the assumptions (1.5), (1.6), and those concerning b made above, we can reduce the Cauchy problem (1.4) to the equivalent form (VE) as follows: define $a : [0,\infty) \to \mathbb{R}^+$ by

$$(1.7) \quad \begin{cases} a(t) = a_\infty + A(t) \;; \quad a_\infty = 1 - c\int_0^\infty b(\tau)\,d\tau > 0 \;; \\ A(t) = c\int_t^\infty b(\tau)d\tau \;; \quad a(0) = 1 \;; \; A(t) > 0 \,, \quad A'(t) < 0 \\ \qquad \text{for} \quad 0 \le t < \infty \;; \quad A(\infty) = 0 \;; \end{cases}$$

define g by (1.3); then (1.4) - (1.7) is equivalent to (VE).

The analysis which follows will be concerned (VE), where a satisfies the physically reasonable assumptions implied by (1.7); for technical reasons based on our analysis we shall require that a satisfy somewhat stronger assumptions.

To motivate our result for (VE) we begin with some general remarks. If $a(t) \equiv 1$, (VE) reduces to the equation of nonlinear elasticity

$$(\mathrm{E}) \qquad u_{tt} = \sigma(u_x)_x + g \,, \quad u(0,x) = u_0(x) \,, \quad u_t(0,x) = u_1(x) \,.$$

If $g \equiv 0$ in (E) and if σ is "genuinely nonlinear" Lax [6], has shown that (E) fails to have global smooth solutions in time, no matter how smooth one takes the initial data u_0, u_1, due to the development of "shocks" (the first derivatives of solutions generally develop singularities when characteristics cross).

Nishida [13] has shown that for the wave equation with "frictional" damping

$$u_{tt} + u_t = \sigma(u_x)_x \quad , \quad u(0,x) = u_0(x) \; , \quad u_t(0,x) = u_1(x)$$

the dissipation precludes the development of shocks if the initial data are sufficiently smooth and "small", resulting in global smooth solutions. The proof rests on the concept of Riemann invariant is restricted to one space dimension.

Mac Camy [10] has recently studied (VE) on $(0,\infty) \times (0,1)$ and homogeneous Dirichlet boundary conditions at $x = 0$, $x = 1$, by combining Nishida's method with certain a priori estimates under suitable assumptions on the kernel $a(t)$ and the forcing term g. His object is to show that the memory term in (VE) induces a dissipative mechanism which guarantees global existence for "small" initial data and forcing term. The problem of obtaining the existence and uniqueness of a suitable local solution of (VE) to be continued with the aid of the derived a priori estimates is not discussed in [10], but this gap can be filled by the method outlined in Nohel [14] where the result of [13] is extended.

The object of this work is to study (VE) by a different approach based entirely on energy estimates and not on Riemann invariants. While the exposition is restricted to the one-dimensional problem (VE) for clarity, the method can be applied to problems in any number of space dimensions, provided estimates on derivatives of sufficiently high order are computed. This approach to (VE) may be regarded as a generalization of recent work of Matsumura [11], [12] who studies multi-space-dimensional nonlinear wave equations with frictional damping for small data by a similar energy method. We are grateful to Professor Nishida for explaining this approach to us during a recent visit.

We remark that the special case of (VE) resulting from the kernel $a(t) = \frac{1}{2}(1 + \exp(-t))$ can easily be shown to be equivalent to the Cauchy problem

$$u_{ttt} + u_{tt} = \sigma(u_x)_{xt} + \tfrac{1}{2}\sigma(u_x)_x + g_t + g \; ,$$

$$u(0,t) = u_0(x) \; , \quad u_t(0,x) = u_1(x) \; , \quad u_{tt}(0,x) = \sigma(u_{0x})_x \; ,$$

which was studied by Greenberg [5]; his result is a special case of ours.

Finally, we note that (VE) is of the abstract form

$$\text{(A)} \qquad \begin{cases} u''(t) + Au(t) + \int_0^t a'(t-\tau)Au(\tau)\,d\tau = F(t) \; , & 0 < t < \infty \; , \\ u(0) = u_0 \; , \quad u'(0) = u_1 & \end{cases}$$

where Au is a nonlinear operator ($Au = -\frac{\partial}{\partial x}\sigma(u_x)$ together with conditions at $\pm\infty$, or boundary conditions at say $x = 0, 1$). Abstract problems of the form (A) have been studied by Londen [7], [8] for a class of kernels $a(\cdot)$ which are positive, decreasing, convex on $[0,\infty)$ and satisfy the crucial assumption $a'(0^+) = -\infty$; unfortunately the latter assumption is not satisfied by most "memory" functions in viscoelasticity.

2. Statement of Results. We make the following assumptions. Concerning σ let

$$(\sigma) \qquad \sigma \in C^3(\mathbb{R}), \quad \sigma(0) = 0, \quad \sigma'(0) > 0,$$

the first for technical reasons and the remaining on physical grounds. Concerning the kernel a assume

$$(a)\quad \begin{cases} \text{(i)} & a \in \mathcal{B}^{(3)}[0,\infty) \\ \text{(ii)} & a(t) = a_\infty + A(t), \quad a_\infty > 0, \quad a(0) = 1, \quad a'(0) < 0, \\ \text{(iii)} & (-1)^j A^{(j)}(t) \geq 0 \qquad (0 \leq t < \infty;\ j = 0,1,2) \\ \text{(iv)} & t^j A^{(m)}(t) \in L^1(0,\infty) \qquad (m, j = 0,1,2,3), \end{cases}$$

where $\mathcal{B}^{(m)}[0,\infty)$ is the set of functions with bounded, continuous derivatives on $[0,\infty)$ up to and including order m. The forcing term g is assumed to satisfy

$$(g) \qquad g, g_t \in L^1([0,\infty); L^2(\mathbb{R})), \quad g_x, g_{tt}, g_{tx} \in L^2([0,\infty); L^2(\mathbb{R})),$$

meaning that g and some of its distributional derivatives decay sufficiently rapidly at infinity. The inital data u_0, u_1 satisfy

$$(u_0) \qquad u_0 \in H^3(\mathbb{R}), \qquad\qquad (u_1) \qquad u_1 \in H^2(\mathbb{R}).$$

Our result concerning (VE) is (see [4; Theorem 5.1]).

<u>Theorem 2.1. Let the assumptions (σ), (a), (g), (u_0), (u_1) hold. If the $H^2(\mathbb{R})$ norms of u_{0x}, u_1, the $L^1([0,\infty); L^2(\mathbb{R}))$ norms of g, g_t, and the $L^2([0,\infty); L^2(\mathbb{R}))$ norms of g_x, g_{tt}, g_{tx} are sufficiently small then (VE) has a unique solution $u \in C^2([0,\infty)\times\mathbb{R})$ having the following</u>

<u>properties</u> :

(2.1) $u_t, u_x, u_{tt}, u_{tx}, u_{xx}, u_{ttt}, u_{ttx}, u_{txx}, u_{xxx} \in L^\infty([0,\infty); L^2(\mathbb{R}))$,

(2.2) $u_{tt}, u_{tx}, u_{xx}, u_{ttt}, u_{ttx}, u_{txx}, u_{xxx} \in L^2([0,\infty); L^2(\mathbb{R}))$,

(2.3) $u_{tt}(t,\cdot), u_{tx}(t,\cdot), u_{xx}(t,\cdot) \to 0$ <u>in</u> $L^2(\mathbb{R})$ <u>as</u> $t \to \infty$,

(2.4) $u_t(t,x), u_x(t,x), u_{tt}(t,x), u_{tx}(t,x), u_{xx}(t,x) \to 0$ <u>uniformly</u> <u>in</u> $\mathbb{R}$ <u>as</u> $t \to \infty$.

We remark that conclusions (2.3), (2.4) are an easy consequence of (2.1), (2.2). It also follows from the proof of the theorem that the solution u has a finite speed of propagation. In addition, we note that the same result holds (and with the same proof) for the following two problems of a viscoelastic rod of unit length:

(i) (VE) on $(0,\infty) \times (0,1)$ with homogeneous Neumann boundary conditions at $x = 0$ and $x = 1$, and with initial data prescribed on $[0,1]$.

(ii) (VE) on $(0,\infty) \times (0,1)$ with homogeneous Dirichlet boundary conditions $u(t,0) = u(t,1) \equiv 0$, and initial data prescribed on $[0,1]$, provided one also assumes that the forcing term g also satisfies $g(t,0) = g(t,1) \equiv 0$. Finally, we observe that a comparison of Theorem 2.1 and its proof with Mac Camy's results in [10] shows that our approach, in addition to being simpler, more direct, and not restricted to one space dimension, yields a more general result.

3. <u>Outline of Proof of Theorem 2.1</u>. To simplify the exposition we shall assume that $g \equiv 0$ in (VE), and we refer the reader to [4] for the technical complication and treatment of terms resulting from $g \neq 0$; no change in the method is involved.

a. <u>Transformation of</u> (VE). Define the resolvent kernel k of a' by the equation

(k) $$k(t) + (a' * k)(t) = -a'(t) \qquad (0 \le t < \infty).$$

By standard harmonic analysis methods, and by a frequency domain argument to obtain the last conclusion (see Nohel and Shea [15, Theorem 1]), the resolvent kernel has the following properties.

Lemma 3.1. Let assumptions (a) be satisfied. Then

(i) $k \in \mathcal{B}^{(2)}[0,\infty)$;

(ii) $k^{(m)} \in L^1(0,\infty)$ $(m = 0, 1, 2)$;

(iii) For every $T > 0$ and for every $v \in L^2(0,T)$ one has

$$\int_0^T v(t)\frac{d}{dt}(k * v)(t)\,dt \geq 0.$$

Let u be a smooth solution of (VE) with $g \equiv 0$, and observe that (VE) is linear in $y = \sigma(u_x)_x$. By the variation of constants formula for linear Volterra equations one has

$$y + a' * y = \varphi \iff y = \varphi + k * \varphi ,$$

where φ is a given function. Applying this to (VE) one sees that u satisfies the equation

$$u_{tt}(t,x) + (k * u_{tt})(t,x) = \sigma(u_x(t,x))_x .$$

Performing an integration by parts shows that (VE) is equivalent to the Cauchy problem

$$(3.1)\quad \begin{cases} u_{tt}(t,x) + \dfrac{\partial}{\partial t}(k * u_t)(t,x) = \sigma(u_x(t,x))_x + \Phi(t,x) & (0<t<\infty, x\in\mathbb{R}) \\ \Phi(t,x) = k(t)\,u_1(x) & \\ u(0,x) = u_0(x), \quad u_t(0,x) = u_1(x) & (x \in \mathbb{R}). \end{cases}$$

Another important equivalent form of (VE) resulting from (3.1) is

$$(3.2)\quad u_{tt}(t,x) + k(0)u_t(t,x) = \sigma(u_x(t,x))_x - (k' * u_t)(t,x) + \Phi(t,x) ;$$

since $k(0) = -a'(0) > 0$, (3.2) suggests the dissipative mechanism induced by the memory term in (VE) and the relationship with Nishida's treatment of the damped nonlinear wave equation. The reader should also note that the above transformation of (VE) rests on the assumption (1.5) which was made to arrive at the model (VE).

The proof of Theorem 2.1 is carried out in two stages: (i) A suitable local existence and uniqueness result is established. (ii) A priori estimates are established to continue the local solution; these will at the same time yield (2.1), (2.2).

b. Local Theory. We shall make the temporary additional assumption concerning σ :

(σ^*) there exists $p_0 > 0$ such that $\sigma'(\xi) \geq p_0 > 0$ $(\xi \in \mathbb{R})$.

Proposition 3.2. Let the assumptions (σ), (σ^*), (u_0), (u_1) hold, and let $k', k'' \in C[0,\infty) \cap L^1(0,\infty)$. Then the Cauchy problem (3.1) (resp. (3.2)) has a unique solution $u \in C^2([0,T_0) \times \mathbb{R})$ on a maximal interval $[0,T_0) \times \mathbb{R}$, $T_0 \leq +\infty$, such that for $T \in [0,T_0)$ one has

(i) all derivatives of u of orders one to three inclusive $\in L^\infty([0,T]; L^2(\mathbb{R}))$;

(ii) if $T_0 < \infty$, then

$$\int_{-\infty}^{\infty} [u_t^2(t,x) + u_x^2(t,x) + u_{tt}^2(t,x) + \dots + u_{xxx}^2(t,x)]\,dx \to \infty \quad \text{as} \quad t \to T_0^- .$$

We remark that the property of finite speed of property of finite speed of propagation of solutions of (VE) is an easy consequence of the proof of Proposition 3.2.

The proof uses the Banach fixed point theorem. Let $X(M,T)$ be the set of functions $u \in C^2([0,T] \times \mathbb{R})$ for any $T > 0$ such that $u(0,x) = u_0(x)$, $u_t(0,x) = u_1(x)$ and such that

(i) $u_t, u_x, u_{tt}, \dots, u_{xxx} \in L^\infty([0,T]; L^2(\mathbb{R}))$ and

(ii) $$\sup_{[0,T]} \int_{-\infty}^{\infty} [u_t^2(t,x) + u_x^2(t,x) + u_{tt}^2(t,x) + \dots + u_{xxx}^2(t,x)]\,dx \leq M^2.$$

Note that $X(M,T)$ is not empty if M is sufficiently large, and that if $u \in X(M,T)$, then

(iii) $$\sup_{[0,T]\times\mathbb{R}} \left\{ |u_t(t,x)|, \ |u_x(t,x)|, \ |u_{tt}(t,x)|, \ |u_{tx}(t,x)|, \ |u_{xx}(t,x)| \right\} \leq M.$$

Let S be the map : $X(M,T) \to C^2([0,T] \times \mathbb{R})$ which carries a function $v \in X(M,T)$ into the solution of the linear Cauchy problem (see (3.2) for motivation)

$$(3.3)\quad \begin{cases} u_{tt}(t,x) + k(0)u_t(t,x) = \sigma'(v_x(t,x))u_{xx}(t,x) + \Phi(t,x) - (k' * v_t)(t,x) \\ \qquad\qquad\qquad\qquad\qquad\qquad\qquad (0 < t < T, \quad x \in \mathbb{R}) \\ u(0,x) = u_0(x), \quad u_t(0,x) = u_1(x). \end{cases}$$

Clearly a fixed point of S will be a solution of (3.1) (respectively (3.2)). To apply the Banach fixed point theorem to the map S one first shows (by an energy argument, for details see [4, Lemma 3.1]) that if M is sufficiently large and if T is sufficiently small, then S maps $X(M,T)$ into itself. One next equips $X(M,T)$ with the metric

$$\rho(u,\bar{u}) = \max_{[0,T]} \left\{ \int_{-\infty}^{\infty} [(u_t(t,x) - \bar{u}_t(t,x))^2 + (u_x(t,x) - \bar{u}_x(t,x))^2]\,dx \right\}^{\frac{1}{2}}.$$

By the lower semicontinuity of norms under weak convergence in Banach space, $X(M,T)$ becomes a complete metric space. One then shows that for M sufficiently large and T sufficiently small the map S is a strict contraction of $X(M,T)$ and the proof of Proposition 3.2 is completed in a standard manner (for details see [4, Lemma 3.2]).

If σ, k, u_0, u_1 are smoother, the solution becomes smoother. A precise regularity result which is needed for the a priori estimates is

<u>Proposition 3.3</u>. <u>Let the assumptions of Proposition 3.2 be satisfied. In addition, assume that</u>

$$(3.4)\qquad \sigma \in C^4(\mathbb{R}), \quad u_{0xxxx}, \; u_{1xxx} \in L^2(\mathbb{R}).$$

Then the solution u of Proposition 3.2 has the addition property

$$(3.5)\qquad u_{tttt},\; u_{tttx},\; u_{ttxx},\; u_{txxx},\; u_{xxxx} \in L^2([0,T]; L^2(\mathbb{R})),$$

<u>for every</u> $T < T_0$, <u>where</u> $[0,T_0) \times \mathbb{R}$ <u>is the maximal interval of existence</u>. For the proof see [4, Theorem 3.2].

c. <u>A Priori Estimates and Continuation</u>. We wish to show that the maximal interval $[0,T_0)$ of Proposition 3.2 is in fact $[0,\infty)$. Recall that the local theory assumes that (σ^*) is satisfied; this assumption will be removed. The a priori estimates will be deduced from equations (3.1) above and (3.9) below. We shall restrict the range of $u_x(t,x)$ for a local solution u to the set on which $\sigma'(\cdot) > 0$; choose $c_0 > 0$ such that

$$(3.6) \qquad \sigma'(w) \geq p_0 > 0 \quad , \quad w \in [-c_0, c_0] \ .$$

We wish to show that there exists a constant $\mu > 0$, $\mu < c_0$, depending on on p_0, $\int_0^\infty |k'(t)|\,dt$, $\max\limits_{[-c_0, c_0]} \{ |\sigma'(\cdot)|, |\sigma''(\cdot)|, |\sigma'''(\cdot)| \}$, but not on $T > 0$ such that if the local solution u of (3.1) satisfies

$$(\mu^*) \qquad \sup_{0 \leq t < T,\ x \in \mathbb{R}} \{ |u_t(t,x)|, |u_x(t,x)|, |u_{tx}(t,x)|, |u_{xx}(t,x)| \} \leq \mu$$

then certain functionals of the solution u are controllably small (i.e. these functionals can be made arbitrarily small by choosing the initial data u_0, u_1 sufficiently small in the appropriate H norms). More precisely the result of the a priori estimates which follow is that if the assumptions of Theorem 2.1 hold (with $g \equiv 0$) and if the $H^2(\mathbb{R})$ norms of u_{0x}, u_1 are sufficiently small, then for as long as the local solution u of (3.1) satisfies the condition (μ^*) for $\mu > 0$ sufficiently small, the condition

$$(\mu^{**}) \qquad \int_{-\infty}^{\infty} [u_t^2(s,x) + u_x^2(s,x) + u_{tt}^2(s,x) + \dots + u_{xxx}^2(s,x)]\,dx$$

$$+ \int_0^s \int_{-\infty}^{\infty} [u_{tt}^2(t,x) + \dots + u_{xxx}^2(t,x)]\,dx\,dt \leq \mu^2 \qquad (0 \leq s < T)$$

is satisfied. The inequality (μ^{**}) in turn implies that condition (μ^*) holds and the cycle closes in a standard manner using Proposition 3.2. Thus the maximal interval of existence of the solution $u(t,x)$ is $[0,\infty) \times \mathbb{R}$ and (μ^{**}) holds for $0 \leq s < \infty$. This proves (2.1), (2.2) and the theorem.

The remainder of this section is devoted to the derivation of the a priori estimates which imply (μ^{**}). Define

$$(3.7) \qquad W(w) = \int_0^w \sigma(\xi)\,d\xi \geq \frac{p_0}{2} w^2 \qquad w \in [-c_0, c_0] \ ,$$

where the inequality follows from (3.6). Let u be a local solution of (3.1) satisfying (μ^*) for some $T > 0$ and $0 < \mu < c_0$. Multiply (3.1) by u_t and integrate over $[0,s] \times \mathbb{R}$. Using (3.7) and Lemma 3.1 (iii) one obtains

$$\frac{1}{2}\int_{-\infty}^{\infty} u_t^2(s,x)\,dx + \int_{-\infty}^{\infty} W(u_x(s,x))dx \le \frac{1}{2}\int_{-\infty}^{\infty} u_1^2(x)\,dx$$

$$+ \int_{-\infty}^{\infty} W(u_{0x}(x))\,dx + \int_0^s \int_{-\infty}^t \Phi\, u_t\, dx\,dt\,.$$

The easy estimate

$$\int_0^s \int_{-\infty}^{\infty} \Phi\, u_t\, dx\,dt \le \frac{1}{4} \max_{[0,s]} \int_{-\infty}^{\infty} u_t^2(t,x)\,dx + \int_0^{\infty} |k(t)|\,dt + \left\{\int_{-\infty}^{\infty} u_1^2(x)\,dx\right\}^{\frac{1}{2}},$$

together with the above inequality and (3.7), yield

$$\text{(3.8)} \quad \frac{1}{2}\int_{-\infty}^{\infty} u_t^2(s,x)\,dx + p_0 \int_{-\infty}^{\infty} u_x^2(s,x)\,dx \le \int_{-\infty}^{\infty} u_1^2(x)dx + 2\int_{-\infty}^{\infty} W(u_{0x}(x))dx$$

$$+ 2\int_{-\infty}^{\infty} |k(t)|\,dt \left\{\int_{-\infty}^{\infty} u_1^2(x)\,dx\right\}^{\frac{1}{2}} \qquad (0 \le s < T).$$

The estimate (3.8) implies that $\int_{-\infty}^{\infty} u_t^2(s,x)\,dx$, $\int_{-\infty}^{\infty} u_x^2(s,x)\,dx$ are controllably small, uniformly on $[0,T)$ provided condition (μ^*) holds. Unfortunately, repetitions of the above simple argument applied to equations obtained by differentiations of (3.1), either with respect to t or x or combinations thereof, do not yield the necessary estimates to obtain (μ^{**}), due to the rather weak nature of the positivity expressed in the inequality of Lemma 3.1 (iii). This is in sharp contrast to what happens in a problem of heat flow in materials with memory also considered by us in [4, Theorem 4.1], see also Mac Camy [9] for a different, less complete approach.

To obtain the further information needed for our estimates we define a modified resolvent $r : [0,\infty) \to \mathbb{R}$ by the equation

$$\text{(r)} \qquad r(t) = \beta + k(t) + \beta \int_0^t k(\tau)\,d\tau \qquad (0 \le t < \infty),$$

where k is the resolvent of Lemma 3.1, and where $\beta > 0$ is a constant to be chosen later (when $\beta = 0$, $r \equiv k$). It is easy to show that solutions of the linear Volterra equation

$$y + a' * y = \varphi$$

can be represented with the aid of r in the form

$$y(t) + \beta \int_0^t y(\tau)\,d\tau = \varphi(t) + (r * \varphi)(t) .$$

Applying this to (VE) with $y = \sigma(u_x)_x$, $\varphi = u_{tt}$, and $g \equiv 0$, we find (after a differentiation with respect to t) that (VE) is equivalent to the Cauchy problem

$$(3.9) \begin{cases} u_{ttt}(t,x) + \frac{\partial}{\partial t}(r * u_{tt})(t,x) = \sigma(u_x(t,x))_{xt} + \beta\sigma(u_x)_x(t,x) & (0 < t < \infty,\ x \in \mathbb{R}) \\ u(0,x) = u_0(x), \quad u_t(0,x) = u_1(x), \quad u_{tt}(0,x) = \sigma(u_{0x}))_x & (x \in \mathbb{R}) . \end{cases}$$

The usefulness of the resolvent kernel r is revealed in the following result summerizing the properties of r (especially part (iii) in which the constant β is suitably adjusted).

Lemma 3.4. Let the assumptions (a) be satisfied. Then the function r defined by equation (r) has the properties:

(i) $r \in B^{[2]}[0,\infty)$;

(ii) $r(t) = r_\infty + R(t)$, $r_\infty = \frac{\beta}{a_\infty}$, $R^{(m)} \in L^1(0,\infty)$ $(m = 0, 1, 2)$;

(iii) for every $T > 0$ and for every $v \in L^2(0,T)$ there exist constants $\gamma, q > 0$ with $\beta_q < 1$, such that

$$q \int_0^T v(t) \frac{d}{dt}(r * v)\ t)\,dt - \int_0^T v(t)(R * v)(t)\,dt \geq (1+\gamma)\int_0^T v^2(t)\,dt.$$

For the harmonic analysis proof of Lemma 3.4 we refer the reader to [10, Lemma 3.2] where a similar, but more complicated result is established. Part (iii) can be proved much more simply by using the frequency domain argument of [15, Theorem 1]. For the benefit of the reader we record here the following corrections in the proof of [10, Proposition 4.1]: Equation (4.16) should read

$$-q\eta \operatorname{Im}\hat{r}(i\eta) - \operatorname{Re}\hat{r}(i\eta) = \Gamma(q,\beta,\eta) + 1, \quad \eta \in \mathbb{R} ;$$

equation (4.26) should read

$$\frac{1}{q} - \beta\gamma = \frac{m(\eta)q^2 + n(\eta)}{q(n(\eta) - q)}$$

and the sentence following this equation should read: "Given $\bar{q}, \bar{Q}$, $0<\bar{q}<\bar{Q}<\infty$, there exists an $\bar{\varepsilon} > 0$, $\bar{\varepsilon} = \bar{\varepsilon}(\bar{q}, \bar{Q})$, such that for every $\bar{q} \le q \le \bar{Q}$ (4.27) holds". Note also that $\frac{1}{q_k} - \beta\gamma(q_k, \eta_k)$ tends to $-\dot{a}(0)$ (and not to $-\dot{a}(0)/q^2$). In the concluding argument of the proposition one needs to choose $0 < \varepsilon < \min(\bar{\varepsilon}(\bar{q}, \bar{Q}), -2\dot{a}(0))$ in order to carry through the proof, since $\gamma(q, \beta, \eta) \to q\beta - 1 - \dot{a}(0)q$ and $0 < \frac{1}{q} - \beta < \frac{\varepsilon}{2}$, by the choice of q and β.

We obtain the next set of estimates from the equivalent form (3.9) of (VE) with the aid of Lemma 3.4. Let u be a local solution of (3.9) (also equivalent to (3.1)) on $[0,T) \times \mathbb{R}$, $T < T_0$. Multiply (3.9) first by qu_{tt} then by u_t, integrate each relation over $[0, s] \times \mathbb{R}$, $0 < s < T < T_0$, and add the resulting equations in which several terms have been simplified by integrations by parts. The result of this tedius calculation is

$$\begin{aligned}
(3.10)\qquad & \frac{\beta}{2a_\infty}\int_{-\infty}^{\infty} u_t^2(s,x)\,dx + \beta\int_{-\infty}^{\infty} W(u_x(s,x))\,dx + \frac{q}{2}\int_{-\infty}^{\infty} u_{tt}^2(s,x)\,dx \\
& + \frac{q}{2}\int_{-\infty}^{\infty} \sigma'(u_x(s,x))\,u_{tx}^2(s,x)\,dx + \gamma\int_0^s\int_{-\infty}^{\infty} u_{tt}^2(t,x)\,dx\,dt \\
& + (1-q\beta)\int_0^s\int_{-\infty}^{\infty} \sigma'(u_x(s,x))\,u_{tx}^2(t,x)\,dx\,dt \le \frac{\beta}{2a_\infty}\int_{-\infty}^{\infty} u_1^2(x)\,dx \\
& + \beta\int_{-\infty}^{\infty} W(u_{0x}(x))\,dx + \frac{q}{2}\int_{-\infty}^{\infty} u_{tt}^2(0,x)\,dx + \frac{q}{2}\int_{-\infty}^{\infty} \sigma'(u_{0x})\,u_{1x}^2(x)\,dx \\
& - \int_{-\infty}^{\infty} u_t(s,x)\,u_{tt}(s,x)\,dx + \int_{-\infty}^{\infty} u_1(x)\,u_{tt}(0,x)\,dx \\
& - q\beta\int_{-\infty}^{\infty} \sigma(u_x(s,x))\,u_{tx}(s,x)\,dx + q\beta\int_{-\infty}^{\infty} \sigma(u_{0x}(x))\,u_{1x}(x)\,dx \\
& - \int_{-\infty}^{\infty} u_t(s,x)\,(R * u_{tt})(s,x)\,dx + \frac{q}{2}\int_0^s\int_{-\infty}^{\infty} \sigma''(u_x(t,x))\,u_{tx}^3(t,x)\,dx\,dt,
\end{aligned}$$

where $1 - q\beta > 0$ by Lemma 3.4, and where $u_{tt}(0,x)$ is the given smooth function in (3.9). The left side of (3.10) is minorized with the aid of (3.6), (3.7). If condition (μ^*) is satisfied for μ sufficiently small, one shows by elementary estimates that each term on the right side of (3.10) is either controllably small, or else it can be majorized by the sum of a controllably small quantity and a quantity which is dominated by the left side of (3.10). Thus, for example, $u_{tt}(0,x) = \sigma'(u_{0x})\,u_{0xx}$ is controllably small, also

$$-\int_{-\infty}^{\infty} u_t(s,x)\,u_{tt}(s,x)\,dx \le \frac{1}{q}\int_{-\infty}^{\infty} u_t^2(s,x)\,dx + \frac{q}{4}\int_{-\infty}^{\infty} u_{tt}^2(s,x)\,dx,$$

$$-\int_{-\infty}^{\infty} u_t(s,x)\,(R*u_{tt})(s,x)\,dx \le \frac{1}{\gamma}\sup_{0\le t<\infty}|R(t)|\int_0^{\infty}|R(t)|\,dt\int_{-\infty}^{\infty}u_t^2(s,x)dx$$
$$+\frac{\gamma}{4}\int_0^s\int_{-\infty}^{\infty}u_{tt}^2(t,x)\,dx\,dt,$$

$$\frac{q}{2}\int_0^s\int_{-\infty}^{\infty}\sigma''(u_x)\,u_{tx}^3(t,x)\,dx\,dt \le \frac{\mu q}{2}\max_{[-c_0,c_0]}|\sigma''(\cdot)|\int_0^s\int_{-\infty}^{\infty}u_{tx}^2(t,x)dxdt.$$

The result of (3.10) and the above estimates is that the quantities $\int_{-\infty}^{\infty}u_{tt}^2(s,x)\,dx$, $\int_{-\infty}^{\infty}u_{tx}^2(s,x)\,dx$, $\int_0^s\int_{-\infty}^{\infty}u_{tt}^2(t,x)\,dx\,dt$, $\int_0^s\int_{-\infty}^{\infty}u_{tx}^2(t,x)\,dx\,dt$, as well as (over again) $\int_{-\infty}^{\infty}u_t^2(s,x)\,dx$, $\int_{-\infty}^{\infty}u_x^2(s,x)\,dx$ are controllably small, uniformly on $[0,T)$, as long as (μ^*) holds for μ sufficiently small (note that $\mu q \max_{[-c_0,c_0]}|\sigma''(\cdot)| \le (1-q\beta)p_0$ for μ sufficiently small). We deduce that $\int_{-\infty}^{\infty}u_{xx}^2(s,x)\,dx$ and $\int_0^s\int_{-\infty}^{\infty}u_{xx}^2(t,x)\,dx\,dt$ are controllably small, uniformly on $[0,T)$, from the equation (see the equation preceeding (3.1))

$$\sigma'(u_x)\,u_{xx} = u_{tt} + k*u_{tt},$$

together with the result just obtained and properties of k.

To obtain the final set of estimates we assume temporarily that the assumptions of Proposition 3.3 hold. We then differentiate (3.9) twice with respect to t and x; we then multiply the resulting equation first by $q\mu_{ttx}$, then by u_{tx}, integrate each over $[0,s]\times\mathbb{R}$, and add the resulting equations By a long calculation involving Lemma 3.4 (iii), similar to that used in arriving at (3.10) and the estimates following, one shows that as long as (μ^*) holds for μ sufficiently small, the quantities $\int_{-\infty}^{\infty} u_{ttx}^2(s,x)\,dx$, $\int_{-\infty}^{\infty} u_{txx}^2(s,x)\,dx$, $\int_0^s \int_{-\infty}^{\infty} u_{ttx}^2(t,x)\,dx\,dt$, $\int_0^s \int_{-\infty}^{\infty} u_{txx}^2(t,x)\,dx\,dt$ are controllably small uniformly on $[0,T)$ (for details see [4, proof of Theorem 5.1]). A careful examination of these estimates enables us to drop the extraneous smoothness assumption of Proposition 3.3 via a density argument. Observing that

$$\sigma'(u_x)\,u_{xxx} = u_{ttx} - \sigma''(u_x)\,u_{xx}^2 + k(0)\,u_{tx} + k' * u_{tx} - k(t)\,u_1(x) ,$$

we use the earlier estimates to conclude that $\int_{-\infty}^{\infty} u_{xxx}^2(s,x)\,dx$ and $\int_0^s \int_{-\infty}^{\infty} u_{xxx}^2(t,x)\,dx\,dt$ are controllably small, uniformly on $[0,T)$, provided (μ^*) holds for μ sufficiently small. Finally, the equation

$$u_{ttt} = \sigma'(u_x)\,u_{txx} + \sigma''(u_x)\,u_{tx}\,u_{xx} - k(0)\,u_{tt} - k' * u_{tt}$$

is used to conclude that $\int_{-\infty}^{\infty} u_{ttt}^2(s,x)\,dx$ and $\int_0^s \int_{-\infty}^{\infty} u_{ttt}^2(t,x)\,dx\,dt$ are controllably small, uniformly on $[0,T)$, provided (μ^*) holds. Combining all of the above estimates yields (μ^{**}) above, provided that (μ^*) holds for μ sufficiently small. This completes the outline of the proof.

References

[1] B.D. Coleman and M.E. Gurtin, Waves in materials with memory. II On the growth and decay of one-dimensional acceleration waves. Arch. Rat. Mech. and Analysis 19 (1965), 239-265.

[2] C.M. Dafermos, An abstract Volterra equation with applications to linear viscoelasticity, J. Differential Equations 7 (1970), 554-569.

[3] C.M. Dafermos, Asymptotic stability in viscoelasticity, Arch. Rat. Mech. and Analysis 37 (1970), 297-308.

[4] C.M. Dafermos and J.A. Nohel, Energy methods for nonlinear hyperbolic Volterra integrodifferential equations (to appear).

[5] J.M. Greenberg, A priori estimates for flows in dissipative materials, J. Math. Anal. and Appl. 60 (1977), 617-630.

[6] P.D. Lax, Development of singularities of solutions of nonlinear hyperbolic partial differential equations, J. Math. Phys. 5 (1964), 611-613.

[7] S.O. Londen, An existence result for a Volterra equation in Banach space, Trans. Amer. Math. Soc. (to appear)

[8] S.O. Londen, An integrodifferential Volterra equation with a maximal monotone mapping, J. Differential Equations (to appear).

[9] R.C. Mac Camy, An integro-differential equation with applications in heat flow, Q. Appl. Math. 35 (1977), 1-19.

[10] R.C. Mac Camy, A model for one-dimensional, nonlinear viscoelasticity, Ibid 35 (1977), 21-33.

[11] A. Matsumura, Global existence and asymptotics of the solutions of the second order quasilinear hyperbolic equations with first order dissipation (to appear).

[12] A. Matsumura, Energy decay of solutions of dissipative wave equations (to appear).

[13] T. Nishida, Global smooth solutions of the second-order quasilinear wave equations with the first-order dissipation (unpublished).

[14] J.A. Nohel, A forced quasilinear wave equation with dissipation, Proceedings of EQUADIFF 4, Lecture Notes, Springer Verlag (to appear).

[15] J.A. Nohel and D.F. Shea, Frequency domain methods for Volterra equations, Advances in Math. 22 (1976), 278-304.

Differential Equations Associated with Continuous and Dissipative Time - Dependent Domain Operators

NICOLAE H. PAVEL and IOAN I. VRABIE (Iaşi)

1. - Introduction .

In the present paper we prove via flow - invariance techniques a necessary and sufficient condition for the existence of a strong solution to a certain nonlinear Cauchy problem with time - dependent right hand side . Our main result is very closely related to a result of Martin [4] (see also Brezis [1]) and the idea of proof is essentialy based on an improvement of Martin's method using similar arguments as those of Pavel and Vrabie [8] and [9] . See also [7].

We begin with the basic notations and definitions .

Let X be a real Banach space with the norm $\| \;\|$ and X^* its dual with the corresponding norm $\| \;\|_*$. If $x \in X$ and $r > 0$, denote by $S(x,r)$ the open ball with center x and radius r and by $B(x,r)$ the closed ball with center x and radius r . If $D \subset X$ is a nonempty set , denote by $d(x,D)$ the usual distance between x and D . Let $D \subset X$ and $E \subset X$ be two nonempty and bounded sets . We recall that the Hausdorff distance between D and E , $\rho(D,E)$ is defined by :

$$(1.1) \quad \rho(D,E) = \inf\left\{ r > 0 \;;\; D \subset \bigcup_{y \in E} S(y,r) \;,\; E \subset \bigcup_{x \in D} S(x,r) \right\} .$$

From (1.1) it follows that $\rho(D,E) = 0$ iff D and E have the same closure .

If $x \in X$ and $x^* \in X^*$, then $\langle x,x^* \rangle$ denotes the value of x^* at x . Let $F : X \longrightarrow 2^{X^*}$ be the duality mapping , i.e.

$F(x) = \{ x^* \in X^* \; ; \langle x,x^* \rangle = \|x\|^2 = \|x^*\|_*^2 \}$, for each $x \in X$.

Let $x \in X$ and $y \in X$ be two arbitrary elements and let us define :

(1.2) $\langle y,x \rangle_s = \sup\{ \langle y,w \rangle ; \; w \in F(x) \}$.

From (1.2) we easily get :

(1.3) $\langle y + z,x \rangle_s \leq \|y\| \|x\| + \langle z,x \rangle_s$.

Let us consider the following nonlinear Cauchy problem :

(1.4) $\dot{u}(t) = A(t)u(t)$, $t_0 \leq t \leq t_0 + T$,

(1.5) $u(t_0) = u_0 \in D(t_0)$,

where for each $t \in [a,b[$, $A(t) : D(t) \longrightarrow X$ is a nonlinear operator whose domain $D(t)$ is a subset of X .

A <u>strong solution</u> to the problem (1.4) , (1.5) is a continuously differentiable (in the norm topology of X) function $u : [t_0,t_0 + T] \longrightarrow X$ such that $u(t) \in D(t)$ for each $t \in [t_0,t_0 + T]$ and which satisfies (1.4) , (1.5) on $[t_0,t_0 + T]$.

In all what follows $\mathcal{D}$ is defined by :

$\mathcal{D} = \{ (t,x) \in [a,b[\times X \; ; \; x \in D(t) \}$.

2. Statement of the main result .

We begin with the hypothesis we need in the sequel .

(H_1) <u>For each</u> $t_0 \in [a,b[$ <u>and</u> $x_0 \in D(t_0)$ <u>there exist</u> $r > o$ <u>and</u> $T > o$ <u>such that the mapping</u> $t \longmapsto B(x_0,r) \cap D(t)$ <u>is nonempty , closed valued and continuous in the Hausdorff metric on</u> $[t_0,t_0 + T]$.

(H_2) <u>For each</u> $t \in [a,b[$, $A(t) : D(t) \longrightarrow X$ <u>and for each</u> x,y <u>from</u> $D(t)$ <u>the following condition</u> :

(2.1) $\langle A(t)x - A(t)y , x - y \rangle_s \leq L \|x - y\|^2$

holds , where L is a positive constant .

(H_2') The mapping $(t,x) \longmapsto A(t)x$ is compact from $\mathcal{D}$ to X .

(H_3) The mapping $(t,x) \longmapsto A(t)x$ is continuous from $\mathcal{D}$ to X .

(H_4) For each $(t,x) \in \mathcal{D}$ the following condition :

(2.2) $$\lim_{h \searrow 0} \frac{1}{h} d(x + hA(t)x , D(t+h)) = 0$$, holds .

Now , we are able to formulate our main results .

THEOREM 2.1.- Assume that (H_1) , (H_2) and (H_3) are satisfied . Then , (H_4) holds if and only if for each $(t_0,x_0) \in \mathcal{D}$ there exists $T > 0$ such that the problem (1.4) , (1.5) has a unique strong solution on $[t_0, t_0 + T]$.

THEOREM 2.2.- Assume that (H_1) , (H_2') and (H_3) are satisfied . Then , (H_4) holds if and only if for each $(t_0,x_0) \in \mathcal{D}$ there exists $T > 0$ such that the problem (1.4) , (1.5) has a strong solution on $[t_0, t_0 + T]$.

Let us remark that in the case in which D(t) does not depend on t , our Theorem 2.1. yields to the well - known result of Martin [4], while Theorem 2.2. improves the results of Nagumo [5] and Crandall [2].

If we assume additional growth conditions on A , we obtain :

THEOREM 2.3.- Assume that (H_1) , (H_2) , (H_3) and (H_4) are satisfied . Assume in addition that for each $c \in [a,b[$ and $r > 0$ there exists $K(c,r) > 0$ such that :

$$\|A(t)x\| \leq K(c,r) \quad \text{for all } t \in [a,c] \text{ and } x \in B(0,r) \cap D(t) .$$

Then , for each $(t_0,x_0) \in \mathcal{D}$ there exists a unique strong solution of (1.4) , (1.5) defined on the whole interval $[t_0, b[$.

THEOREM 2.4.- Assume that (H_1) , (H_2') , (H_3) and (H_4) are satisfied . Assume in addition that for each $c \in [a,b[$ and $r > 0$ there exists $K(c,r) > 0$ such that :

$$\|A(t)x\| \leq K(c,r) \quad \text{for all } t \in [a,c] \text{ and } x \in B(0,r) \cap D(t) .$$

Then , for each $(t_0,x_0) \in \mathcal{D}$ there exists at least a strong solution u *of* (1.4) , (1.5) *defined on a maximal interval of existence* $[t_0, T_{max}[$, *where either* $T_{max} = b$ *or if* $T_{max} < b$, *then* $\limsup_{t \to T_{max}} \|u(t)\| = +\infty$.

Before proceeding to the proof of Theorem 2.1. we recall for easy references the following technical results :

LEMMA 2.1. (Kato [3]) - *Let* $u : [a,b] \longrightarrow X$ *be a weakly differentiable function at* $t \in [a,b]$. *Assume further that* $s \longmapsto \|u(s)\|$ *is also differentiable at* t . *Then , for each* $x^* \in F(u(t))$ *the following relation* :

$$(2.3) \quad \|u(t)\| \frac{d}{dt}\|u(t)\| = \langle \dot{u}(t) , x^* \rangle ,$$

holds .

LEMMA 2.2. (Martin [4]) - *Let* $f: [a,b] \longrightarrow [0,+\infty[$ *be a given function satisfying* :

(i) *there exists* $\{r_k\}_{k=0}^{\infty} \subset [a,b]$ *with* $r_k < r_{k+1}$ *for* $k = 0,1,\dots$ *such that* f *is continuous on each* $[r_k, r_{k+1}[$.

(ii) $f(a) = 0$ *and* $f(r_{k+1}^-) = \lim_{h \searrow 0} f(r_{k+1} - h)$ *exists for* $k = 0,1,\dots$

(iii) $\dot{f}(t)$ *exists for almost all* $t \in [a,b]$ *and satisfies* :

$$(2.4) \quad \dot{f}(t) \leq pf(t) + q$$

where p *and* q *are positive constants* .

Then , for each $t \in [r_k, r_{k+1}[$ *the following inequality* :

$$(2.5) \quad f(t) \leq \left(p^{-1}q + \sum_{i=0}^{k} |f(r_i) - f(r_i^-)|\right) \exp p(t - a) - p^{-1}q$$

holds .

Now , we may proceed to the proof of Theorem 2.1.

3. *Proof of Theorem 2.1. First Part* .

Necessity . Let $(t,x) \in \mathcal{D}$ be arbitrary and assume that there exist $T > 0$ and a continuously differentiable function $u : [t, t + T] \to X$ such that :

$$(3.1)\begin{cases} u(s)\in D(s) & t\leq s\leq t+T\ ,\\ \dot{u}(s)=A(s)u(s) & t\leq s\leq t+T\ ,\\ u(t)=x\ . \end{cases}$$

In view of (3.1) we get :

$$(3.2)\quad \frac{1}{h}\,d(x+hA(t)x\ ,\ D(t+h))\leq \frac{1}{h}\,\|x+hA(t)x-u(t+h)\| =$$

$$= \left\| \frac{u(t+h)-u(t)}{h} - A(t)x \right\| \ ,$$

for all $o<h\leq T$.

But (3.1) and (3.2) imply (2.2) and this completes the proof of the necessity .

Suffiency . The proof of the suffiency is somewhat awkward and follows (with some modifications) the same lines as those of Martin [4] and Pavel and Vrabie [8], [9] .

First step . The construction of the approximate solutions .

Let $(t_o,x_o)\in \mathcal{D}$ and let $M>1$, $T>o$ and $r>o$ be such that :

$$(3.3)\quad \|A(t)y\|\leq M-1$$

for all $t\in[t_o,t_o+T]$ and $y\in B(x_o,r)\cap D(t)$,

$$(3.4)\quad MT\leq r\ .$$

Such M , T and r exist as A is continuous at (t_o,x_o) and therefore locally bounded . The fact that $B(x_o,r)\cap D(t)$ is nonempty has been admitted by the hypothesis (H_1) .

Let n be an arbitrary natural number and define $t_o^n=t_o$, $u_o^n=x_o$ Inductively , starting from t_i^n , $u_i^n\in B(x_o,r)\cap D(t_i^n)$ let us define t_{i+1}^n and u_{i+1}^n as below . If $t_i^n=t_o+T$, set $t_{i+1}^n=t_o+T$, but if $t_i^n<t_o+T$, choose the largest number $d_i^n\in\,]o,1/n]$ such that :

$$(3.5)\quad t_{i+1}^n=t_i^n+d_i^n\leq t_o+T\ ,$$

$$(3.6)\quad \|A(t)y-A(t_i^n)u_i^n\|\leq 1/n\ ,$$

for all $y\in D(t)$ with $t\in[t_i^n,t_i^n+d_i^n]$ and $\|y-u_i^n\|\leq d_iM$,

(3.7) $d(u_i^n + d_i^n A(t_i^n)u_i^n \, , \, D(t_i^n + d_i^n)) \leqslant d_i/2n$.

The number d_i^n is well defined in view of the continuity of the mapping $(t,x) \longmapsto A(t)x$ and of (H_4) .

By (3.7) it follows that there exists $u_{i+1}^n \in D(t_i^n + d_i^n)$ such that :

(3.8) $\|u_i^n + d_i^n A(t_i^n)u_i^n - u_{i+1}^n\| \leqslant d_i^n/n$.

Denoting :

(3.9) $\frac{1}{d_i^n}(u_{i+1}^n - u_i^n - d_i^n A(t_i^n)u_i^n) = p_i^n$,

by (3.8) and (3.9) we have :

(3.1o) $u_{i+1}^n = u_i^n + d_i^n A(t_i^n)u_i^n + d_i^n p_i^n$,

where $\|p_i^n\| \leqslant 1/n$.

Define now the polygonal line :

(3.11) $u_n(t) = u_i^n + (t - t_i^n)A(t_i^n)u_i^n + (t - t_i^n)\, p_i^n$, $t \in [t_i^n, t_{i+1}^n[$.

Clearly , $u_n(t_i^n) = u_i^n$ and is no hard to verify (inductively) that u_n given by (3.11) may be rewritten as :

(3,12) $u_n(t) = x_o + \sum_{j=o}^{i-1} (t_{j+1}^n - t_j^n)(A(t_j^n)u_j^n + p_j^n) + (t - t_i^n)A(t_i^n)u_i^n +$

$+ (t - t_i^n)p_i^n$.

By hypothesis , $u_i^n \in B(x_o,r) \cap D(t_i^n)$ and so from (3.3) and (3.12) one derives:

$$\|u_n(t) - x_o\| \leqslant (M - 1 + 1/n)\left(\sum_{j=o}^{i-1}(t_{j+1}^n - t_j^n) + t - t_i^n\right) \leqslant M(t - t_o) \leqslant$$

$MT \leqslant r$ and therefore $u_n(t) \in B(x_o,r)$ for all t from the domain of u_n . In what follows we shall prove that for each n = 1,2,..., $\lim_{i\to\infty} t_i^n = t_o + T$.

Second step . The convergence of the sequences $\{t_i^n\}_{i=o}^{\infty}$ to $t_o + T$.

Assume by contradiction that there exists a natural number n for which $\lim_{i\to\infty} t_i = t_o + \tilde{T}$, with $o < \tilde{T} < T$.

Let us remark that (3.1o) implies $\|u_{i+1}^n - u_i^n\| \leqslant M(t_{i+1}^n - t_i^n)$, relation which shows that $\lim_{i\to\infty} u_i^n = u^n$ exists . Inasmuch as $u_i^n \in B(x_o,r) \cap D(t_i^n)$, it follows by hypothesis (H_1) that $u^n \in B(x_o,r) \cap D(t_o + \tilde{T})$. Denote $t_o + \tilde{T} = t^n$ and let $c > o$ be such that :

(3.13) $\|A(t)y - A(t^n)u^n\| \leqslant 1/3n$,

for all $|t - t^n| \leqslant 2c$ and $y \in D(t)$ with $\|y - u^n\| \leqslant 2cM$.

Choose $s > o$, small enough such that :

(3.14) $d(u^n + sA(t^n)u^n , D(t^n + s)) \leqslant s/6n$, $o < s < \min\{1/n, c, t_o + T - t^n\}$

Let N_o be a natural number such that $|t^n - t_i^n| \leqslant s$, $\|u^n - u_i^n\| \leqslant rM$, for all $i \geqslant N_o$. Then , for all $i \geqslant N_o$ and $|t - t_i^n| \leqslant s$, $y \in D(t)$ with $\|y - u_i^n\| \leqslant sM$ we have :

$\|A(t)y - A(t_i^n)u_i^n\| \leqslant \|A(t)y - A(t^n)u^n\| + \|A(t^n)u^n - A(t_i^n)u_i^n\| \leqslant 2/3n < 1/n$.

Since from (3.14) we have $t_i^n + s \leqslant t^n + s < t_o + T$ and d_i^n is the greatest number in $]o,1/n]$ satisfying (3.5) , (3.6) , (3.7) , it follows :

$d(u_i^n + sA(t_i^n)u_i^n , D(t_i^n + s)) > s/2n$, for $i \geqslant N_o$.

On the other hand , from (3.14) and (H_1) we get :

$d(u_i^n + sA(t_i^n)u_i^n , D(t_i^n + s)) \leqslant s/3n$, for i large enough .

The contradiction between the last two relations shows that the supposition $o < \tilde{T} < T$ is false and consequently $\lim_{i\to\infty} t_i^n = t_o + T$, as claimed .

Now , we shall point out two properties of u_n which follow directly from (3.11) and from the estimate $\|p_i^n\| \leqslant 1/n$.

(3.15) $\|u_n(t) - u_n(s)\| \leqslant M|t - s|$ for $t,s \in [t_o , t_o + T]$.

(3.16) $\|\dot{u}_n(t) - A(t_i^n)u_i^n\| \leqslant 1/n$ for all $t \in [t_i^n, t_{i+1}^n[$, $i = o,1,\ldots$

Define now the step functions a_n by :

$a_n(s) = t_i^n$ for $s \in [t_i^n, t_{i+1}^n[$, $i = o,1,\ldots$, $a_n(t_o + T) = t_o + T$.

It follows that :

(3.17) $u_n(a_n(s)) = u_i^n$ for all $s \in [t_i^n, t_{i+1}^n[$, $i = 0,1,\ldots$

Set $u_n(t_0 + T) = \lim_{t \to T_0} u_n(t)$ where $T_0 = t_0 + T$. Noting that this limit exists by (3.15) we conclude that u_n is continuous on $[t_0, t_0 + T]$

Denoting :

$\sum_{j=0}^{i-1} (t_{j+1}^n - t_j^n)p_j^n + (t - t_i^n)p_i^n = g_n(t)$ for $t \in [t_i^n, t_{i+1}^n[$ and

taking into account (3.14) , we can easily verify that u_n may be rewritten under the following convenient form :

$$(3.18) \quad u_n(t) = x_0 + \int_{t_0}^{t} A(a_n(s))u_n(a_n(s))ds + g_n(t) ,$$

for all $t \in [t_0, t_0 + T]$, where obviously $\|g_n(t)\| \leqslant T/n$.

Let us assume that $\lim_{n\to\infty} u_n(t) = u(t)$ exists uniformnely on $[t_0, t_0+T]$ (we shall prove this fact in the second part of the suffiency) .

Then , it is easy to check out that $\lim_{n\to\infty} a_n(s) = s$ and $\lim_{n\to\infty} u_n(a_n(s)) = u(s)$ uniformely on $[t_0, t_0 + T]$. One may also verify that for each $t \in [t_0, t_0 + T]$, $u(t) \in B(x_0, r) \cap D(t)$, relation which follows from (H_1) and from the fact that $u_n(t) \in B(x_0, r) \cap D(t)$ for $t = a_n(s)$.

Now , from (3.18) and the last two remarks , we may conclude that u is a strong solution of (1.4) , (1.5) , as claimed .

4. Proof of Theorem 2.1. Second Part .

We shall prove now that the sequence $\{u_n\}_{n=1}^{\infty}$ given by (3.11) (and also by (3.18)) is uniformely convergent on $[t_0, t_0 + T]$ to some continuous function u .

The proof of this fact follows Martin's techniques . See for instance [4] .

With the notations above , set $S_{mn} = \bigcup_{i=1}^{\infty} \{t_i^m, t_i^n\}$ and define

$\{r_k\}_{k=0}^{\infty} \subset [t_o, t_o + T]$ by $r_o = t_o$, $r_{k+1} = \min\{ s \in S_{mn} , s > r_k\}$ for $k = o, 1, \ldots$. We shall construct the sequences v_n and v_m on $[t_o, t_o + T]$ with the following properties :

(4.1) $\|v_b(t) - x_o\| \leq M(t - t_o)$, $\|v_b(t) - v_b(s)\| \leq M|t - s|$.
for $t, s \in [r_k, r_{k+1}[$ where $b \in \{m, n\}$.

(4.2) v_n and v_m are differentiable on $[r_k, r_{k+1}]$ excepting a countable subset and in addition :

$$\langle \dot{v}_n(t) - \dot{v}_m(t) , v_n(t) - v_m(t) \rangle_s \leq L \|v_n(t) - v_m(t)\|^2 +$$
$$+ (1/n + 1/m)(1 + \|v_n(t)\| + \|v_m(t)\|) .$$

(4.3) If i , j , k are such that t_i^n , $t_j^m \leq r_k < r_{k+1} \leq t_{i+1}^n$, t_{j+1}^m , then:

a) $v_n(r_{k+1}) = u_n(r_{k+1})$ if $r_{k+1} = t_{i+1}^n$ and
$v_n(r_{k+1}) = v_n(r_k^-)$ if $r_{k+1} < t_{i+1}^n$ (where $v_n(t^-) = \lim\limits_{s \searrow t} v_n(s)$)

b) $v_m(r_{k+1}) = u_m(r_{k+1})$ if $r_{k+1} = t_{j+1}^m$ and
$v_m(r_{k+1}) = v_m(r_k^-)$ if $r_{k+1} < t_{j+1}^m$,

c) $\|v_b(t_{i+1}^b) - v_b(t_{i+1}^{b-})\| \leq \frac{3}{b}(t_{i+1}^b - t_i^b)$, $b \in \{m, n\}$.

(4.4) $\|v_n(t) - u_n(t)\| \leq \frac{3}{n}(t - t_i^n)$, $\|v_m(t) - u_m(t)\| \leq \frac{3}{m}(t - t_j^m)$,
for all $t \in [r_k, r_{k+1}[$.

(4.5) $v_b(r_{k+1}) \in B(x_o, r) \cap D(r_{k+1})$, $v_b(r_{k+1}^-) \in B(x_o, r) \cap D(r_{k+1})$,
$b \in \{m, n\}$.

Define $v_n(r_o) = v_m(r_o) = x_o$, where $r_o = t_o$ and assume that we have constructed v_n and v_m on $[t_o, r_k]$. The construction of v_n and v_m on $[r_k, r_{k+1}]$ is similar to that of u_n ; namely , take $s_o = r_k$ and define the sequence $\{s_l\}_{l=0}^{\infty} \subset [r_k, r_{k+1}]$ by induction .

If $s_l = r_{k+1}$ set $s_{l+1} = r_{k+1}$ and if $s_l < r_{k+1}$, set $s_{l+1} = s_l + c_l$, where $c_l > o$ is the largest positive number with the properties :

(4.6) $s_l + c_l \leq r_{k+1}$

(4.7) $d(v_b(s_l) + c_l A(s_l)v_b(s_l)\ ,\ D(s_l + c_l)) \leqslant c_l/2b \quad , \ b \in \{m,n\}$,

(4.8) $< A(s_l)v_n(s_l) - A(s_l)v_m(s_l)\ ,\ v_n(s_l) - v_m(s_l) + x - y>_s \leqslant$

$\leqslant L\ \|v_n(s_l) + x - (v_m(s_l) + y)\|^2 + 1/n + 1/m$,

for all x,y with $\|x\| \leqslant c_l M$ and $\|y\| \leqslant c_l M$.

Here we have supposed that v_n and v_m are constructed on $[r_k, s_l]$ with $v_b(s_l) \in B(x_0, r) \cap D(s_l)$, $\|v_b(s_l) - x_0\| \leqslant M(s_l - t_0)$, $b \in \{m,n\}$.

In view of (4.7) there exists an element $v_b(s_{l+1}) \in D(s_{l+1})$ such that:

(4.9) $\|v_b(s_l) + c_l A(s_l)v_b(s_l) - v_b(s_{l+1})\| \leqslant c_l/b \quad , \ b \in \{m,n\}$.

Denoting by :

$q_l^b = \frac{1}{c_l}(v_b(s_{l+1}) - v_b(s_l) - c_l A(s_l)v_b(s_l))$, it follows :

(4.10) $v_b(s_{l+1}) = v_b(s_l) + (s_{l+1} - s_l)(A(s_l)v_b(s_l) + q_l^b)$,

with $\|q_l^b\| \leqslant 1/b$.

Define now v_b on $[s_l, s_{l+1}]$ by :

(4.11) $v_b(t) = v_b(s_l) + (t - s_l)(A(s_l)v_b(s_l) + q_l^b)$,

for $t \in [s_l, s_{l+1}]$.

It is easy to see that :

$\|v_b(t) - v_b(s)\| \leqslant M|t - s|$ for all $t,s \in [r_k, s_{l+1}]$ and also that:

$\|v_b(t) - x_0\| \leqslant M(t - t_0)$ for all $t \in [s_l, s_{l+1}]$.

Therefore , (4.1) is satisfied on $[r_k, r_{k+1}]$ if we show that $\lim_{l\to\infty} s_l = r_{k+1}$. To this end , let us assume by contradiction that $\lim_{l\to\infty} s_l = \bar{r} < r_{k+1}$. From the relation $\|v_b(s_{l+1}) - v_b(s_l)\| \leqslant M|s_l - s_{l+1}|$ it follows that $\lim_{l\to\infty} v_b(s_l) = \bar{v}_b$, exists and in addition that $\bar{v}_b \in B(x_0, r) \cap D(\bar{r})$. As $A(\bar{r})$ is dissipative we get :

(4.12) $< A(\bar{r})\bar{v}_m - A(\bar{r})\bar{v}_n\ ,\ \bar{v}_m - \bar{v}_n>_s \leqslant L\ \|\bar{v}_m - \bar{v}_n\|^2$.

Taking into account (4.12) and the upper semicontinuity of the mapping $<\cdot,\cdot>_s : X \times X \longrightarrow R$, we can easily prove the existence of a sufficiently small number $0 < c < r_{k+1} - \bar{r}$ such that :

(4.13) $\langle A(\bar{r})\bar{v}_m + y_1 - (A(\bar{r})\bar{v}_n + y_2)\ ,\ \bar{v}_m + x_1 - (\bar{v}_n + x_2)\rangle_s \leq$

$\leq L\|\bar{v}_m + x_1 - (\bar{v}_n + x_2)\|^2 + 1/2m + 1/2n$,

for all $\|x_i\| \leq 2cM$, $\|y_i\| \leq c$, $i = 1,2$.

Moreover ,

(4.14) $d(\bar{v}_b + cA(\bar{r})\bar{v}_b\ ,\ D(\bar{r} + c)) \leq c/3b$, $b \in \{m,n\}$.

Now let N_0 be a natural number , large enough , such that :

(4.15) $c_l = s_{l+1} - s_l < c$, $\|v_b(s_l) - \bar{v}_b\| \leq cM$, $\|A(s_l)v_b(s_l) - A(\bar{r})\bar{v}_b\| \leq c$, for $l \geq N_0$.

For $y_1 = A(s_l)v_m(s_l) - A(\bar{r})\bar{v}_m$, $y_2 = A(s_l)v_n(s_l) - A(\bar{r})\bar{v}_n$, $x_1 = x + v_m(s_l) - \bar{v}_m$, $x_2 = y + v_n(s_l) - \bar{v}_n$, with $\|x\| \leq cM$, $\|y\| \leq cM$, it follows from (4.13) that (4.8) holds with c instead of c_l , for all $l \geq N_0$. In addition we have $s_l + c < \bar{r} + c < r_{k+1}$. From the maximality of c_l with the properties (4.6) , (4.8) and from the fact that $c_l < c$ for $l \geq N_0$, it follows that (4.7) cannot hold if we set c instead of c_l . Therefore , for $l \geq N_0$ we have :

(4.16) $d(v_b(s_l) + cA(s_l)v_b(s_l)\ ,\ D(s_l + c)) > c/2b$,

either for $b = m$, or for $b = n$.

Letting $l \to +\infty$ in (4.16) and taking into account the continuity of $(u,G) \longmapsto d(u,G)$, $u \in X$, $G \subset X$, with respect to G (with the Hausdorff topology) we get the contrary of (4.14) . This contradiction shows that $\lim_{l\to\infty} s_l = r_{k+1}$, as claimed .

Now , we define $v_b(r_{k+1})$ as we have indicated in (4.3) .

Taking into account (4.11) and (1.3) we easily get :

(4.17) $\langle v_n(t) - v_m(t)\ \ v_n(t) - v_m(t)\rangle_s = \langle A(s_l)v_n(s_l) + q_l^n - (A(s_l)\cdot v_m(s_l) + q_l^m)\ ,\ v_n(t) - v_m(t)\rangle_s \leq \langle A(s_l)v_n(s_l) - A(s_l)v_m(s_l)\ ,\ v_n(t) - v_m(t)\rangle_s + (1/n + 1/m)\|v_n(t) - v_m(t)\|$.

Let us observe that $v_b(t) = v_b(s_l) + x_b$, with :

$\|x_b\| = (t - s_l)\|A(s_l)v_b(s_l) + q_l^b\| \leq c_1 M$, $l = o,1,\dots$, $b \in \{m,n\}$.

Therefore , if we set $x = x_n$, $y = x_m$, from (4.8) it follows :

(4.18) $\langle A(s_l)v_n(s_l) - A(s_l)v_m(s_l)$, $v_n(t) - v_m(t)\rangle_s \leq L\|v_n(t) - v_m(t)\|^2 +$
$+ 1/n + 1/m$.

Combining (4.17) and (4.18) one may verify (4.2) . Now , let i , j , k be such that t_i^n , $t_j^m \leq r_k < r_{k+1} \leq t_{i+1}^n$, t_{j+1}^m .

Since $\|v_n(s_l) - u_n(t_i^n)\| = \|v_n(s_l) - v_n(t_i^n)\| \leq M(t_{i+1}^n - t_i^n)$, by (3.6) it follows :

(4.19) $\|A(s_l)v_n(s_l) - A(t_i^n)u_n(t_i^n)\| \leq 1/n$, $l = o,1,\dots$.

Now , let $t \in]s_l , s_{l+1}[$. Taking into account (4.10) , (4.11) and (3.11) we have :

$$\|\dot{v}_n(t) - \dot{u}_n(t)\| \leq \|\dot{v}_n(t) - A(s_l)v_n(s_l)\| + \|A(s_l)v_n(s_l) - A(t_i^n)u_n(t_i^n)\| +$$
$$+ \|A(t_i^n)u_n(t_i^n) - \dot{u}_n(t)\| \leq 3/n .$$

Thus we have proved that :

(4.20) $\|\dot{v}_n(t) - \dot{u}_n(t)\| \leq 3/n$,

for all $t \in [t_o, t_o + T] \setminus S_{mn}$.

From (4.20) we derive (4.4) as below :

$$\|v_n(t) - u_n(t)\| \leq \|v_n(t_i^n) - u_n(t_i^n)\| + \int_{t_i^n}^{t} \|v_n(s) - u_n(s)\|ds \leq 3(t - t_i^n)/n$$

for all $t \in [t_i^n, t_o + T]$. For $t = r_k$ we get $\|v_n(r_k) - u_n(r_k)\| \leq 3(r_k - t_i^n)/n$. Similarly , from $v_n(t) - u_n(t) = v_n(r_k) - u_n(r_k) + \int_{r_k}^{t}(v_n(s) - u_n(s))ds$ one obtains : $\|v_n(t) - u_n(t)\| \leq 3(t - t_i^n)/n$, for all $t \in [r_k, r_{k+1}]$. In the same way we may prove the second inequality of (4.4) . Let us remark that (4.5) follows from the continuity of $t \longmapsto B(x_o,r) \cap D(t)$ in the Hausdorff metric .

Finally , it is easy to verify : $\|v_b(t_{i+1}^b) - v_b(t_{i+1}^{b-})\| =$
$\|u_b(t_{i+1}^b) - v_b(t_{i+1}^{b-})\| = \lim_{h \searrow o} \|u_b(t_{i+1}^b - h) - v_b(t_{i+1}^b - h)\| \leq \lim_{h \searrow o} 3(t_{i+1}^b - t_i^b -$

$- h) = 3(t_{i+1}^b - t_i^b)$ and so (4.3) (c) is proved .

Denote by $g(t) = \|v_n(t) - v_m(t)\|^2$, $t \in [t_o, t_o + T]$.
It is no hard to conclude that the real function $t \longmapsto \|v_n(t) - v_m(t)\|$ is Lipschitz continuous on $[t_o, t_o + T]$ and therefore almost everywhere differentiable . By Lemma 2.1. (see also [6] p. 45) we have : $(dg(t)/dt) = 2\|v_n(t) - v_m(t)\| \frac{d}{dt}\|v_n(t) - v_m(t)\| = 2\langle \dot{v}_n(t) - \dot{v}_m(t), v_n(t) - v_m(t)\rangle_s$ a.e. on $[t_o, t_o+T]$.

From (4.2) it follows :

(4.21) $\frac{dg(t)}{dt} \leq 2Lg(t) + 2(1/n + 1/m)(1 + \tilde{k})$, $\tilde{k} > o$,

where $\tilde{k} \geq \|v_n(t) + v_m(t)\|$ for all $m,n \in N$ and $t \in [t_o, t_o + T]$. Such a $\tilde{K}$ exists since $v_b(t)$ satisfies (4.1) . One may choose $\tilde{k} = 2(TM + \|x_o\|)$ since $\|v_b(t)\| \leq \|v_b(t) - x_o\| + \|x_o\| \leq TM + \|x_o\|$. To apply Lemma 2.2. we need an estimation for $|g(r_k) - g(r_k^-)|$. We have :

$|g(r_k) - g(r_k^-)| = \left| \|v_n(r_k) - v_m(r_k)\|^2 - \|v_n(r_k^-) - v_m(r_k^-)\|^2 \right| \leq$

$\leq 2\tilde{k}\left| \|v_n(r_k) - v_m(r_k)\| - \|v_n(r_k^-) - v_m(r_k^-)\| \right| \leq 2\tilde{k}\,\|v_n(r_k) - v_n(r_k^-)\| +$

$+ \ 2\tilde{k}\,\|v_m(r_k) - v_m(r_k^-)\|$.

Taking into account (4.3) (c) we derive :

$$\sum_{k=o}^{\infty} |g(r_k) - g(r_k^-)| \leq 2\tilde{k} \sum_{i=o}^{\infty} \frac{3}{n}(t_{i+1}^n - t_i^n) + 2\tilde{k} \sum_{i=o}^{\infty} \frac{3}{m}(t_{i+1}^m - t_i^m) \leq$$

$6\tilde{k}T(1/n + 1/m)$.

Now , applying Lemma 2.2. (whose proof may be found in [6] p. 171) with $p = 2L$, $q = 2(1/n + 1/m)(1 + \tilde{k})$, we get :

(4.22) $\|v_n(t) - v_m(t)\|^2 \leq k_1(1/n + 1/m)$, for all $t \in [t_o, t_o + T]$,

where $k_1 = (\frac{1 + \tilde{k}}{L} + 6\tilde{k}T)\exp 2TL - \frac{\tilde{k} + 1}{L}$. Therefore , we have proved that $v_n(t) \longrightarrow u(t)$ as $n \to +\infty$ uniformly on $[t_o, t_o + T]$. On the other hand , by (4.4) (i.e. from $\|u_n(t) - v_n(t)\| \leq 3T/n$) it follows that $u_n(t) \longrightarrow u(t)$ as $n \longrightarrow +\infty$, uniformly on $[t_o, t_o + T]$.

We have already proved that if $\lim_{n \to \infty} u_n(t) = u(t)$ exists , then u is a solution of (1.4) , (1.5) . The uniqueness of the solution follows from the hypothesis (H_2) in a standard manner and this completes the proof of the Theorem 2.1.

We conclude with the following remarks .

(a) The proof of Theorem 2.2. is similar to the preceding one . We have only to use the compactness assumption (H_2') to obtain a convergent subsequence of u_n .

(b) Theorem 2.3. and also Theorem 2.4. follows using standard arguments and therefore we do not give details .

(c) Finally we would like to point out a case in which (H_4) is automatically satisfied .

Take X = H a real Hilbert space , D : $[a,b[\longrightarrow 2^H$, a nonempty , bounded , open and convex valued mapping such that $\bar{D}$ is continuous in the Hausdorff metric on $[a,b[$.

Let $A(t) : D(t) \longrightarrow H$ be a given operator , $t \in [a,b[$.

Then , (H_3) implies (H_4) . A simple proof of this fact may be obtained by using Lemma 2.1. and Lemma 3.1. from [10] .

REFERENCES

1. H. Brezis , On a characterization of flow - invariance sets , Comm. Pure Appl. Math. 23(1970) , 261 - 263 .

2. M. G. Crandall , A generalization of Peano's existence theorem and flow invariance , Proc. A. M. S. 36(1972) , 151 - 155 .

3. T. Kato , Nonlinear semigroups and evolution equations , J. Math. Soc. Japan . 19(1967) , 508 - 520 .

4. R. H. Martin Jr. , Differential equations on closed subsets of a Banach space , Trans. A. M. S. 179(1973) , 399 - 414 .

5. M. Nagumo , Über die Lage der Integralkurven gewöhnlicher Differentialgleichnungen , Proc. Phys. Math. Soc. Japan (3)24(1942) , 551-559 .

6. N. Pavel , Ecuaţii diferenţiale ataşate unor operatori neliniari pe spaţii Banach , Editura Academiei R. S. R. , Bucureşti 1977 .

7. N. Pavel and F. Iacob , Invariant sets for a class of perturbed differential equations of retarded type , Israel J. Math. 28(1977) , 254 - 264 .

8. N. Pavel and I. I. Vrabie , Semi linear evolution equations with multivalued right hand side in Banach spaces (to appear in An. St. Univ. "Al. I. Cuza" Iaşi)

9. N. Pavel and I. I. Vrabie , Flow invariance for differential equations associated to nonlinear operators , Proc. of the First Romanian-American Seminar on Operators Theory and Applications held at Iaşi and Suceava March 20-24 , 1978 , (to appear)

10. I. I. Vrabie , Time optimal control for contingent equations in Hilbert spaces , (to appear in An. St. Univ. "Al. I. Cuza" Iaşi) .

Seminarul Matematic "Al. Myller"
Universitatea "Al. I. Cuza" Iaşi - 6600
The Socialist Republic of Romania

ON SOME NONLINEAR PROBLEMS OF DIFFUSION

by Marie Lise RAYNAL

Université de Bordeaux I
351, Cours de la Libération
33405 Talence FRANCE

***

In this report, we are concerned with the diffusion equation :

$$\frac{\partial u}{\partial t} - Au + \int_0^t \gamma(t-\sigma)\, Bu(\sigma)\, d\sigma = f$$

which may occur in the heat theory, in some gas diffusion problems, in some fluid flow in porous medium problems and so on ... That equation arises from two laws :

1. A conservation law, which is, according to the cases, a conservation of heat, gas or fluid :

$$\frac{\partial u}{\partial t} + \operatorname{Div} \vec{q} = f$$

where $\vec{q}$ is the flux vector, f is a function representing the source term.

2. A diffusion law, which may be nonlinear and gives the flux vector according to the gradient of u, let :

$$\vec{q} = \vec{H}(\overrightarrow{\operatorname{grad}}\, u) + \int_0^t \gamma(t-\sigma)\, \vec{G}(u(\sigma), \overrightarrow{\operatorname{grad}}\, u(\sigma))\, d\sigma \ .$$

A first section (see [7] for a detailed information) is devoted to the enunciation of results obtained for problems involving homogeneous and isotropic materials, while an example of problems which occur with composite media is given in the second section.

Section I :

The results of this first section deal with the equation :

$$\frac{\partial u}{\partial t} - \Delta u - \sum_{i=1}^{i=n} \frac{\partial}{\partial x_i} \int_o^t \gamma(t-\sigma)\, G(u(\sigma), \frac{\partial u(\sigma)}{\partial x_i})\, d\sigma = f$$

with as follows :

a) $$G(u, \frac{\partial u}{\partial x_i}) = g(u)$$

b) $$G(u, \frac{\partial u}{\partial x_i}) = g(u) \frac{\partial u}{\partial x_i}$$

c) $$G(u, \frac{\partial u}{\partial x_i}) = g(\frac{\partial u}{\partial x_i}) .$$

Our aim is to find the solutions $u(t) : x \longrightarrow u(t,x)$ of that equation, in a bounded open set of R^n, Ω, with sufficiently smooth boundary, by joining :

- the initial condition $u(o,x) = u_o(x)$
- the boundary conditions :

$$u = 0 \quad \text{on} \quad \Gamma_1$$

$$-\sum_{i=1}^{i=n} \left(\frac{\partial u}{\partial x_i} + \int_o^t \gamma(t-\sigma)\, G(u(\sigma), \frac{\partial u(\sigma)}{\partial x_i}) d\sigma\right) \cos(\vec{n}, \vec{x}_i) = \Phi(u) \quad \text{on} \quad \Gamma_2$$

where $\vec{n}$ is the outward normal, and $\{\Gamma_1, \Gamma_2\}$ is a partition of Γ.
One notes the problems listed above, successively P_a, P_b, P_c.
In every case our work proceeds as follows :

1) Variational formulation of the problem. In that order one gives the following set up :

$$V = \{ v \in H^1(\Omega) \;/\; v/\Gamma_1 = 0 \}$$

$$H = L^2(\Omega)$$

V' topological dual space of V.

2) Galerkin approximation (see [4]).

3) Passage to the limit in the approximate equation by methods which make use of monotonicity or compacity results (cf. LIONS [5]). One obtains the following statements :

Case (a) :

Assume :

a.1 - Φ monotone nondecreasing, belonging to $C^{o}(R, R)$ such that :

$$|\Phi(\lambda)| < C_1 |\lambda|^{\frac{n-1}{n-2}} + C_2 \quad n \neq 1,\ n \neq 2 \quad C_1, C_2 \text{ positive constants}$$

Φ increasing as any polynomial if $n=2$.

Any Φ if $n=1$.

a.2 - γ function defined on $(0, T) \subset R$, real valued, differentiable with a bounded derivative.

Let us give :

a.3 - $f \in L^2(0, T ; V)$ $f' \in L^2(0, T ; V')$ $f(0) \in H ; u_o \in V / \exists k \in H$ with $a(u_o, v) + \int_{\Gamma_2} \Phi(u_o)\, v\, d\Gamma = (k, v)$ $\forall v \in V$ where $a(u, v)$ is the bilinear form associated to the Laplacian.

One then arrives at :

THEOREM 1-a :

Let g be a real hölder function, defined on R ; there exists a unique solution u to the problem P_a satisfying :

$$u \in L^{\infty}(0, T ; V)$$
$$u' \in L^2(0, T ; V) \cap L^{\infty}(0, T ; H).$$

Proof :

Variational formulation and Galerkin approximation of P_a, lead to the approximated problem :

$$\mathrm{Pa}_m \begin{cases} (u'_m(t), w_i) + a(u_m(t), w_i) + \sum_{j=1}^{j=n} \int_0^t \gamma(t-\sigma)\left(g(u_m(\sigma)), \frac{\partial w_i}{\partial x_j}\right) d\sigma \\ \quad + \int_{\Gamma_2} \Phi(u_m(t))\, w_i \, d\Gamma = (f(t), w_i) \quad i = 1, 2, \dots, m \qquad (E_{am}) \\ g_{im}(0) = \alpha_{im} \quad \text{with} \quad \sum_{i=1}^{i=m} \alpha_{im} w_i \longrightarrow u_0 \quad \text{strongly in } V, \end{cases}$$

where $(w_i)_{i \in \mathbb{N}}$ is a basis for the separable space V, $u_m(t) = \sum_{i=1}^{i=m} g_{im}(t)\, w_i$, and $(\,,\,)$ denotes both the scalar product in H and the duality $< V, V' >$. It is well known that Pa_m has a unique solution $u_m(t)$ defined for $t \in [0, t_m[$, which could be extended to $t \in [0, T[$ provided a priori estimates on $u_m(t)$.

To find these estimates, one multiplies the equation above, $(E a_m)$, by $g_{im}(t)$ and sums on the index i. Then, using (a.1), (a.2), (a.3) assumptions, g hölder, and Gronwall lemma one has :

$$\begin{cases} |u_m(t)| < C \\ \int_0^t \|u_m(s)\|^2 \, ds < C \end{cases} \qquad \forall t \in [0, T]$$

where $|\ |$, $\|\ \|$ denotes respectively the norms in H and V. The C's here and in what follows are constants which do not depend on m. However to pass to the limit in $E a_m$ one needs estimates on the derivative $u'_m(t)$, which could be easily obtained from equation $E a_m$ by making use of the method of differential quotients (see LIONS [4] ch. V. 7 for a detailed information). One has :

$$\begin{cases} |u'_m(t)| \leq C \\ \\ \int_0^t \|u'_m(s)\|^2 \, ds < C \end{cases} \qquad \forall t \in [0, T] \quad \text{and it follows} \quad \begin{cases} \|u_m(t)\| < C \\ \|\Phi(u_m(t))\|_* < C \end{cases}$$

Therefore one can extract subsequences, still denoted by u_m such that :

$u_m \longrightarrow u$ weakly in $H^1(Q)$ $(Q = \Omega \times]0, T[)$ and in $L^\infty(0, T; V)$ weak star

$u'_m \longrightarrow u$ weakly in $L^2(0, T; V)$ and in $L^\infty(0, T; V)$ weak star

$\Phi(u_m) \longrightarrow \gamma$ in $L^\infty(0, T; V')$ weak star.

Then, involving the compacity of the injection of $H^1(Q)$ in $L^2(Q)$ and the Lebesgue theorem, we conclude :

$$\begin{cases} (u'(t),v)+a(u(t),v)+\sum_{j=1}^{j=n}\int_0^t \gamma(t-\sigma)\left(g(u(\sigma)),\frac{\partial v}{\partial x_j}\right)d\sigma+\int_{\Gamma_2} v\,d\Gamma=(f(t),v) & \forall v\in V \\ & \text{pp } t\in[0,T] \\ u(0)=u_o . \end{cases}$$

It remains to prove $\gamma = \Phi(u)$. It is done without any difficulty by a method of monotonicity.

To show the unicity of the solution, on the understanding that g is hölder with power $p=1$, we write the equation above with u and u^*, then we take the difference and form the inner product with $v = u-u^*$. The proof follows from assumptions (a.1), (a.2) and Gronwall lemma.

Case (b) :

Suppose :

b.1 - The function Φ monotone, belonging to $C^o(R,R)$ suitably increasing (cf. (a.1)).

b.2 - The function γ differentiable with a bounded derivative such that :

$$\operatorname{Re} \mathcal{F}(\tilde{\gamma}(\tau)) \geq 0 \qquad \forall \tau \in R$$

where $\mathcal{F}$ is the Fourier operator and $\tilde{\gamma}$ the extension of γ to R, by defining $\tilde{\gamma}=0$ out of $(0,T)$.

b.3 - The given functions f, u_o, as in a.3

One has the following results :

THEOREM 1-b :

If we assume the real function g to be a positive constant α, it exists only one solution to the problem P_b such that :

$$u \in L^\infty(0,T;V)$$
$$u' \in L^2(0,T;V) \cap L^\infty(0,T;H) .$$

If, in addition $f \in L^2(0,T;H)$ then $\Delta u \in L^2(0,T;L^2(\Omega))$.

Proof :

By the same process as in theorem 1-a (approximated equation, a priori estimates on $u_m(t)$, then on $u'_m(t)$, passage to the limit) we arrive at :

$$\begin{cases} (u'(t), v)+a(u(t), v)+\alpha \int_0^t \gamma(t-\sigma)\, a(u(\sigma), v)\, d\sigma + \int_{\Gamma_2} \gamma(t)\, v\, d\Gamma = (f(t), v) \\ \qquad\qquad \forall v \in V \quad \text{pp} \quad t \in [0, T] \\ u(0) = u_0 \, . \end{cases}$$

In order to prove $\gamma = \Phi(u)$, we have to show :

$$\text{(1.1)} \qquad \underline{\lim}\, \alpha \int_0^T \int_0^t \gamma(t-\sigma)\, a(u_m(\sigma), u_m(t))\, d\sigma\, dt \geqslant \int_0^T \int_0^t \gamma(t-\sigma)\, a(u(\sigma), u(t))\, d\sigma\, dt \, ,$$

and it is a typical difficulty of that case, because of the weak convergence of the space derivatives $\frac{\partial u_m}{\partial x_i}$. So, we introduce the functional Q_γ defined on $L^2(0, T; V)$ by :

$$Q_\gamma(u) = \int_0^T \int_0^t \gamma(t-\sigma)\, a(u(\sigma)\, u(t))\, d\sigma\, dt \, .$$

Then, making use of the Fourier transformation, we can show that :

$$Q_\gamma(u) = \int_{-\infty}^{+\infty} \operatorname{Re} \mathfrak{F}(\widetilde{\gamma}(\tau)) \sum_{i=1}^{i=n} |\mathfrak{F}(\frac{\partial \widetilde{u}}{\partial x_i}(\tau))|^2\, d\tau$$

where u and γ are equal to zero out of $(0, T)$. Moreover one has easily that Q_γ is convex and continue and (2.1) follows. To prove the unicity we proceed as in 1.a, and use the positivity of Q_γ.

When the function g is no longer a constant, the most appropriate method seems to involve compacity in order to obtain strong convergences, but such a method does not admit any Φ. So we assume Φ lipschitz. As a counterpart this enables us to impose only the function γ to be differentiable with a bounded derivative. Let the given functions f and u_0 be as in a.3.

One obtains the :

THEOREM 2-b :

Suppose the function g to be a continuous bounded real function defined on R ; there exists at least a solution u to the problem P_b satisfying

$$u \in L^{\infty}(0, T ; V)$$
$$u' \in L^{\infty}(0, T ; V) \cap L^2(0, T ; H) .$$

Proof :

Let $W = \{ v \in H^2(\Omega) \, / \, v_{|\Gamma_1} = 0 \}$.

Because of the regularity of Ω, the embedding of W in H is compact ; then if we denote by $(\, , \,)_W$ the scalar product in W, the spectral problem :

$$(w, v)_W = \lambda \, (w, v) \qquad \forall \, v \in W$$

admits a sequence of non-zero solutions $(w_j)_{j \in \mathbb{N}}$, complete and orthogonal in W, associated to a sequence of positive eigenvalues λ_j. In Galerkin-approximation of the problem we make use of these functions and we obtain the approximated equation :

$$(u'_m(t), w_j) + a(u_m(t), w_j) + \sum_{i=1}^{i=n} \int_0^t \gamma(t-\sigma) \left(g\left(\frac{\partial u_m(\sigma)}{\partial x_i}\right), \frac{\partial w_j}{\partial x_i} \right) d\sigma \, +$$

$$+ \int_{\Gamma_2} \Phi(u_m(t)) \, w_j \, d\Gamma = (f(t), w_j) .$$

The same estimates as in 1-a hold and it follows that one can extract subsequences, still denoted by u_m, satisfying the same properties.

Provided the regularity of the functions w_j and the Sobolev embedding of $H^1(\Omega)$ in $L^p(\Omega)$ for a suitably p, one can show that :

$$\int_0^t \gamma\,(t-\sigma) \left(g\left(\frac{\partial u_m(\sigma)}{\partial x_i}\right), \frac{\partial w_j}{\partial x_i} \right) d\sigma$$

converges to :

$$\int_0^t \gamma\,(t-\sigma) \left(g\left(\frac{\partial u(\sigma)}{\partial x_i}\right), \frac{\partial w_j}{\partial x_i} \right) d\sigma .$$

To take the limit in the term $\int_{\Gamma_2} \Phi(u_m(t))\, w_j\, d\Gamma$ we set :

$$X = \{ v \,/\, v \in L^2(0,T;H^1(\Omega)),\quad v' \in L^2(0,T;H) \}$$

normed with :

$$\|v\|_X = \|v\|_{L^2(0,T;H^1(\Omega))} + \|v'\|_{L^2(0,T;H)} .$$

Given ε, $0 < \varepsilon < \frac{1}{2}$, the injection of X in $L^2(0,T;H^{1-\varepsilon}(\Omega))$ is compact ; then it results from the estimates that u_m is bounded in X and hence strongly converges in $L^2(0,T;H^{1-\varepsilon}(\Omega))$. Moreover, $\varepsilon < \frac{1}{2}$ implies the continuity of the injection of $H^{1-\varepsilon}(\Omega)$ in $L^2(\Gamma)$ and consequently the strongly convergence in $L^2(0,T;\Gamma_2)$ of $u_{m|\Gamma_2}$ to $u_{|\Gamma_2}$. The assumption Φ lipschitz ends the proof of the existence.

Note that in this case we have only partial results for unicity.

Case (c) :

Let Ω be now an open bounded set of R with sufficiently smooth boundary ; let the boundary conditions be of Dirichlet type, or Neumann type, or mixed (i. e. $\Phi = 0$).

1. - Under the hypothesis :

1-c-1 g belonging to $C^0(R,R)$, bounded, differentiable with a bounded derivative.

1-c-2 γ real function defined on $(0,T)$ differentiable with a bounded derivative.

1-c-3 $f \in L^2(0,T;V)$, $f' \in L^2(0,T;V')$, $f(0) \in H$, $u_0 \in V \cap H^2(\Omega)$.

One has the :

THEOREM 1-c :

There exists only one solution u to the problem P_c with :

$$u \in L^2(0,T;H^2(\Omega)) \cap L^\infty(0,T;V)$$

$$u' \in L^2(0,T;V) \cap L^\infty(0,T;H) .$$

Proof :

Let $(w_j)_{j\in N}$ be the sequence of the eigenfunctions defined by :

$$\begin{cases} -\Delta w_j = \lambda_j w_j \\ w_{j|\Gamma} = 0 \quad (\text{i.e. } w_j(0) = w_j(1) = 0) . \end{cases}$$

It is well known that the functions w_j belong to $H^2(\Omega) \cap H^1_o(\Omega)$ and form an orthogonal complete sequence in H. The use of these functions in Galerkin approximation enables us to obtain additional estimates on $u_m(t)$. Indeed, let us write the approximated equation in the form :

$$(u'_m(t), w_j) - (\Delta u_m(t), w_j) - \int_0^t \gamma(t-\sigma)(g'(\frac{\partial u_m(\sigma)}{\partial x}) \Delta u_m(\sigma), w_j)\, d\sigma = (f(t), w_j)$$

which we multiply by $\lambda_j\, g_{jm}(t)$ and sum on j. Then we obtain :

$$a(u'_m(t), u_m(t)) + |\Delta u_m(t)|^2 + \int_0^t \gamma(t-\sigma)(g'(\frac{\partial u_m(\sigma)}{\partial x}) \Delta u_m(\sigma), \Delta u_m(t))\, d\sigma$$
$$= -(f(t), \Delta u_m(t)) ,$$

and hence by integration over $]0, t[$, use of Schwartz inequality and iteration it results that $\|\Delta u_m\|_{L^2(0,T\,;\,H)} < C$ and $\|u_m\|_{L^2(0,T\,;\,H^2(\Omega))} < C$ by Niremberg regularity.

From that estimate and the usual ones follows the existence of the solution by taking limit in the approximated equation.

It is easy to prove the unicity when one remarks that assumptions on g give g lipschitz.

2. - Under the hypothesis :

2-c-1 g belonging to $C^o(R, R)$, defined by $g(\lambda) = |\lambda|^{p-2}\lambda \quad p \geq 2$.

2-c-2 γ real function defined on $(0, T)$ of positive type (i.e. $\gamma \in C^2(0, T)$), $(-1)^k\, \gamma^{(k)} \geq 0 \quad k \in \{0, 1, 2\}$, with $\gamma^{(2)}$ bounded on $(0, T)$.

2-c-3 $f \in L^2(0, T\,;\, W)$, $f' \in L^2(0, T\,;\, W')$, $u_o \in W \cap H^2(\Omega)$ where W is the usual Sobolev space $W^{1,p}_o(\Omega)$.

One has the :

THEOREM 2-c :

There exists one solution u to the problem P_c satisfying :

$$u \in L^{\infty}(0,T;W)$$

$$u' \in L^{\infty}(0,T;H) \cap L^2(0,T;W)$$

$$\left|\frac{\partial u}{\partial x}\right|^{\frac{p-2}{2}} \frac{\partial^2 u}{\partial x^2} \in L^2(0,T;W).$$

Moreover $2 \leq p \leq 4$ the solution is unique.

Proof :

Let $(w_j)_{j\in\mathbb{N}}$ be the sequence of the eigenfunctions given in (c-1), we note as a consequence of Sobolev's embedding theorems, that these functions belong to W. Let P_m be the orthogonal projection of H on W_m, generated by $(w_1, \ldots, w_m)$, we define as follows the approximated problem :

$$P_{cm} : \begin{cases} (u'_m(t), w_j) - (\Delta u_m(t), w_j) + \int_0^t \gamma(t-\sigma)(P_m A u_m(\sigma), w_j)\, d\sigma = (P_m f(t), w_j) & E_{cm} \\ u_{om} = P_m u_o \end{cases}$$

where Au is the operator $-\frac{\partial}{\partial x}\left(\left|\frac{\partial u}{\partial x}\right|^{p-2}\frac{\partial u}{\partial x}\right)$.

Denote now $P_m f = fm$, $P_m Au_m(t) = V_m(t) = \sum_{k=1}^{k=m} \gamma_{km}(t)\, w_k$ and multiply E_{cm} by $\gamma_{jm}(t)$. After summation on the index j and manipulations on the terms $(u'_m(t), v_m(t))$ and $(-\Delta u_m(t), v_m(t))$, we obtain :

$$\frac{1}{p}\frac{\partial}{\partial t}\|u_m(t)\|_W^p + (p-1)\int_\Omega \left|\frac{\partial u_m(t)}{\partial x}\right|^{p-2}\left(\frac{\partial^2 u_m(t)}{\partial x}\right)^2 dx$$

$$= (f_m(t), v_m(t)) - \int_0^t \gamma(t-\sigma)\,(v_m(\sigma), v_m(t))\, d\sigma$$

where $\|\ \|_W$ denotes the norm in W.

If we introduce now, the function :

$$t \longrightarrow N_m(t) = \frac{1}{p}\|u_m(t)\|_W^p + \frac{\gamma(t)}{2}\left|\int_0^t v_m(\sigma)\, d\sigma\right|^2 - \frac{1}{2}\int_0^t \gamma'(t-\sigma)\left|\int_0^t v_m(s)\, ds\right|^2 d\sigma$$

an integration by part, and the assumption γ of positive type give :

$$\frac{\partial}{\partial t} N_m(t)+(p-1)\int_\Omega \left|\frac{\partial u_m(t)}{\partial x}\right|^{p-2} \left(\frac{\partial^2 u_m(t)}{\partial x^2}\right)^2 dx \leq (f_m(t),\ v_m(t)).$$

Integrating then over $(0, t)$, we deduce consequently to (2-c-3) that u_m is bounded independently of m in $L^\infty(0, T ; W)$ and $\left|\frac{\partial u_m}{\partial x}\right|^{p-2/2} \frac{\partial^2 u_m}{\partial x^2}$ in $L^2(0, T ; H)$. Hence follows that Au_m is bounded in $L^\infty(0, T ; W')$ and $\left|\frac{\partial u_m}{\partial x}\right|^{p-2/2} \frac{\partial u_m}{\partial x}$ in $L^2(0, T ; V)$.

It is now a standard process to obtain by derivation of E_{cm}, the additionnal estimates :

$$\|u'_m\|_{L^\infty(0, T ; H)} < C \ , \quad \|u'_m\|_{L^2(0, T ; V)} < C \ .$$

Consider now the set $S = \{v \mid |v|^{p-2/2}\ v \in V\}$ and the function M defined on S by $M(v) = \left(\int_\Omega |v|^{p-2} \left(\frac{\partial v}{\partial x}\right)^2 dx\right)^{1/p}$.

It is easy to prove :

$$\begin{cases} S \hookrightarrow L^p(\Omega) \hookrightarrow L^2(\Omega) \\ M(v) \geq 0 \quad \text{on} \quad S \\ M(\lambda v) = |\lambda|\, M(v) \\ \{v / v \in S \quad M(v) \leq 1\} \ \text{has a compact closure in} \ L^p(\Omega) \end{cases}$$

(see LIONS [5] ch. I. 12 for a detailed information).

Invoking then a result due to DUBINSKII [3], one has that :

$$E = \{v / v \in L^1_{loc}(0, T ; H),\ \int_0^T (Mv(t))^p\, dt < C\ ,\ v' \ \text{bounded in}\ L^2(Q)\}$$

has a compact closure in $L^p(Q)$.

Thus one can extract a subsequence of u_m, still denoted u_m, such that $\frac{\partial u_m}{\partial x}$ strongly converge in $L^p(Q)$ and almost everywhere in Q. That enables us to prove, provided a lemma due to LIONS (Lemme 1. 3, LIONS [5]), that Au_m weakly converges to Au in $L^p(0, T ; V')$ and consequently the existence of a solution by taking limit in P_{cm}.

Unicity can be shown in the case $2 \leq p \leq 4$ by a standard process.

Note that this problem is a generalization of [1] distinct from above in :

- The investigation of a priori estimates, which is not standard (use of an additional functional).

- The passage to the limit which involves a compacity result due to DUBINSKY [3].

Section II :

We suppose in this section that the material is composite and fills an open set Ω of R^n, with sufficiently smooth boundary and that Ω can be covered with a set of periods, each of them being homothetic with ratio ε, $\varepsilon > 0$ of a basic period Y. The problem we have to solve is then to find the limit law of comportment when the structure of the material becomes finer and finer. We study as an example of such situations the homogeneization of the following problem of diffusion.

Find a function $u_\varepsilon = u_\varepsilon(t, x)$ $(t, x) \in]0, T[\times \Omega$ satisfying in a generalized sense :

$$P_\varepsilon \quad \begin{cases} \dfrac{\partial u_\varepsilon}{\partial t} - \displaystyle\sum_{i,j=1}^{n} \frac{\partial}{\partial x_i}\left(a_{ij}\left(\frac{x}{\varepsilon}\right)\frac{\partial u_\varepsilon}{\partial x_j}\right) - \sum_{i=1}^{n} \frac{\partial}{\partial x_i}\left(\int_0^t \gamma(t-\sigma)\, g(u_\varepsilon(\sigma))\, d\sigma\right. \\ u_\varepsilon(0, x) = u_0(x) \\ u_\varepsilon = 0 \quad \text{sur } \Gamma . \end{cases}$$

Where the functions a_{ij} belong to $L^{\infty}(R^n)$ are Y-periodics and satisfy :

H_1 : $a_{ij} = a_{ji}$

H_2 : there exists a constant α positive, such that :

$$\sum_{i,j=1}^{n} a_{ij} \xi_i \xi_j \geq \alpha \sum_{i=1}^{n} \xi_i^2 \quad \forall \xi , \xi = (\xi_1, \ldots, \xi_n) \in R^n .$$

Let us give :

$V = H_o^1(\Omega)$

$H = L^2(\Omega)$ where we note (,) the scalar product

$$a_\varepsilon(u, v) = \sum_{i,j=1}^{n} \int_\Omega a_{ij}\left(\frac{x}{\varepsilon}\right) \frac{\partial u}{\partial x_j} \frac{\partial v}{\partial x_i} dx$$

$$b(u, v) = \sum_{i=1}^{n} \int_\Omega g(u) \frac{\partial v}{\partial x_i} dx .$$

We then write P_ε :

$$\left[\begin{array}{l} \text{Find } u_\varepsilon \in V \text{ such that :} \\ (u'_\varepsilon, v) + a_\varepsilon(u_\varepsilon, v) + \int_0^t \gamma(t-\sigma)\, b(u_\varepsilon(\sigma), v)\, d\sigma = (f, v) \\ u_\varepsilon(0, x) = u_o(x) . \end{array}\right.$$

Assume moreover that :

H_3 : g hölder real function defined on R

H_4 : γ real function defined on $(0, T)$, bounded

H_5 : $f \in L^2(0, T ; V)$, $u_o \in V$.

One has the :

<u>LEMMA 2-1</u> :

For a given ε, $\varepsilon > 0$, the solution u_ε of the problem P_ε satisfies

$$u_\varepsilon \in L^{\infty}(0, T ; V) \quad u'_\varepsilon \in L^2(0, T ; V') .$$

Remark : One can show that lemma by Galerkin approximation and passage to the limit by making use of the "uniform" coercitivity of A_ε, with $A_\varepsilon = -\frac{\partial}{\partial x_j}(a_{ij}(\frac{x}{\varepsilon})\frac{\partial}{\partial x_i})$. Moreover the standard estimates on $u_{\varepsilon m}(t)$ lead to u_ε bounded in $L^2(0,T;V)$ and u'_ε bounded in $L^2(0,T;V')$ independently of ε.

If the structure of the material becomes finer and finer i.e. if $\varepsilon \to 0$, one can show by LIONS-TARTAR [6, 8], the following result of homogeneization :

THEOREM 2-1 :

Assume the hypothesis H_1, H_2, H_3, H_4, H_5 hold, the solution u_ε to the problem P_ε weakly converges in V to u **solution of the problem** :

$$(u', v) + A(u,v) + \int_0^t \gamma(t-\sigma)\, b(u(\sigma), v)\, d\sigma = (f, v)$$
$$u(0,x) = u_0(x)$$

where :

$$A(u,v) = \sum_{i,j=1}^{n} \int_\Omega q_{ij} \frac{\partial u}{\partial x_j} \frac{\partial v}{\partial x_i} dx .$$

The coefficients q_{ij} are given by the following process ; we set :

$W(Y) = \{v \in H^1(Y)$ such that the values of V are equal on the opposite faces of $Y\}$

$a_Y(\varphi,\psi) = \sum_{i,j=1}^{n} \int_\Omega a_{ij}(y) \frac{\partial \varphi}{\partial y_j} \frac{\partial \psi}{\partial y_i} dy$ which is a coercive form on

$\dot{W}(Y) = W(Y)/R$.

Then Y_i being the function $y = (y_1, \ldots, y_n) \to y_i$, we define $X_i \in \dot{W}(Y)$ by $a_Y(X_i, \psi) = a_Y(y_i, \psi) \quad \forall \psi \in W(Y)$

and one has :

$$q_{ij} = \frac{1}{\textbf{meas } Y} a_Y(X_i - y_i, X_j - y_j) .$$

Proof :

Provided the remark following the lemma 2.1, one can extract a subsequence of u_ε still denote u_ε, such that u_ε weakly converges to u in

$L^2(0,T;V)$ and u'_ε weakly converges to u' in $L^2(0,T;V')$; then making use of the compacity of the embedding of V into H, we have that u'_ε strongly converges to u in $L^2(0,T;H)$. Now, the assumption g hölder enables us to show that $Bu_\varepsilon \underset{def}{=} -\sum_{i=1}^{i=n} \frac{\partial}{\partial x_i}\left(\int_0^t \gamma(t-\sigma)\, g(u_\varepsilon(\sigma))\, d\sigma\right)$ strongly converges to $Bu = -\sum_{i=1}^{i=n} \frac{\partial}{\partial x_i}\left(\int_0^t \gamma(t-\sigma)\, g(u(\sigma))\, d\sigma\right)$ in V'.

Let us denote $v(\Phi) = \int_0^T v(t)\,\Phi(t)\,dt$, with $\Phi \in C_0^\infty(]0,T[)$, $v \in L^2(0,T;V')$. We have :

$$\begin{cases} u'_\varepsilon(\Phi) = -u_\varepsilon(\Phi') \\ (A_\varepsilon u_\varepsilon)(\Phi) = A_\varepsilon(u_\varepsilon \Phi). \end{cases}$$

where A_ε is the operator associated to a_ε. Then, from lemma 2.1, we deduce :

$$(A_\varepsilon u_\varepsilon)(\Phi) = -u'_\varepsilon(\Phi) - (Bu_\varepsilon)(\Phi) + f(\Phi)$$

and :

$$A_\varepsilon(u_\varepsilon \Phi) = u_\varepsilon(\Phi) - (Bu_\varepsilon)(\Phi) + f(\Phi) \underset{def}{=} \psi_\varepsilon(\Phi).$$

Since u_ε and Bu_ε strongly converge in V' respectively to u and Bu, we have that $\psi_\varepsilon(\Phi)$ strongly converges in V' to $\psi(\Phi)$ with $\psi(\Phi) = u(\Phi) - (Bu)(\Phi) + f(\Phi)$. Let us remark now that $u_\varepsilon(\Phi)$ weakly converges to $u(\Phi)$ in V, we have, provided an homogeneization result due to LIONS-TARTAR, that $\mathcal{Q}(u(\Phi)) = \psi(\Phi)$, where $\mathcal{Q}$ is the operator associated to the bilinear form A and the theorem 2.1 as a consequence.

Remarks.

2-1 It seems possible by using recent results due to ARTOLA-DUVAUT [2] to find the homogeneized problems of problems of the following type :

$$u' - \sum_{i,j=1}^{n} \frac{\partial}{\partial x_i}\left(a_{ij}\left(\frac{x}{\varepsilon}\right)\frac{\partial u_\varepsilon}{\partial x_j}\right) - \sum_{i=1}^{n} \frac{\partial}{\partial x_i} \int_0^t \gamma\left(t-\sigma, \frac{x}{\varepsilon}, u_\varepsilon(\sigma)\right) d\sigma = f.$$

2-2 In problems with "bad" nonlinearities (as in (b) or (c)), it seems that the results are more difficult to obtain and involve asymptotical methods.

BIBLIOGRAPHY

[1] ARTOLA (M.). - Sur les perturbations des équations d'évolution. Application à des problèmes de retard. Annales E. N. S. 1969, t. 2, p. 137-253.

[2] ARTOLA (M.), DUVAUT (G.). - Sur l'homogénéisation de quelques problèmes non linéaires. EVANSTON, Juillet 1978.

[3] DUBINSKII (J. A.). - Convergence faible dans les équations elliptiques paraboliques non linéaires. Math. Sbornick 67 (109), 1965, p. 609-612.

DUVAUT (G.), ARTOLA (M.). - Cf. [2].

[4] LIONS (J. L.). - Equations différentielles opérationnelles et problèmes aux limites (Springer 1961).

[5] LIONS (J. L.). - Quelques méthodes de résolution de problèmes aux limites non linéaires (Dunod 1969).

[6] LIONS (J. L.). - Cours au Collège de France (1975-1976).

[7] RAYNAL (M. L.). - Thèse 1975.

[8] TARTAR (L.). - Cours au Collège de France (1976-1977).

ON CERTAIN BOUNDED SOLUTIONS OF A VOLTERRA INTEGRAL EQUATION

George Seifert
(Iowa State University, Ames, Iowa 50011)

We denote by R^n real Euclidean n-space, and if x and y are in R^n, define

$$\langle x,y\rangle = \sum_{i=1}^{n} x_i y_i, \quad \text{and} \quad |x| = \langle x,x\rangle^{1/2}.$$

Let $M \subset R^n$ be closed and convex. For each $x \in \partial M$, the boundary of M, we denote by $N(x)$ the set of all $u \in R^n$ such that

$$\lim_{k\to\infty} |x_k - y(x_k)|^{-1}(x_k - y(x_k)) = u$$

for some sequence $x_k \to x$ as $k \to \infty$, $x_k \notin M$; here $y(x_k)$ is the unique point in ∂M such that $\operatorname{dist}(x_k, M) = |x_k - y(x_k)|$.

Consider the integral equation in R^n:

$$x(t) = \int_0^t K(t,s,x(s))ds + f(t) \tag{1}$$

where K is continuous in (t,s,x) for $t \geq s$, $x \in R^n$, and f is continuous for $t \geq 0$.

The following theorem is related to a similar result, Corollary 1, in [1], and can be proved similarly.

THEOREM 1. Let $M \subset R^n$ be closed and convex, and $M_0 \subset M$ be closed and such that $M_0 \cap \partial M$ is nonempty. Suppose that

$$\langle K(t,t,y(t)),u\rangle + \limsup_{h\to 0+} h^{-1}\Big\{\int_0^t \langle K(t+h,s,y(s)) - K(t,s,y(s)),u\rangle ds + \langle f(t+h)-f(t),u\rangle\Big\} < 0 \tag{2}$$

for $t \geq 0$, $y(s)$ any continuous function on $0 \leq s \leq t$ to R^n with $y(s) \in M_0$, $y(t) \in \partial M \cap M_0$, and $u \in N(y(t))$. Then if $x(t)$ is a solution of (1) for $t \geq t_0 \geq 0$ such that $x(s) \in M_0$ for $0 \leq s \leq t_0$, then $x(t) \in M$ for $t \geq t_0$ as long as it exists. In particular if $f(0) \in M_0$, then any solution of (1) satisfies $x(t) \in M$ for $t \geq 0$ as long as it exists.

The following theorem is an easy consequence of Theorem 1.

THEOREM 2. Suppose M and M_0 are as in Theorem 1 and in addition to the hypotheses of this theorem, suppose that

(i) K is locally Lipschitz in x;

(ii) There exists a function $h(x)$ continuous on a neighborhood of M_0 to R^n such that $\langle h(x),u\rangle < 0$ for $x \in \partial M \cap M_0$, $u \in N(x)$; and

(iii) (2) holds with the strict inequality $<$ replaced by the non-strict in equality $\leq$.

Then the conclusion of Theorem 1 holds.

PROOF. We consider

$$(1\varepsilon) \quad x(t) = \int_0^t K(t,s,x(s))ds + f(t) + \varepsilon \int_0^t h(x(s))ds.$$

By standard methods, cf. [2], given $x_0(s)$ continuous on $0 \leq s \leq t_0$, there exists a $t_1 > t_0$ and a unique solution $x(t)$ of (1) on $t_0 \leq t \leq t_1$ satisfying $x(s) = x_0(s)$ on $0 \leq s \leq t_0$. Also it follows that for all $\varepsilon > 0$ and suffiently small, (1ε) has a solution $x(t,\varepsilon)$ on $t_0 \leq t \leq t_1$ such that $x(s,\varepsilon) = x_0(s)$ on $0 \leq s \leq t_0$ and $x(t,\varepsilon) \to x(t)$ as $\varepsilon \to 0+$ for $t_0 \leq t \leq t_1$. The hypotheses of Theorem 1 with K replaced by $K+\varepsilon h$ now hold since for $x \varepsilon \partial M$ and $u \in N(x)$, we have

$<K(t,t,x) + \varepsilon h(x),u> < <K(t,t,x),u>$, and hence (2) holds with K replaced by $K+\varepsilon h$.

Now if there exists a solution $x(t)$ of (1) such that $x(s) \varepsilon M_0$ for $0 \leq s \leq t_0$ and $x(t_1) \notin M$ for $t_1 > t_0$, then there exists an $\varepsilon > 0$ and a solution $x(t,\varepsilon)$ of (1ε) on $t_0 \leq t \leq t_1$ such that $x(t,\varepsilon) = x(t)$ on $[0,t_0]$ and $x(t_1,\varepsilon) \notin M$. But this contradicts Theorem 1 and proves Theorem 2.

The following corollary can be used in cases where ∂M is sufficiently smooth.

COROLLARY 1. Suppose all the hypotheses of Theorem 2 hold except (ii). Suppose for each $x \varepsilon \partial M \cap M_0$, the set $N(x)$ consists of a single vector $u(x)$ continuous on $\partial M \cap M_0$. Then the conclusion of Theorems 1 and 2 holds.

PROOF. By Dugundji's extension of Tietze's theorem [3], there exists a continuous function $h(x)$ on R^n to R^n such that $h(x) = -u(x)$ for $x \varepsilon \partial M \cap M_0$. Using this h as the h in Theorem 2, the proof is complete.

We now consider a special case of (1) of interest in certain applications to heat conduction problems; cf. [4]. Somewhat related results are due to Gripenberg [5] and Levin [6], [7], for the case where $n = 1$, i.e., (1) is a scalar equation.

In particular, we consider

$$(3) \qquad x(t) = \int_0^t B(t-s)g(x(s))ds + f(t);$$

here $B(t) = (b_{ij}(t))$ is an $n \times n$ matrix, $g(x)$ is a function on R^n to R^n, $f(t)$ is a function on $[0,\infty)$ to R^n, and the following properties are assumed.

(iv) $b_{ij}(0) \geq 0$ for $i \neq j$, $0 \leq f_i(0) \leq 1$; $g_i(x) = 0$ if $x_i = 0$, and $g_i(x) > 0$ if $x_i > 0$; here $x = (x_1,\ldots,x_n)$;

(v) there exist constants L_1, L_2, L_3 such that for $h > 0$ and small,

a) $0 \leq b_{ij}(t+h) - b_{ij}(t) \leq L_1 h$,

b) $0 \leq f_i(t+h) - f_i(t) \leq L_2 h$,

c) $|g_i(x) - g_i(y)| \le L_3|x-y|$, $i = 1,\dots,n$, and $x,y\varepsilon M_0 = \{x\varepsilon R^n : 0 \le x_i \le 1\}$.

We define $M_{0i} = \{x\varepsilon M_0 : x_i = 1\}$, (note that $M_{0i} \subset \partial M_0$)

$m_i = \min\{g_i(x) : x\varepsilon M_{0i}\}$, $m_{ij} = \max\{g_j(x) : x\varepsilon M_{0i}\}$, and
$b_0 = \max\{|g(x)| : x\varepsilon M_0\}$.

THEOREM 3. With the conditions and notation as above, assume that

(4) $$b_{ii}(0)m_i + \sum_{j\neq i} b_{ij}(0)m_{ij} + L_1 b_0 + L_2 \le 0, \quad i = 1,\dots,n.$$

Then for $x(t)$ a solution of (3), $x(t)\varepsilon M_0$ for all $t \ge 0$.

PROOF. We first note that $m_j > 0$ for $j = 1,\dots,n$, and that (iv) and (4) imply $b_{ii}(0) < 0$ for $i = 1,\dots,n$.

The first part of the proof consists of showing that the conclusion of Theorem 1 or 2 holds for the case where $M = M^+ = \{x\varepsilon R^n : x_i \ge 0, i = 1,\dots,n\}$ and M_0 is as in c) of (v). The smoothness conditions of Theorem 2 are easily verified. To satisfy (ii), we choose $h(x) = e = (1,\dots,1)$; note that for $x\varepsilon\partial M^+$, $N(x)$ consists of unit vectors u such that $u_i \le 0$ for $i = 1,\dots,n$. To check that (iii) holds, we observe that if $x\varepsilon\partial M^+, x_i > 0$, and $u\varepsilon N(x)$ then $u_i = 0$. From this and the fact that $g_j(x) = 0$ for $x_j = 0$ it follows that

(5) $$\langle B(0)g(x), u\rangle = \sum_{i=1}^{n}\sum_{j=1}^{n} b_{ij}(0)g_j(x)u_i = \sum_{i\neq j} b_{ij}(0)g_j(x)u_i \le 0$$

for $x\varepsilon\partial M^+$ and $u\varepsilon N(x)$. Thus the first term on the left in (2) is nonpositive. To see that the other terms are also, we use properties a) and b), and the conclusion of Theorem 1 holds for (3) with M^+, and $M_0 \subset M^+$.

The next part of the proof consists of showing that for each $i = 1,\dots,n$, the conclusion of Theorem 1 holds for $M = M_i = \{x\varepsilon R^n : x_i \le 1\}$ and our particular M_0. With M_{0i} as previously defined, we note that $M_{0i} = \partial M_i \cap M_0$. Also it follows easily that if $x\varepsilon\partial M_i$, then $N(x) = \{e_i\}$, $e_i = (0,\dots,1,0,\dots0)$, the 1 in the ith place. We may now use Corollary 1, provided (iii) is satisfied. But the left side of (2) is now just

(6) $$\sum_{j=1}^{n} b_{ij}(0)g_j(y(t)) + \limsup_{h\to 0+} h^{-1}\{\int_0^t \sum_{j=1}^{n}(b_{ij}(t+h-s) - b_{ij}(t-s))g_j(y(s))ds + f_i(t+h) - f_i(t)\};$$

using the facts that $y(t)\varepsilon M_{0i} = \partial M_i \cap M_0$ and $y(s)\varepsilon M_0$ for $0 \le s \le t$, properties a) and b) and hypothesis (4), it follows easily that (6) is nonpositive, and using Theorem 2 the conclusion of this theorem holds for $M = M_i$, $i = 1,\dots,n$.

To complete the proof we note that $M_0 = \bigcap_{i=1}^{n}(M_i \cap M^+)$, and that if a solution

$x(t)$ of (3) leaves M_0 at say $t_0 \geq 0$, it must therefore leave M^+ or some M_i at $t = t_0$; since this would contradict at least one of the two conclusions verified above, and since M_0 is a bounded set, our theorem is proved.

It is possible to obtain weaker conditions on $B(t)$, $f(t)$ and $g(x)$ under which the conclusion of Theorem 3 will hold. For the scalar case of (3) one need not suppose $B(t) = b_{11}(t)$ is defined at $t = 0$; i.e., one can have $b_{11}(t) = -t^{-1/2}$, a case of interest in certain heat conduction problems; cf. [5], [6]. Referring to the general result in [1] which can be used to prove our Theorem 1, we need to consider the inner product of

$$(7) \quad \limsup_{h \to 0+} h^{-1} \{\int_0^t (B(t+h-s) - B(t-s))g(y(s))ds + \int_t^{t+h} B(t+h-s)g(y(s))ds$$
$$+ f(t+h) - f(t)\}$$

with $u \varepsilon N(y(t))$ for $y(t) \varepsilon \partial M$ and $y(s) \varepsilon M$ for $0 \leq s \leq t$. It is easy to see that if M is bounded as in our case, g is continuous, and $B(t) = -t^{-1/2}$, (7) can certainly be finite, or in any case, its inner product with u could certainly be nonnegative.

Even if (3) is not scalar, the local integrability of the matrix $B(t)$ on $[0,\infty)$ and the monotonicity of its entries are the really crucial requirements.

Finally, results on the asymptotic behavior of solutions of (3) are obtained in [8] and [9].

REFERENCES

[1] G. Seifert, Positive Invariance for Closed Sets for Generalized Volterra Equations, (submitted for publication).

[2] R. K. Miller, Nonlinear Volterra Integral Equations, Benjamin, Inc. (1971).

[3] J. Dugundji, An Extension of Tietze's Theorem, Pac. J. Math. 1 (1972), 161-170.

[4] H. S. Carslaw and J. C. Jaeger, Conduction of Heat in Solids, 2nd ed., Oxford at Clarendon Press, New York (1959).

[5] G. Gripenberg, Bounded Solutions of a Volterra Equation, J. Diff. Eq. 28 (1), (1978), 18-22.

[6] J. J. Levin, On a Nonlinear Volterra Equation, J. Math. Analysis and Appl. 39 (1972), 458-476.

[7] ___________, A Bound on the Solutions of a Volterra Equation, Arch. Rat. Mech. Anal. 52 (1973), 339-349.

[8] D. G. Weis, Asymptotic Behavior of Some Nonlinear Volterra Integral Equations, J. Math. Anal. Appl. 49 (1975), 59-87.

[9] R. K. Miller, Almost-Periodic Behavior of Solutions of a Nonlinear Volterra System, Q. Appl. Math. (1971), 553-570.

HYPERBOLIC STRUCTURES FOR LINEAR VOLTERRA DIFFERENTIAL EQUATIONS

By

George R. Sell*
National Science Foundation
Washington, D.C. 20550

I. Introduction. One of the basic principals underlying much of the recent research in the general theory of differential equations is that "small" perturbations do not substantially alter the behavior of solutions. This has taken on many forms. One of the most interesting variants of this principal is a recent theorem of Kurzwei concerning differential-delay equations that are close to ordinary differential equations, [4]. With this in mind, it would seem reasonable to expect that the solutions of the linear Volterra differential equation

$$\text{(VDE)} \qquad x'(t) = A(t)\,x(t) + f(t) + \int_0^t B(t-s)x(s)ds$$

remain close to the solutions of the associated ordinary differential equation

$$\text{(ODE)} \qquad x'(t) = A(t)\,x(t) + f(t)$$

provided the kernel $B(\xi)$ is small.

As we shall see, there are significant differences between the flows generated by (VDE) and (ODE) for any kernel $B \neq 0$. However in the special case that the homogeneous ordinary differential equation

$$\text{(HODE)} \qquad x'(t) = A(t)\,x(t)$$

admits on exponential dichotomy one can show that in an appropriate sense these two

*Current address: School of Mathematics, University of Minnesota, Minneapolis, Minnesota 55455 USA

flows are close together provided the L_1-norm of B is sufficiently small.

The framework in which we shall study this problem is that of a linear skew-product flow and the corresponding spectral theory as developed in [11, 12, 14]. The basic theory of flows is reviewed in Section II. In Section III we shall show that (VDE) gives rise to a suitable flow, which is an extension of the flows generated by (ODE) and (HODE). The consequences of our assumption that (HODE) admit an exponential dichotomy will be analyzed in Sections IV and V. Finally in Section VI we shall indicate several directions for further research.

II. Flows. Let W be a topological space and R denote the real line. A flow on W is a continuous mapping $\pi : W \times R \to W$ that satisfies $\pi(w,o) = w$ and $\pi(\pi(w,s),t) = \pi(w,s+t)$ for all $w \in W$ and $s,t \in R$. If $W = X \times Y$ is a product space then π is a skew-product flow if, in addition, π has the form

(1) $$\pi(x,y,t) = (\varphi(x,y,t), \sigma(y,t))$$

where σ is a flow on Y. If $W = X \times Y$, where X is a topological vector space, then π is a linear skew-product flow (LSPF) if it π is a skew-product flow given by (1) where $\varphi(x,y,t)$ is linear in x. Notice that if π is a LSPF on $X \times Y$ then the mapping $\Phi(y,t)x = \varphi(x,y,t)$ is a continuous linear transformation of X onto X. The inverse of $\Phi(y,t)$ is given by

$$\Phi^{-1}(y,t) = \Phi(\sigma(y,t),-t) .$$

A very useful concept for the study of flows is the notion of a homomorphism. Let π_i be a flow on W_i for $i=1,2$. A mapping h between W_1 and W_2 is said to be a homomorphism if (i) h is a homeomorphism of W_1 onto W_2 and (ii) h satisfies

(2) $$h(\pi(w_1, t)) = \pi_2(h(w_1),t)$$

for all $w_1 \in W_1$ and $t \in R$. Statement (2) says that the mapping h (and h^{-1}) commutes with the flows.

We shall use the notation (W,π) to denote the space W with the flow π.

III. Volterra Differential Equations. Let us consider now the initial value problem

(3) $$x'(t) = A(t)\,x(t) + f(t) + \int_0^t B(t-s)x(s)ds$$
$$x(0) = x_0 ,$$

where $x_0 \in R^n$, Euclidean n-space. Let M^n denote the space of $(n \times n)$ matrices with real coefficients, and let $BUC = BUC(R,R^n)$ and $BUC(R,M^n)$ denote the spaces of bounded uniformly continuous functions from R to R^n and M^n, respectively. We assume that both of the BUC-spaces have the topology of uniform convergence on compact sets.

Let $\mathcal{C}$ denote a subset of $BUC(R,M^n)$. We shall say that $\mathcal{C}$ is translation-invariant if for every $A \in \mathcal{A}$ and $\tau \in R$ one has $A_\tau \in \mathcal{A}$ where $A_\tau(t) = A(\tau + t)$. We will now study (VDE) under the following three hypotheses:

(H1) $f \in BUC(R, R^n)$

(H2) $A \in \mathcal{A}$ where $\mathcal{A}$ is a compact subset of $BUC(R, M^n)$ that is translation-invariant.

(H3) B is continuous and $B \in L_1(-\infty, \infty)$.

An example of a set $\mathcal{A}$ which satisfies (H2) is

$$\mathcal{A} = \text{Hull}(B) = \text{Cl}\{B_\tau : \tau \in R\}$$

where Cl is the closure operation in $BUC(R, M^n)$, cf. [8, 11, 13].

Assume that (H1), (H2) and (H3) are satisfied. Let $\phi(t) = \phi(x_0, f, A, t)$ denote the unique solution of (VDE) that satisfies $\phi(0) = x_0$. With ϕ so determine we define $T_\tau f = T_\tau(x_0, f, A)$ by

$$T_\tau f(t) = f(\tau + t) + \int_0^\tau B(\tau + t - s)\phi(s)ds$$

The following theorem can now be proven by using the methods of [2, 6, 8]:

1 Theorem. The mapping

(4) $$\pi(x_0, f, A, \tau) = (\phi(\tau), T_\tau f, A_\tau)$$

defines a LSPF on $R^n \times BUC \times \mathcal{A}$ where $X = R^n \times BUC$ and $Y = \mathcal{A}$.

Notice that the vector coordinate space X is infinite dimensional for this flow π eventhough the initial vector x_0 belongs to a finite dimensional space. Let us now look at two special cases of (4). The first case describes the flow generated by (ODE) and the second is the flow generated by (HODE).

First Case: $B \equiv 0$. In this case the translation operator $T_\tau f$ becomes

$$T_\tau f(t) = f(\tau + t) = f_\tau(t)$$

i.e. $T_\tau f$ is independent of x_0 and A. Consequently the flow π reduces to

(5) $$\pi(x_0, f, A, \tau) = (\phi(\tau), f_\tau, A_\tau).$$

In this case, the flow π can be viewed as a LSPF in two ways:

(i) $X = R^n \times BUC$, $Y = \mathcal{A}$

(ii) $X = R^n$, $Y = BUC \times \mathcal{A}$

In the second way, the vector coordinate space X is finite dimensional. Notice that for $B \not\equiv 0$, the translate operator $T_\tau f$ depends on x_0, and consequently the second option with $X = R^n$ is not available under this formulation. We see therefore an essential difference between the flows generated by (VDE) and (ODE) when $B \not\equiv 0$.

Second Case: $B \equiv 0$, $f \equiv 0$. For $A \in \mathcal{A}$ let $\Phi(A, t)$ denote the fundamental matrix solution of (HODE) that satisfies $\Phi(A, 0) = I$. Since $B \equiv 0$ one has $\phi(x_0, 0, A, t) = \Phi(A, t)x_0$. Consequently

(6) $\pi(x_o,A,\tau) = (\Phi(A,\tau)x_o,A_\tau)$

is a LSPF on $R^n \times \mathcal{A}$, and the vector coordinate space $X = R^n$ is finite dimensional.

Our interest in difference between LSPF s with finite dimensional and infinite dimensional vector coordinate spaces goes beyond the superficial level. In the case of a "finite dimensional" LSPF, one has a highly developed spectral theory [12], which gives significant information about the qualitative structure of the flow. At the present time, it is not known whether this spectral theory admits a meaningful extension to an "infinite dimensional" LSPF. However a partial extension to the flow generated by (VDE) is sometimes possible, as we shall soon see.

IV. The ODE Flow. We return to the study of

(ODE) $x'(t) = A(t)x(t) + f(t)$

where $f \in BUC$, $A \in \mathcal{A}$ and Hypotheses (H1) and (H2) are valid. Let $M \subseteq \mathcal{A}$. A *projector* (over M) is a continuous mapping $\hat{P} : R^n \times M \to R^n \times M$ that has the form $\hat{P}(x,A) = (P(A)x,A)$ for all $(s,A) \in R^n \times M$, where $P(A)$ is a linear projection on R^n. We shall say that (HODE) admits an *exponential dichotomy over* M if there is a projector $\hat{P}$ over M and positive constants K and α such that

$$|\Phi(A,t)\,P(A)\,\Phi^{-1}(A,s)| \le Ke^{-\alpha(t-s)},\quad s \le t,$$

(7)

$$|\Phi(A,t)[I-P(A)]\Phi^{-1}(A,s)| \le Ke^{-\alpha(s-t)},\quad t \le s,$$

for all $A \in M$.

In addition to the above hypotheses we shall assume the following :

(H4) (HODE) *admits an exponential dichotomy over* $\mathcal{A}$.

Remark. (H4) means that (7) is valid for every $A \in \mathcal{A}$. The fact that the projector $\hat{P}$ is continuous means that the exponential dichotomy (7) varies continuously over $\mathcal{A}$, cf. [11]. Also we recall that if $\mathcal{A} = \mathrm{Hull}(B)$ for some $B \in BUC(R,M^n)$ and if (HODE) admits an exponential dichotomy over the one point set $M = \{B\}$, then (HODE) admits an exponential dichotomy over $\mathcal{A}$, cf. [11].

Define the stable and unstable manifolds for (HODE) to be respectively

$$\mathcal{S} = \{(x,A) \in R^n \times \mathcal{A} : \Phi(A,t)x \to o \text{ as } t \to +\infty\}$$

$$\mathcal{U} = \{(x,A) \in R^n \times \mathcal{A} : \Phi(A,t)x \to o \text{ as } t \to -\infty\}$$

and let

$$\mathcal{S}(A) = \{x \in R^n : (x,A) \in \mathcal{S}\}$$

$$\mathcal{U}(A) = \{x \in R^n : (x,A) \in \mathcal{U}\}$$

Clearly $\mathcal{S}$ and $\mathcal{U}$ are invariant sets for the flow π generated by (HODE). Under hypothesis (H4), $\mathcal{S}(A)$ and $\mathcal{U}(A)$ vary continuous for $A \in \mathcal{A}$, and the decay rate in $\mathcal{S}$ and $\mathcal{U}$ is exponential. Furthermore one has $R^n = \mathcal{S}(A) + \mathcal{U}(A)$ for all $A \in \mathcal{A}$. We will now show that this hyperbolic structure is inherited by (ODE).

The following result is well-known and easily verified:

2. Proposition. Assume that (H2) and (H4) are satisfied. Then for every $f \in BUC$ and $A \in \mathcal{C}$, there is a unique bounded solution $\varphi(f,A,t)$ of (ODE) which is given by

$$\varphi(f,A,t) = \int_{-\infty}^{t} \Phi(A,t)\, P(A)\, \Phi^{-1}(A,s) f(s) ds - \int_{t}^{\infty} \Phi(A,t)\, [I - P(A)]\, \Phi^{-1}(A,s) f(s) ds$$

Now define the set $M \subseteq R^n \times BUC \times \mathcal{C}$ by

$$M = \{(\varphi(f,A,0),f,A) : f \in BUC,\, A \in \mathcal{C}\}$$

Let $p : R^n \times BUC \times \mathcal{C} \to BUC \times \mathcal{C}$ be the natural projection and define $h : M \to BUC \times \mathcal{C}$ to the restriction of p to M. Because of Proposition 1, we see that h is a one-to-one mapping of M onto $BUC \times \mathcal{C}$. Let σ denote the flow on $BUC \times \mathcal{C}$ given by

$$\sigma(f,A,\tau) = (f_\tau, A_\tau) \quad .$$

We shall call this the σ - flow on $BUC \times \mathcal{C}$. The following result is now valid.

3. Proposition. The set M is an invariant set in the flow π generated by (ODE). In addition, the mapping $h : M \to BUC \times \mathcal{C}$ describes a homomorphism between the flow π on M and the σ - flow on $BUC \times \mathcal{C}$.

The flow π on M takes the form

$$\pi(\varphi(f,A,0),f,A,\tau) = (\varphi(f,A,\tau),f_\tau,A_\tau) \; .$$

Since M is invariant this means that one has

$$\varphi(f,A,\tau) = \varphi(f_\tau,A_\tau,0) \tag{8}$$

for all $(f,A) \in BUC \times \mathcal{C}$ and $\tau \in R$.

Let us now introduce new coordinate system on $R^n \times BUC \times \mathcal{C}$ by

$$x_o = x_n + \varphi(f,A,0)\,,$$
$$f = f \;,\; A = A \;.$$

Only the vector coordinate changes, but the change depends on f and A. One should think of x_o as the old coordinates and x_n as the new coordinates on R^n. Let q be the mapping $q\,(x_o,f,A) = (x_n,f,A)$ defined by this coordinate change. Clearly q is a homeomorphism.

Next consider the mapping $\hat{\pi}$ on $R^n \times B\ \ C \times \mathcal{C}$ given by

$$\hat{\pi}(x_n,f,A,\tau) = (\Phi(A,\tau)x_n, f_\tau, A_\tau)$$

It is easy to verify that $\hat{\pi}$ is a LSPF with vector coordinate $X = R^n$ and base space $Y = BUC \times \mathcal{C}$. We now have the following result :

4. Theorem. The mapping

$$q : (R^n \times BUC \times \mathcal{C}, \pi) \to (R^n \times BUC \times \mathcal{C}, \hat{\pi})$$

is a homomorphism.

Proof. We have already observed that q is a homeomorphism. In order to show that q commutes with the flows π and $\hat{\pi}$, we recall the superposition principle for (ODE), which in our notation becomes

$$(9) \qquad \phi(x_0,f,A,t) = \Phi(A,t)x_n + \varphi(f,A,t)$$

where $x_0 = x_n + \varphi(f,A,0)$. By combining (8) and (9) we have

$$q(\pi(x_0,f,A,t)) = q(\Phi(A,t)x_n + \varphi(f_\tau,A_\tau,0),f_t,A_t)$$
$$= (\Phi(A,t)x_n,f_t,A_t) = \hat{\pi}(x_n,f,A,t)$$
$$= \hat{\pi}(q(x_0,f,A),t) .$$

The flow $\hat{\pi}$ is especially simple. It is precisely the product of the (HODE) flow

$$\pi(x_n,A,\tau) = (\Phi(A,\tau)x_n,A_\tau)$$

on $R^n \times \mathcal{G}$, and the translational flow $\sigma(f,\tau) = f_\tau$ on BUC. Thus the homomorphism q decouples the flow generated by (ODE) into two smaller components.

The invariant set M in $(R^n \times BUC \times \mathcal{G})$ has a hyperbolic structure. More precisely, define S and U by

$$S = \{(x_0,f,A) \in R^n \times BUC \times \mathcal{G} : x_n \in \mathcal{S}(A)\}$$
$$U = \{(x_0,f,A) \in R^n \times BUC \times \mathcal{G} : x_n \in \mathcal{U}(a)\}$$

where $x_0 = x_n + \varphi(f,A,0)$. Then q maps S and U onto

$$\hat{S} = \{(x_n,f,A) \in R^n \times BUC \times \mathcal{G} : x_n \in \mathcal{S}(A)\}$$
$$\hat{U} = \{(x_n,f,A) \in R^n \times BUC \times \mathcal{G} : x_n \in \mathcal{U}(A)\}$$

Clearly $\hat{S}$ and $\hat{U}$ are invariant sets for the flow $\hat{\pi}$ on $R^n \times BUC \times \mathcal{G}$. Since q is a homomorphism, the sets S and U are invariant for the flow π. For $(f,A) \in BUC \times \mathcal{G}$ let

$$S(f,A) = \{x_0 \in R^n : (x_0,f,A) \in S\}$$
$$U(f,A) = \{x_0 \in R^n : (x_0,f,A) \in U\} .$$

Then (9) implies that

$$S(f,A) = \varphi(f,A,0) + \mathcal{S}(A)$$
$$U(f,A) = \varphi(f,A,0) + \mathcal{U}(A)$$

for all $(f,A) \in BUC \times \mathcal{G}$. In addition (9) and (H4) imply that for $x_0 \in S(f,A)$ one has

$$|\phi(x_0,f,A,t) - \varphi(f,A,t)| \leq K|x_n| e^{-\alpha t} , \quad t \geq 0$$

where $x_0 = x_n + \varphi(f,A,0)$, and likewise for $x_0 \in U(f,A)$ one has

$$|\phi(x_0,f,A,t) - \varphi(f,A,t)| \leq K|x_n| e^{\alpha t} , \quad t \leq 0 .$$

V. <u>The VDE Flow</u>. We now consider (VDE) where the four hypotheses (H1) - (H4) are satisfied. We assume that K and α have been chosen so that (7) is valid for all $A \in \mathcal{G}$. Next we let

$$\beta = \int_{-\infty}^{\infty} |B(\xi)| d\xi$$

and we assume that

$$\text{(H5)} \quad \beta < \alpha(4K)^{-1} .$$

5. Theorem. Assume that (H1)-(H5) are satisfied. Then for every $(f,A) \in BUC \times \mathcal{G}$, there is a unique bounded solution $\varphi^B(f,A,t)$ of (VDE).

We emphasize that the solution φ^B is uniformly bounded for all t, $-\infty < t < \infty$. The proof of this theorem is accomplished by showing that the operator

$$\mathcal{J}\varphi(t) = \int_{-\infty}^{t} \Phi(A,t)\, P(A)\, \Phi^{-1}(A,s)[f(s) + \int_0^s B(s-r)\,\varphi(r)\,dr]\,ds$$

$$- \int_t^{\infty} \Phi(A,t)[I - P(A)]\Phi^{-1}(A,s)[f(s) + \int_0^s B(s-r)\,\varphi(r)\,dr]\,ds$$

is a contraction mapping on the Banach space of bounded continuous functions on R. Therefore $\mathcal{J}$ has a fixed point, which in turn is a solution of (VDE).

With φ^B given by Theorem 3, we next define M^B as

$$M^B = \{(\varphi^B(f,A,0),f,A) : (f,A) \in BUC \times \mathcal{G}\}$$

The uniqueness of φ^B assures us that the mapping $h : M^B \to BUC \times \mathcal{G}$ given by

$$h : (\varphi^B(f,A,0),f,A) \to (f,A)$$

is a one-to-one mapping of M^B onto $BUC \times \mathcal{G}$. Moreover one can show that h is a homeomorphism.

Define a mapping $\sigma^B : BUC \times \mathcal{G} \times R \to BUC \times \mathcal{G}$ by

$$\sigma^B(f,A,\tau) = (T^B_\tau f, A_\tau)$$

where

$$T^B_\tau f(t) = f(\tau + t) + \int_0^\tau B(\tau + t - s)\,\varphi^B(f,A,s)\,ds \quad .$$

Because of the uniqueness of φ^B and the fact that $\varphi^B(f,A,t)$ depends continuous on (f,A,t) it is not difficult to verify that σ^B is a flow on $BUC \times \mathcal{G}$. We now have the following extension of Proposition 2.

6. Theorem. Assume that (H1)-(H5) are valid. Then the set M^B is an invariant set for the flow π generated by (VDE). Furthermore the mapping $h : (M^B,\pi) \to (BUC \times \mathcal{G}, \sigma^B)$ is a homomorphism.

The flow π on M^B takes the form

$$\pi(\varphi^B(f,A,0),f,A,\tau) = (\varphi^B(f,A,\tau), T^B_\tau f, A_\tau)$$

Since M^B is invariant this means that one has

$$\varphi^B(f,A,\tau) = \varphi^B(T^B_\tau f, A_\tau, 0)$$

for all $(f,A) \in BUC \times \mathcal{G}$ and $\tau \in R$.

Now define the following subsets of $R^n \times BUC \times \mathcal{G}$ and $R^n \times \mathcal{G}$:

$$S^B = \{(x,f,A) : |\phi(x,f,A,t) - \varphi^B(f,A,t)| \to 0 \text{ as } t \to +\infty\}$$

$$U^B = \{(x,f,A) : |\phi(x,f,A,t) - \varphi^B(f,A,t)| \to 0 \text{ as } t \to -\infty\}$$

$$\mathcal{S}^B = \{(x,A) : \phi(x,o,A,t) \to o \text{ as } t \to +\infty\}$$

$$\mathcal{U}^B = \{(x,A) : \phi(x,o,A,t) \to o \text{ as } t \to -\infty\} ,$$

as well as the following subsets of R^n:

$$S^B(f,A) = \{x : (x,f,A) \in S^B\}$$
$$U^B(f,A) = \{x : (x,f,A) \in U^B\}$$
$$\mathcal{S}^B(A) = \{x : (x,A) \in \mathcal{S}^B\}$$
$$\mathcal{U}^B(A) = \{x : (x,A) \in \mathcal{U}^B\}$$

By using the change of variables

$$x_o = x_n + \varphi^B(f,A,0)$$
$$f = f, \ A = A,$$

together with the superposition principle for (VDE), one can easily verify the following:

7. *Proposition. For all* $(f,A) \in BUC \times \mathcal{G}$ *one has*

$$S^B(f,A) = \varphi^B(f,A,0) + \mathcal{S}^B(A)$$
$$U^B(f,A) = \varphi^B(f,A,0) + \mathcal{U}^B(A)$$

In addition one has the following result which describes the hyperbolic structure of M^B :

8. *Theorem. Assume that* (H1) - (H5) *are valid. Let* $\mathcal{S}$ *and* $\mathcal{U}$ *be the stable and unstable manifolds generated by* (HODE). *Then for all* $A \in \mathcal{G}$ *one has the following facts*:

(i) $\mathcal{S}^B(A)$ *and* $\mathcal{U}^B(A)$ *are linear subspaces of* R^n .

(ii) $\mathcal{S}^B(A) \cap \mathcal{U}^B(A) = \{0\}$

(iii) $R^n = \mathcal{S}^B(A) + \mathcal{U}^B(A)$

(iv) $\dim \mathcal{S}^B(A) = \dim \mathcal{S}(A)$

(v) $\dim \mathcal{U}^B(A) = \dim \mathcal{U}(A)$

(vi) *There are constants* $\hat{K} > 0$ *and* $\beta > 0$ *(independent of* A *) such that*

$$|\phi(x,o,A,t)| \leq \hat{K}\,|x|\,e^{-\beta t} \qquad , t \geq 0$$

for all $x \in \mathcal{S}^B(A)$ *and*

$$|\phi(x,o,A,t)| \leq \hat{K}\,|x|\,e^{\beta t} \qquad , t \leq 0$$

for all $x \in \mathcal{U}^B(A)$.

The proof is this is accomplished by using the methods of [3, Chapter 5] and [14, Theorem 1]. We shall give a brief outline of the proof here.

For $A \in \mathcal{G}$ we let $Z = Z(A)$ denote the space of all bounded linear transformations

$$\psi(t) : \mathcal{S}(A) \to R^n \qquad , t \geq 0 ,$$

that are continuous in t and satisfy

$$\sup_{t \geq 0} \|\psi(t)\| < +\infty .$$

Now define a mapping $\mathcal{J}$ on Z formally by

$$\mathcal{J}\psi(t)\xi = \Phi(A,t)\xi + \Phi(A,t)\int_o^t P(A)\Phi^{-1}(A,s)[\int_o^s B(s-r)\,\psi(r)\xi\,dr]ds$$
$$- \Phi(A,t)\int_t^\infty [I - P(A)]\,\Phi^{-1}(A,s)[\int_o^s B(s-r)\,\psi(r)\,\xi\,dr]\,ds$$

where $\xi \in S(A)$ and $t \geq 0$. One shows that $\mathcal{J}$ is well-defined and that it maps Z into itself. Also one shows that the contraction mapping theorem is applicable and therefore $\mathcal{J}$ has a unique fixed point, say $\psi(t)$ satisfies $\mathcal{J}\psi = \psi$. It is easily seen that for each $\xi \in S(A)$, the function $\psi(t)\xi$ is a solution of (VDE) with $f \equiv 0$. Next one verifies that

$$P(A)\,\psi(o)\,\xi = \xi$$

for all $\xi \in S(A)$. Therefore the linear subspace

$$\{\psi(o)\xi : \xi \in S(A)\}$$

is isomorphic to $\mathcal{S}(A)$. Finally by deriving appropriate growth estimates one verifies that

$$\mathcal{S}^B(A) = \{\varphi(0)\xi : \xi \in \mathcal{S}(A)\} .$$

and that $\|\psi(t)\| \leq \hat{K} e^{-\beta t}$ for $t \geq o$, where $\hat{K}$ and β are appropriately chosen positive numbers. The proof of this last assertion is a straight-forward adaptation of [14, pp. 372-373].

Remark 1. We have studied (VDE) for the case where the "input" function $f(t)$ lies in BUC. Proposition 1 and Theorem 3 are variations on the characterization of Perron-stability, which says that "bounded input implies bounded output". By using the admissibility theory of Massera and Schaffer [5] one can replace BUC with other Banach spaces, for example the L_p - spaces. Also see [1,10] for recent related developments.

2. The spectral theory of [12] can be extended to (VDE) for small values of $\beta = \|B\|_1$. More precisely, the stable and unstable manifolds $\mathcal{S}^B(A)$ and $\mathcal{U}^B(A)$ can be decomposed into linear subspaces which are isomorphic to the spectral subspaces described in [12]. Furthermore the Lyapunov exponents for the solutions $\phi(x_o, o, A, t)$ can be estimated in terms of the associated spectral intervals and the number β.

VI. Directions for Further Research. One important difference between the flows generated by (ODE) and (VDE) when $B \neq 0$ is that the subspace

$$\Gamma = \{(x,f,A) \in R^n \times BUC \times \mathcal{G} : f = 0\}$$

is an invariant set for the flow generated by (ODE). It is not an invariant set for the flow generated by (VDE) when $B \neq 0$. The set Γ is, of course, homeomorphic to $R^n \times \mathcal{G}$. An interesting problem is to determine whether the (VDE) flow has an invariant set $\Gamma(B)$, which is homeomorphic to $R \times \mathcal{G}$ and which satisfies $\Gamma(B) \to \Gamma$ as $\beta = \|B\|_1 \to 0$.

2. Can the results of Proposition 3 and Theorem 8 be extended to systems where (HODE) does not admit an exponential dichotomy? Perhaps the spectral theory methods of [12, 14] can be used for this analysis.

3. Can the results of Proposition 3 and Theorem 8 be extended to kernels B which are not in $L_1(R,R)$ or which are not of convolution - type? (The dynamical theory developed in [8] can be used to study (VDE) with general kernels $B = B(t,s)$.)

Bibliography

1. H.A. Antosiewicz, A general approach to linear problems for nonlinear ordinary differential equations, "Ordinary Differential Equations," pp. 3 - 9, Academic Press, New York, 1972.

2. V. Barbu and S.I. Grossman, Asymptotic behavior of linear integrodifferential systems, Trans. Amer. Math. Soc. 171 (1972), 277 - 288.

3. W.A. Coppel, "Stability and Asymptotic Behavior of Differential Equations," Heath, Boston, 1965.

4. J. Kurzweil, Exponentially stable integral manifolds, averaging principle and continuous dependence on a parameter, Czechoslovak. Math. J. 16(91)(1966), 380-423 and 16(91)(1966), 463 - 492.

5. J.L. Massera and J.J. Schäffer, "Linear Differential Equations and Function Spaces," Academic Press, New York, 1966.

6. R.K. Miller, Linear Volterra integrodifferential equations as semigroups, Funk. Ekvac. 17(1974), 35 - 51.

7. R.K. Miller and G.R. Sell, Existence, uniqueness and continuity of solutions of integral equations, Ann. Mat. Pura Appl. 80 (1968), 135 - 152.

8. R.K. Miller and G.R. Sell, "Volterra Integral Equations and Topological Dynamics," Memoir No. 102, Amer. Math. Soc., Providence, R.I., 1970.

9. R.K. Miller and G.R. Sell, Topological dynamics and its relation to integral equations and nonautonomous systems, "Dynamical Systems. An International Symposium Vol. I," pp. 223 - 249, Academic Press, New York, 1976.

10. R.J. Sacker, The splitting index for linear differential systems, (to appear).

11. R.J. Sacker and G.R. Sell, Existence of dichotomies and invariant splittings for linear differential systems I, II and II , J. Differential Equations 15(1974), 429-548; 22(1976), 478-496; 22(1976), 497-522.

12. R.J. Sacker and G.R. Sell, A spectral theory for linear differential systems, J. Differential Equations 27(1978), 320-358.

13. G.R. Sell, "Topological Dynamics and Ordinary Differential Equations," Van Nostrand - Reinhold, London, 1971.

14. G.R. Sell, The structure of a flow in the vicinity of an almost periodic motion, J. Differential Equations 27(1978), 359-393.

A NONLINEAR VOLTERRA INTEGRAL EQUATION WITH SQUARE INTEGRABLE SOLUTIONS

Olof J. Staffans
Helsinki University of Technology
Institute of Mathematics
SF-02150 Espoo 15, Finland

1. Introduction and Summary of Results

We study the asymptotic behavior of the solutions of the integro-differential equation

$$x'(t) + \int_0^t a(t-s)g(x(s))ds = f(t) \quad (t \in \mathbb{R}^+). \tag{1.1}$$

Here $\mathbb{R}^+ = [0,\infty)$, a, g and f are given real continuous functions, and x is the unknown solution. In particular, we give sufficient conditions which imply that x is square integrable.

Our assumptions are the following (the notations are explained below):

(1.2) a *is continuous and strongly positive definite on* $\mathbb{R}^+$,

(1.3) $a-a(\infty) \in L^1(\mathbb{R}^+)$, $a' \in L^1 \cap BV(\mathbb{R}^+)$,

(1.4) f, $f' \in L^2(\mathbb{R}^+)$,

(1.5) $g \in C(\mathbb{R})$, $\xi g(\xi) > 0 \quad (\xi \neq 0)$,

(1.6) $-\int_{-\infty}^0 g(\xi)d\xi = \int_0^\infty g(\xi)d\xi = \infty$,

(1.7) $\liminf_{\xi\to 0} g(\xi)/\xi > 0$,

(1.8) $\limsup_{\xi\to 0} g(\xi)/\xi < \infty$.

Here the strong positive definiteness of a means that there should exist $\varepsilon > 0$ such that $a(|t|) - \varepsilon e^{-|t|}$ $(t \in \mathbb{R})$ is a (Bochner) positive definite function. The statements concerning a' and f' should be interpreted as requirements that a, f be locally absolutely continuous, together with the additional conditions on the derivatives.

BV stands for functions of bounded variation.

Theorem 1. *(i) Let* (1.1) - (1.5) *hold. Define*

(1.9) $\psi(t) = a(\infty) \int_0^t g(x(s))ds \quad (t \in \mathbb{R}^+)$.

Then

(1.10) $x', \psi \in L^\infty(\mathbb{R}^+)$

and

(1.11) $x' + \psi \in L^2(\mathbb{R}^+)$.

(ii) In addition, let (1.6) *hold. Then*

(1.12) $x \in L^\infty(\mathbb{R}^+)$.

(iii) If moreover (1.7) *holds, then*

(1.13) $x, x', \psi \in L^2(\mathbb{R}^+)$

and

(1.14) $x(t), x'(t), g(x(t)), \psi(t) \to 0 \quad (t \to \infty)$.

(iv) Finally, let (1.1) - (1.8) *hold. Then*

(1.15) $x'', \varphi \in L^2(\mathbb{R}^+)$,

where $\varphi(t) = g(x(t)) \quad (t \in \mathbb{R}^+)$.

In Theorem 1 the statements (i) and (ii) are not as sharp as they could be. More precisely, the hypotheses could be weakened slightly, and one could make (1.14) a part of (ii) by using results e.g. from [3]. The most interesting part of Theorem 1 is the claim (1.13). When $a(\infty) = 0$, then Theorem 1 is essentially a special case of [6, Theorem 4.3]. The proof of the case $a(\infty) \neq 0$ is similar to the proof in [6] of the case $a(\infty) = 0$, but an additional step is needed. This additional step has been inspired by MacCamy's [1, Thm II] (although the actual argument is quite different from the argument in [1]). MacCamy studies an abstract integrodifferential Volterra equation, and our proof of Theorem 1 can be generalized to the abstract situation. Generally speaking, our hypothesis is much weaker that MacCamy's hypothesis, but on the other hand, our conclusion is also weaker (e.g., MacCamy gets $x(t) = O(t^{-m}) \quad (t \to \infty)$ for some integer $m > 0$). We shall return elsewhere [7] to the question of sufficient conditions for MacCamy's stronger conclusion.

2. Proof of Theorem 1

Throughout in the proof of Theorem 1 we write $b = a - a(\infty)$. The strong positive definiteness of a implies that $a(\infty) \geq 0$, and that also b is strongly positive definite (cf. [2, Cor. 2.1] and [5, Remark 3.3]).

Clearly, (1.1) is equivalent to

$$(2.1) \qquad x'(t) + \psi(t) + w(t) = f(t) \quad (t \in \mathbb{R}^+),$$

where

$$(2.2) \qquad w(t) = \int_0^t b(t-s)g(x(s))ds \quad (t \in \mathbb{R}^+).$$

(i) It follows from (1.1), (1.2), (1.4), (1.5) and [6, Thm 1.1 and Prop 4.1] (cf. the proof of [6, Thm 4.3]) that

$$(2.3) \qquad \sup_{T \in \mathbb{R}^+} \int_0^T g(x(t)) \int_0^t a(t-s)g(x(s))ds\, dt < \infty,$$

$$(2.4) \qquad \sup_{T \in \mathbb{R}^+} \int_0^{x(T)} g(\eta)d\eta < \infty.$$

Substitute $a = a(\infty)+b$ in (2.3), and use the positive definiteness of b to get

$$(2.5) \qquad \sup_{T \in \mathbb{R}^+} a(\infty)[\int_0^T g(x(t))dt]^2 < \infty,$$

$$(2.6) \qquad \sup_{T \in \mathbb{R}^+} \int_0^T g(x(t)) \int_0^t b(t-s)g(x(s))ds\, dt < \infty.$$

By (1.2), (1.3), (2.2), (2.6) and two inequalities for positive definite functions (see [3, Lemma 6.1] and [4, Lemma 1 and Thm 2(ii)]),

$$(2.7) \qquad w \in L^\infty \cap L^2(\mathbb{R}^+).$$

Observe that (1.4) implies $f \in L^\infty(\mathbb{R}^+)$. Hence (1.10), (1.11) follow from (1.4), (1.9), (2.1), (2.5) and (2.7).

(ii) Obviously, (1.6) and (2.4) imply (1.12).

(iii) Differentiating (1.1) we get for almost all $t \in \mathbb{R}^+$,

$$(2.8) \qquad x''(t) + a(0)g(x(t)) + v(t) = f'(t),$$

where

$$(2.9) \qquad v(t) = \int_0^t a'(t-s)g(x(s))ds \quad (t \in \mathbb{R}^+).$$

The same argument as in [6, p. 85] yields

$$(2.10) \qquad v \in L^2(\mathbb{R}^+)$$

(one bounds the L^2-norm of v by using (2.6), the strong positive definiteness of b, and (1.3) which implies that the Fourier transform $(a')^\wedge$ of a' satisfies $(a')^\wedge(\omega) \le C(1+|\omega|)^{-1}$ for some constant C and all $\omega \in \mathbb{R}$). Multiply (2.8) by $x(t)$, and integrate over $(0,t)$ to get

$$a(0) \int_0^T x(t)g(x(t))dt - \int_0^T (x'(t))^2 dt = x(0)x'(0)$$
$$- x(T)x'(T) + \int_0^T x(t)(f'(t)-v(t))dt.$$

Hence by (1.4), (1.10), (1.12), (2.10) and the Schwarz inequality,

$$(2.11) \quad a(0) \int_0^T x(t)g(x(t))dt - \int_0^T (x'(t))^2 dt \le C(1 + [\int_0^T (x(t))^2 dt]^{1/2}),$$

where C is some constant independent of T.

To simplify the notations we shall below use the letter C in the same way as in (2.11), namely C represents a constant independent of T. The actual value of C may change from one line to the next.

Multiply (2.1) by $x'(t)$, integrate over $(0,T)$, and use (1.9) to get

$$\int_0^T (x'(t))^2 dt - a(\infty) \int_0^T x(t)g(x(t))dt = -a(\infty)x(T) \int_0^T g(x(s))ds$$
$$+ \int_0^T x'(t)(f(t)-w(t))dt.$$

Hence by (1.4), (1.9), (1.10), (1.12), (2.7) and the Schwarz inequality,

$$(2.12) \quad \int_0^T (x'(t))^2 dt - a(\infty) \int_0^T x(t)g(x(t))dt \le C(1 + [\int_0^T (x'(t))^2 dt]^{1/2}).$$

The strong positive definiteness of b implies $b(0) > 0$, hence $a(0) > a(\infty)$. Choose any number $\lambda > 1$ such that $a(0) > \lambda a(\infty)$, multiply (2.12) by λ, and add the result to (2.11). This gives

$$(a(0)-a(\infty))\int_0^T x(t)g(x(t))dt + (\lambda-1)\int_0^T (x'(t))^2dt$$

$$\leq C(1+\lambda+[\int_0^T (x(t))^2dt]^{1/2} + [\int_0^T (x'(t))^2dt]^{1/2}),$$

or equivalently (redefine C),

$$\int_0^T x(t)g(x(t))dt + \int_0^T (x'(t))^2dt$$

$$\leq C(1+[\int_0^T x(t))^2dt]^{1/2} + [\int_0^T (x'(t))^2dt]^{1/2}).$$

It follows from (1.5), (1.7) and (1.12) that

$$(x(t))^2 \leq Cx(t)g(x(t)) \quad (t \in \mathbb{R}^+),$$

and so we finally get

$$\int_0^T (x(t))^2dt + \int_0^T (x'(t))^2dt$$

$$\leq C(1+[\int_0^T (x(t))^2dt]^{1/2} + [\int_0^T (x'(t))^2dt]^{1/2}).$$

This implies

(2.13) $x\ ,\ x' \in L^2(\mathbb{R}^+).$

Combining (1.4), (2.1), (2.7) and (2.13) one gets (1.13).

That $x(t) \to 0$, hence also $g(x(t)) \to 0$ $(t \to \infty)$ follows from (1.5), (1.13). Combining (2.8) with (1.4), (1.5), (1.12) and (2.10) one observes that x' is uniformly continuous, which together with (1.13) yields $x'(t) \to 0$ $(t \to \infty)$. To show that also $\psi(t) \to 0$ $(t \to \infty)$ one finally uses (1.4) (which implies $f(t) \to 0$ $(t \to \infty)$), (2.1), (2.7) and the fact that w is uniformly continuous ($w' = b(0)g(x(t)) + v(t) \in L^\infty(\mathbb{R}^+) + L^2(\mathbb{R}^+)$). This completes the proof of (iii).

(iv) Clearly (1.5), (1.8) and (1.12) imply

$$|g(x(t))| \leq C\,|x(t)| \quad (t \in \mathbb{R}^+),$$

which combined with (1.13) yields the second half of (1.15). This combined with (1.4), (2.8) and (2.10) gives our last conclusion $x'' \in L^2(\mathbb{R}^+)$, and completes the proof of Theorem 1.

References

[1] R. C. MacCamy, A model for one-dimensional, nonlinear viscoelasticity, Quart. Appl. Math. 35 (1977), 21-33.

[2] J. A. Nohel and D. F. Shea, Frequency domain methods for Volterra equations, Advances in Math. 22 (1976), 278-304.

[3] O. J. Staffans, Positive definite measures with applications to a Volterra equation, Trans. Amer. Math. Soc. 218 (1976), 219-237.

[4] O. J. Staffans, An inequality for positive definite Volterra kernels, Proc. Amer. Math. Soc. 58 (1976), 205-210.

[5] O. J. Staffans, Systems of nonlinear Volterra equations with positive definite kernels, Trans. Amer. Math. Soc. 228 (1977), 99-116.

[6] O. J. Staffans, Boundedness and asymptotic behavior of solutions of a Volterra equation, Michigan Math. J. 24 (1977), 77-95.

[7] O. J. Staffans, A nonlinear Volterra equation with rapidly decaying solutions (to appear).

AN ABSTRACT VOLTERRA STIELTJES-INTEGRAL EQUATION

C. C. Travis
Health and Safety Research Division
Oak Ridge National Laboratory
Oak Ridge, Tennessee 37830

1.0 Introduction. Our objective in this paper is to establish local existence of solutions to the abstract Volterra Stieltjes-integral equation

(1.1) $$u(t) = f(t) + \int_0^t d_s K(t-s) \int_0^s g(s-r, u(r))dr.$$

In equation (1.1) $K(t)$, $t > 0$, is a semigroup of bounded linear operators in a Banach space X and g is, in general, a nonlinear operator from $R^+ \times X \to X$.

Many equations with physical applications can be written in the form (1.1). In particular, we will use equation (1.1) to establish local existence of solutions to a class of partial differential equations arising in the theory of heat flow in materials with memory. As a model for this class one may take the equation

(1.2) $$w_t(x,t) = w_{xx}(x,t) + \int_0^t a(t-s)\sigma(w_{xx}(x,s))ds + h(x,t),\ 0 \le x \le \pi,\ t \ge 0,$$
$$w(0,t) = w(\pi,t) = 0,\ t \ge 0,$$
$$w(x,0) = \phi(x),\ 0 \le x \le \pi,$$

where a: $R^+ \to R$ is of local bounded variation, σ: $R \to R$ is continuous, and h: $[0,\pi] \times R^+ \to R$ is continuous in the first variable and of local bounded variation in the second variable.

2.1 Local Existence. In this section we establish local existence of a solution to the Volterra Stieltjes-integral equation (1.1). We make the following assumptions concerning this equation:

(2.1) $K(t)$, $t \ge 0$, is a strongly continuous semigroup of bounded linear operators in X with infinitesimal generator A satisfying $||K(t)\,x|| \le Me^{\omega t}||x||$ for $t \ge 0$, $x \in X$, where M and ε are real constants;

(2.2) $K(t)X \subset D(A)$;

(2.3) $K(t)$: $X \to X$ is compact for each $t > 0$;

(2.4) f: $R^+ \to X$ is continuous;

(2.5) g: $R^+ \times D \to X$ is continuous where D is an open subset of X, and is of local strong bounded variation in its first variable;

(2.6) given $\varepsilon > 0$ and $T > 0$ there exists $\delta > 0$ such that $x, y \in D$ and $||x-y|| < \delta$ implies $\nu(T; x,y) < \varepsilon$ where $\nu(T; x,y)$ is the total variation of $g(\cdot,x)-g(\cdot,y)$ on the interval $[0,T]$.

(2.7) g is locally equicontinous on bounded subsets of D.

Remark. Conditions (2.6) and (2.7) are satisfied if either of the following conditions is satisfied:

(2.8) g is differentiable with respect to its first variable and the derivative g_1: $R^+ \times D \to X$ is continuous;

(2.9) $g(t,x) = a(t)\ f(x)$ where a: $R^+ \to R$ is continuous and of local bounded variation and f: $D \to X$ is continuous.

Proposition 2.1. Suppose (2.1)-(2.7) hold and $f(0) \varepsilon D$. Then there exists $T > 0$ and a continuous function u: $[0,T] \to X$ such that u satisfies (1.1) on $[0,T]$.

Proof. For $\sigma > 0$, let $N_\sigma = \{\psi \ \varepsilon\ X: \ ||g(s,\psi)|| \leq \sigma\}$. Choose $\sigma > 0$ and $T > 0$ such that

(2.10) $N_\sigma \subset D$;

(2.11) $||g(s,\psi)|| \leq N$ for $s \ \varepsilon\ [0,T]$ and $\psi \ \varepsilon\ N_\sigma$;

(2.12) $||f(t)-f(0)|| < \sigma/2$ for $t \ \varepsilon\ [0,T]$;

(2.13) $||\int_0^t d_sK(t-s)\int_0^s g(s-r,\ \psi)dr|| < \sigma/2$ for $t \ \varepsilon\ [0,T]$ and $\psi \varepsilon N_\sigma$.

Let C be the Banach space $C([0,T];X)$ with the norm $||\eta||_C = \sup\limits_{0 \leq t \leq T} ||\eta(t)||$, and let K be the closed bounded convex subset of C defined by

$$K = \{\eta \ \varepsilon\ C: \ ||\eta - f||_C \leq \sigma/2\}\ .$$

Consider the transformation G: $K \to C$ defined by

$$(G\eta)(t) = f(t) + \int_0^t d_sK(t-s)\ \int_0^s g(s-r,\eta(r))dr\ .$$

Notice that G: $K \to K$ since $G\eta$ is continuous as a function from $[0,T]$ to X, and for $t\varepsilon\ [0,T]$,

$$||(G\eta)(t)-f(t)|| = ||\ \int_0^t d_sK(t-s)\ \int_0^s g(s-r,u(r))\ || < \frac{\sigma}{2}.$$

We now show that G is continuous. Since g: $R^+ \times D \to X$ is continuous, given $\varepsilon > 0$, there is a $\delta > 0$ such that for $\eta_1,\eta_2 \varepsilon K$, $||\eta_1-\eta_2||_C < \delta$, and $s \ \varepsilon\ [0,T]$, we have

$$\sup_{0 \leq s \leq T} ||\int_0^s [g(s-r,\eta_1(r))-g(s-r,\eta_2(r))]dr|| < \varepsilon,$$

and

$$\sup_{0 \leq s \leq T} ||d \int_0^s [g(s-r,\eta_1(r))-g(s-r,\eta_2(r))]dr|| < \varepsilon.$$

To establish the second of these inequalities, notice that as a result of condition (2.6), there exists a $\delta > 0$ such that for $\eta_1, \eta_2 \varepsilon K$, $||\eta_1 - \eta_2|| < \delta$, we have $\sup_{0 \le r \le T} \nu(T; \eta_1(r), \eta_2(r)) < \varepsilon$. Thus if we let $\{s_i\}_{i=1}^{n}$ be a chain from 0 to T.

$$\sup_{0 \le s \le T} || d \int_0^s [g(s-r, \eta_1(r)) - g(s-r, \eta_2(r))] dr ||$$

$$\underset{=}{\text{def}} \sup_n \sum_{i=1}^{n} || \int_0^{s_i} [g(s_i - r, \eta_1(r)) - g(s_i - r, \eta_2(r))] dr$$

$$- \int_0^{s_{i-1}} g(s_{i-1} - r, \eta_1(r)) - g(s_{i-1} - r, \eta_2(r))] dr ||$$

$$\le \int_0^T \sup_n \sum_{i=1}^{n} || g(s_i - r, \eta_1(r)) - g(s_i - r, \eta_2(r))$$

$$- g(s_{i-1} - r, \eta_1(r)) + g(s_{i-1} - r, \eta_2(r)) || dr$$

$$+ \sup_n \sum_{i=1}^{n} \int_{s_{i-1}}^{s_i} || g(s_i - r\ \ \eta_1(r)) - g(s_i - r, \eta_2(r)) || dr$$

$$\le \int_0^T \nu(T; \eta_1(r), \eta_2(r)) dr + T\varepsilon$$

$$\le 2T\varepsilon$$

Now for η_1, $\eta_2 \varepsilon K$ and $t \varepsilon [0,T]$

$$||(G\eta_1)(t) - (G\eta_2)(t)||$$

$$= || \int_0^t d_s K(t-s) \int_0^s [g(s-r, \eta_1(r)) - g(s-r, \eta_2(r))] dr ||$$

$$\le || \int_0^t K(t-s) d \int_0^s [g(s-r, \eta_1(r)) - g(s-r, \eta_2(r))] dr ||$$

$$+ || \int_0^t [g(t-r, \eta_1(r)) - g(t-r, \eta_2(r))] dr ||$$

$$\le \varepsilon [M \int_0^t e^{\omega s} ds + 1] \quad ,$$

and this yields the continuity of G.

We now show that the set $\{G\eta : \eta \varepsilon K\}$ is equicontinuous as a collection of functions in C. For $\eta \varepsilon K$ and $0 \le t \le \hat{t} \le T$, define $V(\alpha, \beta)$ to be the total variation of $\int_0^s g(s-r, \eta(r)) dr$ on the interval $[\alpha, \beta] \subset [0,T]$, and notice that

$$||(G\eta)(t) - (G\eta)(\hat{t})||$$

$$= || \int_0^t d_s K(t-s) \int_0^s g(s-r, \eta(r)) dr - \int_0^{\hat{t}} d_s K(\hat{t}-s) \int_0^s g(s-r, \eta(r)) dr ||$$

$$= \|-\int_0^t K(t-s)d\int_0^s g(s-r,\ n(r))\,dr + \int_0^t g(t-r,\ n(r))\,dr$$

$$+ \int_0^{\hat t} K(\hat t-s)d\int_0^s g(s-r,\ n(r))dr - \int_0^{\hat t} g(\hat t-r,\ n(r))dr\ \|$$

$$\le \|\int_0^t [K(t-s) - K(\hat t-s)]d\int_0^s g(s-r,\ n(r))dr\ \|$$

$$\|\int_t^{\hat t} K(\hat t-s)d\int_0^s g(s-r,\ n(r))dr\ \|$$

$$+ \|\int_0^t [g(t-r,\ n(r)) - g(\hat t-r,\ n(r))]dr\|$$

$$+ \|\int_t^{\hat t} g(\hat t-r,\ n(r))dr\ \|$$

$$\le \|\ [K(\varepsilon) - K(\hat t-t+\varepsilon)]\int_0^{t-\varepsilon} K(t-s-\varepsilon)d\int_0^s g(s-r,\ n(r))dr\ \|$$

$$+ \|\int_{t-\varepsilon}^t [K(t-s) - K(\hat t-s)]d\int_0^s g(s-r,\ n(r))dr\ \|$$

$$+ Me^{|\omega|(\hat t-t)}\ V(t,\hat t)$$

$$+ T\sup_{0\le r\le T}\|\ g(t-r,\ n(r)) - g(\hat t-r,\ n(r))\ \|$$

$$+ (\hat t-t)\ N$$

$$\le \|\ K(\varepsilon) - K(\hat t-t+\varepsilon)\ \|\ Me^{|\omega|\varepsilon}\ V(0,t-\varepsilon)$$

$$+ M[\int_0^{\varepsilon} e^{\omega s}\ ds + \int_0^{\hat t-t+\varepsilon} e^{\omega s}\ ds]\ V(t-\varepsilon,t)$$

$$+ M\int_0^{\hat t-t} e^{\omega s}\ ds\ V(t,\hat t)$$

$$+ T\sup_{0\le r\le T}\|\ g(t-r,\ n(r)) - g(\hat t-r,\ n(r))\ \|$$

$$+ (\hat t-t)N\ .$$

The equicontinuity now follows from the continuity of $K(t)$ for $t > 0$ in the uniform operator topology and the fact that g is locally equicontinuous on bounded subsets of D.

Lastly we show that for each $t\varepsilon[0,T]$ the set $\{(Gn)(t):n\varepsilon K\}$ is precompact in X. Recall that for any bounded set $E\subset X$, the Kuratowski measure $\alpha(E)$ of noncompactness of E is defined as

$$\alpha(E) = \inf\ \{d > 0\colon E \text{ has a finite cover of diameter } < d\}$$

The number $\alpha(E)$ satisfies the following properties:

(i) $\alpha(E) = 0$ if and only if E is precompact,

(ii) $E_1\subset E_2$ implies $\alpha(E_1)\le\alpha(E_2)$,

(iii) $\alpha(E_1\cup E_2) = \max\ \{\alpha(E_1),\ \alpha(E_2)\}$,

(iv) $\alpha(E_1+E_2) \le \alpha(E_1) + \alpha(E_2)$.

Now given $\varepsilon > 0$, notice that

$$\int_0^t d_s\, K(t-s) \int_0^s g(s-r,\, n(r))dr$$

$$= K(\varepsilon) \int_0^{t-\varepsilon} d_s\, K(t-s-\varepsilon) \int_0^s g(s-r,\, n(r))dr$$

$$+ \int_{t-\varepsilon}^t d_s\, K(t-s) \int_0^s g(s-r,\, n(r))dr \quad .$$

Since the set

$$\{K(\varepsilon) \int_0^{t-\varepsilon} d_s\, K(t-s-\varepsilon) \int_0^s g(s-r,\, n(r))dr:\ n\varepsilon K\}$$

is precompact and the diameter of the set

$$\{\int_{t-\varepsilon}^t d_s K(t-s) \int_0^s g(s-r,n(r))dr:\quad n \varepsilon K\}$$

can be made arbitrarily small,

$$\alpha\{(Gn)(t):\ n\varepsilon K\} \leq \varepsilon\ ,$$

demonstrating the precompactness in X of this set.

By Schauder's fixed point theorem, G has a fixed point in K and the proof of the theorem is complete.

Proposition 2.2 Suppose (2.1), (2.2) and

(2.14) $f: R^+ \to X$ is continuous,

(2.15) $g: R^+ \times D \to X$, where D is an open subset of X, g is continuous and of local strong bounded variation in its first variable, and for each $x\varepsilon D$ there exists a neighborhood D_x and a continuous function $a: R^+ \to R^+$ such that $||g(s,x_1) - g(s,x_2)|| \leq a(s)||x_1-x_2||$ for all $s\varepsilon R^+$, $x_1,x_2\ \varepsilon D_x$,

(2.16) there exists a continuous function $b:R^+ \to R^+$ such that $\nu(t;x,y) \leq b(t)$ $||\ x-y\ ||$, where $\nu(t;x,y)$ is the total variation of $g(\cdot,x) - g(\cdot,y)$ on the interval $[0,t]$.

If $f(0)\varepsilon D$, then there exists $T > 0$ and a continous function $u:[0,T] \to X$ such that u satisfies (1.1) on $[0,T]$.

Proof Let N be a neighborhood of f(0) such that $\overline{N} \subset D$. Let $T > 0$ be undefined for the moment and consider the mapping

$$G:\ C([0,T];\ \overline{N}) \to C([0,T];\ \overline{N})$$

defined by

$$(Gh)(t) = f(t) + \int_0^t d_s\, K(t-s) \int_0^s g(s-r,\, h(r))dr.$$

Notice that G is continuous from $[0,T] \to X$ and for T sufficiently small, G is a contraction on C, since

$$||(Gn_1)(t) - (Gn_2)(t)\ ||$$

$$= ||\ \int_0^t dK(t-s) \int_0^s [g(s-r,\, n_1(r)) - g(s-r,\, n_2(r))]dr\ ||$$

$$\leq \int_0^t a(s-r) \| \eta_1(r) - \eta_2(r) \| dr$$

$$+ \| \int_0^t K(t-s)\, d \int_0^s [g(s-r, \eta_1(r)) - g(s-r, \eta_2(r))] dr \|$$

$$\leq [\int_0^t [a(s-r) + b(r)] dr + t \sup_{0 \leq s \leq t} a(s) \| \eta_1 - \eta_2 \|_C .$$

By the Contraction Mapping Theorem, G has a unique fixed point and the proof is complete.

3.0 Applications

In [3] local existence is established for equation (1.2) by treating it as an abstract Volterra integrodifferential equation in a Banach space of the form

(3.1) $$du(t)/dt + Au(t) = \int_0^t g(t-s, \mu(s))ds + f(t).$$

It is assumed that -A is the infinitesimal generator of a semigroup of bounded linear operators in a Banach space X, $f: R^+ \to X$ is continuously differentiable and $g: R^+ \times D \to X$, where D is an open subset of the domain of A with the graph norm (denoted [D(A)]) is continuous, has a continuous derivative g_1, with respect to its first variable, and both g and g_1 are locally Lipschitz continuous with respect to their second variable.

In the present study, we will establish local existence of solutions to this equation under somewhat milder assumptions. In particular, we make the following assumptions on the operator A and the functions g and f:

(3.1) A is a closed densely defined operator in a Banach space X and -A is the infinitesimal generator of a strongly continuous semigroup T(t), $t \leq 0$, in X satisfying $\|T(t)\, x \| \leq Me^{wt} \| x \|$ for $t \geq 0$, $x \in X$, where M and w are real constants;

(3.2) $T(t)\, X \subset D(A)$;

(3.3) $f: R^+ \to X$ is locally of strong bounded variation;

(3.4) $g: R^+ \times D \to X$, where D is an open subset of [D(A)]. g is locally of strong bounded variation in its first variable, and for each $x \in D$ there exist a neighborhood D_x and a continuous function $a: R^+ \to R^+$ such that $\|g(s,x_1) - g(s,x_2)\| \leq a(s) \| x_1 - x_2 \|_A$ for $s \in R^+$ and $x_1, x_2 \in D_x$.

(3.5) There exists a continuous function $b: R^+ \to$ such that $\nu(T;x,y) \leq b(T) \|x - y\|$, where $\nu(T;x,y)$ is the total variation of $g(\cdot,x) - g(\cdot,y)$ on the interval [0,T].

<u>Proposition 3.1</u> Suppose (3.1)-(3.5) hold. For each $x \in D$, there exists $T_x > 0$ and a continuous function $\mu:[0,T_x] \to [D(A)]$ satisfying

(3.6) $$\mu(t) = T(t)x + \int_0^t T(t-s)\, f(s)\, ds + \int_0^t T(t-s) \int_0^s g(s-r, \mu(r))\, dr\, ds \quad .$$

Proof In [4] G. F. Webb establishes that if $f: [0,r] \to X$ is of strong bounded variation on $[0,r]$ and $T(t)\,X \subset D(A)$ for $t > 0$, then $\int_0^t T(t-s)\, f(s)\, ds \in D(A)$ and $A \int_0^t T(t-s)\, f(s)\, ds = \int_0^t d_s\, T(t-s)\, f(s)$. To establish the validity of Proposition 3.1, consider the equation

(3.7) $$Z(t) = T(t)\, AX + \int_0^t d_s\, T(t-s)\, f(s) + \int_0^t ds\, T(t-s) \int_0^s g(s-r, A^{-1}Z(r))\, dr.$$

Notice that $G: R^+ \times A^{-1}(D) \to X$ defined by $G(s,x) = g(s,A^{-1}x)$ is continuous and locally of bounded variation in its first variable, and for each $x \in A^{-1}(D)$ there exists a neighborhood N of x such that $||G(s,x_1) - G(s-x_2)|| \le a(s)\, ||x_1 - x_2||$ for $s \in R^+$ and $x_1, x_2 \in N$. Also $F(t) = T(t)\, AX + \int_0^t d_s T(t-s)\, f(s)$ is continuous. It thus follows from Proposition 2.2 that equation (3.7) has a unique solution $Z(t)$. The function $\mu(t) = A^{-1}Z(t)$ is a solution of equation (3.6).

We now make slightly different assumptions on the operator A and the functions g and f.

(3.8) For $0 \le \alpha < 1$, the fractional power A^α exists. $A^{-\alpha} \in B(X;X)$ so that $D(A^\alpha)$ is a Banach space when endowed with the norm $||\, x\, ||_\alpha = ||A^\alpha\, x\, ||$ for $x \in D(A^\alpha)$. We denote this space by $[D(A^\alpha)]$;

(3.9) $T(t): X \to X$ is compact for each $t > 0$;

(3.10) $g: R^+ \times D \to X$ is continuous, where D is an open subset of $[D(A^\alpha)]$, and is locally of strong bounded variation in its first variable.

Proposition 3.2 Suppose (3.1)-(3.3) and (3.8)-(3.10) hold. For each $x \in D$, there exists $T_x > 0$ and a continuous function $\mu: [0,T_x] \to D(A^\alpha)$ satisfying equation (3.6).

Proof The validity of the Proposition follows by applying Proposition 2.1 to the equation

$$Z(t) = T(t)A^\alpha\, X + A^{\alpha-1} \int_0^t d_s\, T(t-s)\, f(s) + A^{\alpha-1} \int_0^t d_s\, T(t-s) \int_0^s g(s-r, A^{-\alpha}\, Z(r))\, dr\, .$$

Remarks

Local existence of solutions has been established in [1], [2] for abstract equations of the form

$$\mu'(t) = A\mu(t) + f(\mu(t)),$$

and

$$\mu''(t) = A\mu(t) + f(\mu(t)),$$

under the assumption that f is continuous with respect to a fractional power A^α $(0 \le \alpha < 1)$ of the semigroup generator A. For these equations and equation (3.6), existence of solutions in the case $\alpha = 1$ remains open.

References

1. C. Travis and G. Webb, Existence, stability, and compactness in the α-norm for partial functional differential equations, Trans. Am. Math. Soc. 240(1978), 129-143.
2. C. Travis and G. Webb, Cosine families and abstract nonlinear second order differential equations, to appear.
3. G. Webb, An abstract semilinear Volterra integrodifferential equation, to appear.
4. G. Webb, "Regularity of solutions to an abstract inhomogeneous linear differential equation," Proc. Am. Math. Soc. 62(2) (1977), 271-277.

ABSTRACT VOLTERRA INTEGRODIFFERENTIAL EQUATIONS AND A CLASS OF REACTION-DIFFUSION EQUATIONS

G.F. Webb
Mathematics Department
Vanderbilt University
Nashville, Tennessee 37235/USA

1. Introduction. Our objective is to study the semilinear Volterra integrodifferential equation

$$(1.1)\quad \begin{aligned} u'(t) &= Au(t) + f(t,u(t)) + \int_0^t g(t,s,u(s))ds \\ u(0) &= u_0 \end{aligned}$$

The equation (1.1) serves as an abstract formulation of many partial integrodifferential equations and we will discuss one such equation as an application of our theory. There has been considerable recent interest in abstract nonlinear integrodifferential equations and a number of recent contributions to this subject are listed in our references. The approach in most of these treatments is to require that $g(t,s,u(s))$ has the form $a(t-s)A(u(s))$, where a is a scalar-valued function and $-A$ is nonlinear and accretive.

Our approach takes advantage of the special semilinear form of (1.1) in that we require A to be linear and treat (1.1) as a nonlinear perturbation of the linear equation $u'(t) = Au(t)$. We will require that the linear operator A is the generator of an analytic semigroup, the nonlinear operator $f(t,.)$ is Lipschitz continuous with respect to a fractional power of $-A$, and the nonlinear operator $g(t,s,.)$ is Lipschitz continuous on the domain of A with respect to the graph norm of A. We will use techniques familiar to the theory of ordinary differential equations to obtain local existence and uniqueness results for (1.1) (the existence of global solutions, as in the theory of ordinary differential equations, depends upon additional conditions on f and g). These results extend earlier results of the author for the case in which the term $f(t,u(t))$ did not depend on u ([20]).

2. Abstract Results. Let X be a Banach space with norm $\|\ \|$ and let A be the infinitesimal generator of an analytic semigroup of linear operators $T(t)$, $t\geq 0$, in X. It is known that there must exist constants $M\geq 1$, $\omega \in R$, $C>0$, and $\rho \in R$ such that

$$|T(t)| \leq Me^{\omega t},\ t\geq 0$$

$$|(-A)^{\alpha}T(t)| \leq Ce^{\rho t}t^{-\alpha},\ 0<\alpha<1,\ t>0$$

(see, e.g., [5], Part 2). We require that A^{-1} is a bounded linear operator in X, which implies that $(-A)^{-\alpha}$ is also a bounded linear operator in X for $0 < \alpha < 1$. Let D_α, $0 < \alpha \leq 1$, denote the Banach space which is the domain of $(-A)^\alpha$ with the norm $||x||_\alpha = ||(-A)^\alpha||$, $x \in D((-A)^\alpha)$. Let g map $R^+ \times R^+ \times D_1$ into X and let f map $R^+ \times D_1$ into D_α, where $\alpha \in (0,1)$ is fixed.

We will require the following lemma from linear semigroup theory (a proof may be found in [8] and [18]):

Lemma 2.1. Let $T(t)$, $t \geq 0$, be a strongly continuous semigroup of linear operators with infinitesimal generator A. Let k map $[0,t_1]$ into X continuously and let $q(t) = \int_0^t T(t-s)\, k(s)\, ds$, $0 \leq t \leq t_1$. If k is continuously differentiable, then q is continuously differentiable and for $0 \leq t \leq t_1$, $q(t) \in D_1$ and $q'(t) = Aq(t) + k(t) = \int_0^t T(t-s)\, k'(s)\, ds + T(t)\, k(0)$. If $q(t) \in D_1$ for $0 \leq t \leq t_1$ and $Aq(t)$ is continuous on $[0,t_1]$, then q is continuously differentiable and $q'(t) = Aq(t) + k(t)$ on $[0,t_1]$.

We also require the following lemma (which is a generalization of Lemma 1.1 of [21]):

Lemma 2.2. Let w be a continuous function from $[0,t_1]$ into R^+, let $a,b,c \geq 0$, $\omega,\rho \in R$, $0 < \alpha < 1$, and let

$$w(t) \leq ae^{\omega t} + b\int_0^t e^{\rho(t-s)}(t-s)^{-\alpha} w(s)\, ds$$

$$+ c\int_0^t e^{\omega(t-s)}\, w(s)\, ds, \quad 0 \leq t \leq t_1.$$

Then for every real γ such that $\gamma > \max\{\omega,\rho\}$ and $k = b\Gamma(1-\alpha)(\gamma-\rho)^{\alpha-1} + c(\gamma-\omega)^{-1} < 1$, we have that

$$(2.1) \quad w(t) \leq a(1-k)^{-1}e^{\gamma t}, \quad 0 \leq t \leq t_1.$$

Proof. We use the gamma function formula

$$\Gamma(z) = \beta^z \int_0^\infty e^{-\beta s}\, s^{z-1}\, ds \text{ for } z > 0,\ \beta > 0$$

(see [22], p. 265). Let γ, k be as above and let $S = \sup_{0 \leq t \leq t_0} e^{-\gamma t} w(t)$, where $0 \leq t_0 \leq t_1$. Then, for $0 \leq t \leq t_0$,

$$e^{-\gamma t}w(t) \leq ae^{(\omega-\gamma)t} + b\int_0^t e^{(\rho-\gamma)(t-s)}(t-s)^{-\alpha}e^{-\gamma s}w(s)\, ds$$

$$+ c\int_0^t e^{(\omega-\gamma)(t-s)}e^{-\gamma s}w(s)\, ds$$

$$\leq a + bS\Gamma(1-\alpha)(\gamma-\rho)^{\alpha-1} + cS(\gamma-\omega)^{-1},$$

which implies $S \leq a + kS$, which implies (2.1).

Theorem 2.1. Let g be continuous, let g be continuously differentiable with respect to its first place, and let $g(t,s,x)$ and $g_1(t,s,x)$ be locally Lipschitz continuous in x uniformly in any bounded interval containing t and s (i.e., if $x \in D_1$ and I is a bounded interval in R^+, then there exists a neighborhood U_x of x in D_1 and a constant c such that if $x_1, x_2 \in U_x$ then $||g(t,s,x_1) - g(t,s,x_2)|| \leq c||x_1-x_2||_1$ for all $t,s \in I$, and similarly for g_1). Let f be continuous and let $f(t,x)$ be locally Lipschitz continuous in x uniformly in bounded intervals of t. For each $u_0 \in D_1$ there exists $t_1 > 0$ and a unique function u from $[0,t_1)$ to X such that u is continuous from $[0,t_1)$ to D_1 and u is continuously differentiable from $[0,t_1)$ to X and satisfies (1.1).

Proof. Let $u_0 \in D_1$ and let $t_1 > 0$ (we will restrict t_1 later). Let U_0 be a neighborhood of u_0 in D_1 and c a positive constant such that if $x_1, x_2 \in \overline{U}_0$, $t \in [0,t_1]$, and $0 \leq s \leq t$, then

$$||g(t,s,x_1) - g(t,s,x_2)|| \leq c||x_1 - x_2||_1,$$

$$||g_1(t,s,x_1) - g_1(t,s,x_2)|| \leq c||x_1 - x_2||_1,$$

$$||f(s,x_1) - f(s,x_2)||_\alpha \leq c||x_1 - x_2||_1.$$

Let $C_0 = C(0,t_1; \overline{U}_0)$. Define the mapping K on C_0 by

$$(2.2)\quad (Ku)(t) = T(t)x + \int_0^t T(t-s) \int_0^s g(s,r,u(r))\, dr\, ds$$
$$+\int_0^t T(t-s)\, f(s,u(s))\, ds,\quad u \in C_0,\ 0 \leq t \leq t_1$$

Using Lemma 2.1 we see that if $u \in C_0$ and $0 \leq t \leq t_1$, then $(Ku)(t) \in D_1$ and

$$(2.3)\quad (AKu)(t) = T(t)Ax + \int_0^t T(t-s)\, [g(s,s,u(s)) + \int_0^s g_1(s,r,u(r))dr]ds$$
$$- \int_0^t g(t,s,u(s))ds - \int_0^t (-A)^{1-\alpha} T(t-s)(-A)^\alpha f(s,u(s))ds.$$

Thus, for $u \in C_0$, Ku and AKu are both continuous from $[0,t_1]$ to X. If t_1 is chosen sufficiently small then K maps C_0 into C_0. Further, if t_1 is chosen sufficiently small then K is a contraction on C_0, since

$$||(AKu)(t) - (AKv)(t)|| \leq \int_0^t Me^{\omega(t-s)}[c||u(r) - v(r)||_1$$

$$+\int_0^s c\,||u(r) - v(r)||_1 dr]\, ds$$

$$+\int_0^t c\,||u(s) - v(s)||_1 ds$$

$$+\int_0^t Ce^{\rho(t-s)} c(t-s)^{\alpha-1} ||u(s) - v(s)||_1 ds.$$

By the Contraction Mapping Theorem there exists a unique $u \in C_0$ such that $Ku = u$. From Lemma 1.1 we see that $u: [0,t_1] \to D_1$ is continuous and $u: [0,t_1] \to X$ is continuously differentiable and satisfies (1.1) on $[0,t_1]$.

Theorem 2.2. Let g be continuous, let g be continuously differentiable with respect to its first place, and let g and g_1 be Lipschitz continuous on bounded sets of $R^+ \times R^+ \times D_1$ (i.e., if U is a bounded set in D_1 and I is a bounded interval in R^+ then there exists a constant c such that if x_1, $x_2 \in U$ and t_1, t_2, s_1, $s_2 \in I$ then $||g(t_1,s_1,x_1,) - g(t_2,s_2,x_2)|| \le c(||x_1 - x_2||_1 + |t_1 - t_2| + |s_1 - s_2|)$, and similarly for g_1). Let f be continuous and let $f(t,x)$ be Lipschitz continuous on bounded sets of $R^+ \times D_1$ (i.e., if U is a bounded set of D_1 and I is a bounded interval then there exists a constant c such that if x_1, $x_2 \in U$ and t_1, $t_2 \in I$ then $||f(t_1,x_1) - f(t_2,x_2)||_\alpha \le c(||x_1 - x_2||_1 + |t_1 - t_2|))$. If u is a solution of (1.1) on $[0,t_1)$ and t_1 is maximal (i.e., there exists no solution of (1.1) on $[0,t_2)$ if $t_2 > t_1$), then either $t_1 = +\infty$ or there exists a sequence t_n such that $t_n \to t_1^-$ and $||u(t_n)||_1 \to +\infty$.

Proof. Assume $t_1 < +\infty$ and $\sup_{0\le t<t_1} ||u(t)||_1 < b < +\infty$. Let $U = \{x \in D_1 : ||x||_1 \le b\}$, let $I = [0,t_1]$, and let c be a constant as in the statement of the theorem. For $0 \le t < t+h < t_1$,

$$||Au(t+h) - Au(t)|| \le ||T(t+h)Ax - T(t)Ax||$$

$$+ ||\int_{-h}^0 T(t-s)\, [g(s+h,s+h,u(s+h)) + \int_0^{s+h} g_1(s+h,r,u(r))dr]ds||$$

$$+ ||\int_0^t T(t-s)\, [g(s+h,s+h,u(s+h)) - g(s,s,u(s))]ds||$$

$$+ ||\int_0^t T(t-s)\, [\int_{-h}^0 g_1(s+h,r+h,u(r+h))dr$$

$$+\int_0^s (g_1(s+h,r+h,u(r+h)) - g_1(s,r,u(r)))dr]ds||$$

$$+ ||\int_{-h}^0 g(t+h,s+h,u(s+h))ds||$$

$$+ ||\int_0^t g(t+h,s+h,u(s+h))ds - \int_0^t g(t,s,u(s))ds||$$

$$+ \left\|\int_{-h}^{0}(-A)^{1-\alpha}T(t-s)(-A)^{\alpha}f(s+h,u(s+h))ds\right\|$$

$$+ \left\|\int_{0}^{t}(-A)^{1-\alpha}T(t-s)(-A)^{\alpha}[f(s+h,u(s+h)) - f(s,u(s))]ds\right\|$$

Let c_1 be a constant such that $||g(t,s,x)|| \le c_1$, $||g_1(t,s,x)|| \le c_1$, and $||f(t,x)||_\alpha \le c_1$ for $x \in U$, t, $s \in I$. Then, for $0 \le t < t+h < t_1$

$$||Au(t+h) - Au(t)|| \le ||T(t)(T(h)Ax - Ax)||$$

$$+ \int_{-h}^{0} Me^{\omega(t-s)}\,[c_1 + \int_{0}^{s+h} c_1\, dr]ds$$

$$+ \int_{0}^{t} Me^{\omega(t-s)}\,[c\, ||u(s+h) - u(s)||_1 + 2h]ds$$

$$+ \int_{0}^{t} Me^{\omega(t-s)}\,[\int_{-h}^{0} c_1\, dr + \int_{0}^{s} c(||u(r+h) - u(r)||_1 + 2h)dr]ds$$

$$+ \int_{-h}^{0} c_1\, ds + \int_{0}^{t} c(||u(s+h) - u(s)||_1 + 2h)\, ds$$

$$+ \int_{-h}^{0} Ce^{\rho(t-s)}(t-s)^{\alpha-1}\, c_1\, ds$$

$$+ \int_{0}^{t} Ce^{\rho(t-s)}(t-s)^{\alpha-1} c(||u(s+h) - u(s)||_1 + h)\, ds.$$

Thus, we see that there exists a continuous function $L:R^+ \to R^+$ such that $L(0) = 0$ and a constant M_1 such that for $0 \le t < t+h < t_1$,

$$||u(t+h)-u(t)||_1 \le L(h) + M_1\int_{0}^{t}||u(s+h)-u(s)||_1 ds + \int_{0}^{t} Ce^{\rho(t-s)}(t-s)^{\alpha-1}c||u(s+h)-u(s)||_1 ds.$$

By Lemma 1.2 there exist constants k and γ such that $||u(t+h) - u(t)||_1 \le L(h)(1-k)^{-1}e^{\gamma t}$ for $0 \le t < t+h < t_1$. Thus, $\lim_{t\to t_1^-} u(t)$ exists in D_1 and by Theorem 2.1 $u(t)$ can be continued past t_1, which contradicts the assumption that t_1 is maximal.

<u>Corollary 2.3</u>. Let the hypothesis of Theorem 2.2 be satisfied and, in addition, let

$$||g(t,s,x)|| \le k_1(t,s)(1+||x||_1),\ x \in D_1,\ t,\ s \ge 0,$$

$$||g_1(t,s,x)|| \le k_2(t,s)(1+||x||_1),\ x \in D_1,\ t,\ s \ge 0,$$

$$||f(t,x)||_\alpha \le k_3(t)(1+||x||_1),\ x \in D_1,\ t \ge 0,$$

where k_1, k_2 are continuous functions from $R^+ \times R^+$ to R^+ and k_3 is a continuous function from R^+ to R^+. Then for each $x \in D_1$ the solution of (1.1) exists for all $t \geq 0$.

Proof. The conclusion follows from Lemma 1.2 and Theorem 2.2, since from (2.3),

$$||u(t)||_1 \leq Me^{\omega t} ||Ax||$$

$$+ \int_0^t Me^{\omega(t-s)} [k_1(s,s)(1 + ||u(s)||_1) ds$$

$$+ \int_0^s k_2(s,r)(1 + ||u(r)||_1) dr] ds$$

$$+ \int_0^t k_1(t,s)(1 + ||u(s)||_1) ds$$

$$+ \int_0^t Ce^{\rho(t-s)}(t-s)^{\alpha-1} k_3(s) (1 + ||u(s)||_1) ds.$$

3. An Example. We will use our development to treat the partial integrodifferential equation

(3.1) $w_t(x,t) = w_{xx}(x,t) + h(w(x,t)) + \int_0^t a(t-s)\, \sigma(w_x(x,s))_x\, ds, \quad 0 \leq x \leq \pi, \quad t \geq 0$

$w(x,0) = u_0(x), \quad 0 \leq x \leq \pi$

$w(0,t) = w(\pi,t) = 0, \quad t \geq 0.$

We require the following conditions on h, a, and σ:

(3.2) $h: R \to R$ is continuously differentiable, $h(0) = 0$, and $h'(x) \leq 0$ for all $x \in R$;

(3.3) $a: R^+ \to R^+$ is continuously differentiable and $\int_0^T b(t) \int_0^t a(t-s)\, b(s)\, ds\, dt \geq 0$ for all $T > 0$ and all continuous functions b from R^+ to R^+;

(3.4) σ: $R \to R$ is continuously differentiable, $x\sigma(x) \geq 0$ for all $x \in R$ (which implies $\sigma(0) = 0$), and there exist constants σ_1 and σ_2 such that $0 < \sigma_1 \leq \sigma'(x) \leq \sigma_2$ for all $x \in R$.

A function a satisfying (3.3) is sometimes called a positive kernel and it is known that if $(-1)^k a^{(k)}(t) \geq 0$ for $k = 0, 1, 2$, then a is a positive kernel (see [9], p. 217).

We will formulate equation (3.1) abstractly as equation (1.1) in the Hilbert space $X = L^2(0,\pi)$. Let $A: X \to X$ with $Az = z''$ and $D(A) = \{z \in X$: z and z' are absolutely continuous, $z' \in X$, and $z(0) = z(\pi) = 0\} = D_1$. Then A has the representation

$Az = \Sigma_{n=1}^{\infty} -n^2 <z,z_n> z_n$, where $z_n(x) = (2/\pi)^{1/2} \sin nx$, $n = 1,2,\ldots$, is the orthonormal set of eigenvectors of A. Further, A is the infinitesimal generator of an analytic semigroup of operators $T(t)$, $t \geq 0$ given by $T(t)z = \Sigma_{n=1}^{\infty} e^{-n^2 t} <z,z_n>z_n$, $z \in X$, $t \geq 0$, and $(-A)^{1/2}T(t)z = \Sigma_{n=1}^{\infty} ne^{-n^2 t}<z,z_n>z_n$, $z \in X$, $t>0$. The following estimates can be shown:

$$|T(t)| \leq e^{-t}, \; t \geq 0$$

$$|(-A)^{1/2}T(t)| \leq C_\rho t^{-1/2} e^{\rho t}, \; t > 0,$$

where $C_\rho = (\sqrt{2e}\ (1 + \rho))^{-1}$, $-1 < \rho < 0$.

Define $g: R^+ \times R^+ \times D_1 \to X$ by

$$(3.5) \quad g(t,s,z) = a(t-s)\sigma(z')', \; z \in D_1, \; t,s \geq 0.$$

Notice that $\sigma(z')' = \sigma'(z')z''$ for $z \in D_1$ since σ' is continuously differentiable. Also, $\sigma(z')' \in X$, since $\sup_{0<x<\pi}|z'(x)| \leq \sqrt{\pi}||z||_1$ for $z \in D_1$. Using the facts that a is continuously differentiable, σ is continuously differentiable, and $\sup_{0<x<\pi}$ $|z'(x)| \leq \sqrt{\pi}||z||_1$ for all $z \in D_1$, one can show that g satisfies the hypothesis of Theorem 2.2.

Define $f: D_1 \to D_{1/2}$ by

$$(3.6) \quad f(z) = h(z(.)) \text{ for all } z \in D_1.$$

If $z \in D_1$ then $z(0) = z(\pi) = 0$ and hence $h(z(0)) = h(z(\pi)) = 0$. Further, $h(z(x))$ is once continuously differentiable with respect to x. Since $||z||_{1/2} = ||z'||$ for $z \in D_1$, we see that f is well-defined from D_1 to $D_{1/2}$ and f satisfies the hypothesis of Theorem 2.2.

Thus, the hypothesis of Theorem 2.2 (and hence of Theorem 2.1) is satisfied for this choice of X, A, α, f, and g. Then for each $u_0 \in D_1$ there exists a unique function u from some maximal interval of existence $[0,t_1)$ to X such that $u: [0,t_1) \to D_1$ is continuous and $u: [0,t_1) \to X$ is continuously differentiable and satisfies (1.1) with $u(0) = u_0$.

We now show that the solutions must exist globally, that is, $t_1 = \infty$. Our techniques are similar to those used in [15]. From (3.4) we have

$$(3.7) \quad \sigma_1 x^2/2 \leq \int_0^x \sigma(\tau)d\tau \leq \sigma_2 x^2/2, \; x \in R.$$

Let $u(t)$ be a solution of (1.1) on $[0,t_1)$ with $u(0) = u_0 \in D_1$. If we multiply (1.1) by $\sigma(u(t)_x)_x$, we obtain

(3.8) $\langle u'(t), \sigma(u(t)_x)_x\rangle = \langle Au(t), \sigma(u(t)_x)_x\rangle$

$$+ \langle f(u(t)), \sigma(u(t)_x)_x\rangle + \int_0^t a(t-s) \langle\sigma(u(s)_x)_x, \sigma(u(t)_x)_x\rangle\, ds$$

From (3.8) we obtain

(3.9) $- \langle u'(t)_x, \sigma(u(t)_x)\rangle = \langle Au(t), \sigma'(u(t)_x)\, Au(t)\rangle$

$$- \langle h'(u(t))u(t)_x, \sigma(u(t)_x)\rangle$$

$$+ \int_0^t a(t-s) \langle\sigma(u(s)_x)_x, \sigma(u(t)_x)_x\rangle\, ds.$$

Define $J(x) = \int_0^x \sigma(s)\, ds$ and integrate (3.9) from 0 to $T < t_1$ to obtain

(3.10) $- \int_0^T [d/dt \int_0^\pi J(u(t)_x)\, dx]\, dt = \int_0^T \langle Au(t), \sigma'(u(t)_x)\, Au(t)\rangle\, dt$

$$- \int_0^T \langle h'(u(t))u(t)_x, \sigma(u(t)_x)\rangle\, dt$$

$$+ \int_0^T\int_0^t a(t-s) \langle\sigma(u(s)_x)_x, \sigma(u(t)_x)_x\rangle\, ds\, dt.$$

Now use (3.2), (3.3), and (3.4) to obtain

(3.11) $- \int_0^\pi J(u(T)_x)dx + \int_0^\pi J(u(0)_x)\, dx \geq \sigma_1 \int_0^T ||Au(t)||^2\, dt.$

From (3.7) and (3.11) we obtain

(3.12) $(\sigma_1/2)\, ||u(T)||_{1/2}^2 + \sigma_1 \int_0^T ||Au(t)||^2\, dt \leq (\sigma_2/2)\, ||u(0)||_{1/2}^2.$

For each positive integer N let $h_N: R \to R$ be a truncation of h such that $h_N(x) = h(x)$ for $|x| \leq N$, h_N is continuously differentiable, and $- c_N \leq h_N'(x) \leq 0$ for all $x \in R$ and some constant $c_N > 0$. With X, A, and g as before and f replaced by $f_N(z) = h_N(z(.))$, $z \in D_1$, we have that the hypothesis of Corollary 2.3 is satisfied. Thus, the solution $u_N(t)$ of (1.1), with f replaced by f_N, is defined for all $t \geq 0$. But $u_N(T)$ must also satisfy the estimate of (3.12) for all $T \geq 0$. Then $\sup_{0\leq x\leq\pi} |u_N(T)(x)| \leq \sqrt{\pi}||u_N(T)||_{1/2} \leq \sqrt{\pi}(\sigma_2/\sigma_1)^{1/2}||u_0||_{1/2}$ for all $T \geq 0$ and all N = 1, 2,... . Thus, we must have $f_N(u_N(t)) = f(u_N(t))$, $t \geq 0$, for N sufficiently large. Hence, for N sufficiently large $u_N(t) = u(t)$ and the maximal interval $[0,t_1)$ of existence must have $t_1 = +\infty$.

In conclusion we observe that the estimate (3.12) implies that the solutions of (3.1) are stable in the 1/2 -norm in the sense that $||u(t)||_{1/2} \leq (\sigma_2/\sigma_1)^{1/2}||u_0||_{1/2}$, $t \geq 0$, and square integrable in the A-norm in the sense that $\int_0^\infty ||u(t)||_1^2\, dt < \infty$.

REFERENCES

1. V. Barbu, *Nonlinear Volterra equations in a Hilbert space*. SIAM J. Math. Anal. 6 (1975), 728-741.

2. ________, *Nonlinear semigroups and differential equations in Banach spaces*, Noordhoff, Leyden, 1976.

3. M. G. Crandall, S.-O. Londen, and J. A. Nohel, *An abstract nonlinear Volterra integrodifferential equation*, J. Math. Anal. Appl. (to appear).

4. M. G. Crandall and J. A. Nohel, *An abstract functional differential equation and a related nonlinear Volterra equation*, Math. Res. Center, Univ. of Wisconsin, Tech. Summary Report #1765, 1977.

5. A. Friedman, *Partial differential equations*, Holt, Rinehart, and Winston, New york, 1969.

6. ___________, *Monotonicity of solutions of Volterra integral equations in Banach space*, Trans. Amer. Math. Soc. 138 (1969), 129-148.

7. A. Friedman and M. Shinbrot, *Volterra integral equations in Banach space*, Trans. Amer. Math. Soc. 126 (1967), 131-179.

8. T. Kato, *Perturbation theory for linear operators*, Springer-Verlag, New York, 1966.

9. M. Loève, *Probability theory. Foundations. Random sequences*, 2nd rev. ed., University Series in Higher Math., Van Nostrand, Princeton, N. J., 1960.

10. S.-O. Londen, *An existence result on a Volterra equation in a Banach space*, Trans. Amer. Math. Soc. (to appear).

11. ___________, *On an integral equation in a Hilbert space*, SIAM J. Math. Anal. (to appear).

12. S.-O. Londen and O. J. Staffans, *A note on Volterra equations in a Hilbert space*, Helsinki Univ. of Tech. Report - HTKK - MAT - A90 (1976).

13. R. C. MacCamy, *Stability theorems for a class of functional differential equations*, SIAM J. Math. Anal. (to appear).

14. ___________, *An integro-differential equation with applications in heat flow*, Quart. Appl. Math. 35 (1977), 1-19.

15. R. C. MacCamy and J. S. W. Wong, *Stability theorems for some functional differential equations*, Trans. Amer. Math. Soc. 164 (1972), 1-37.

16. R. K. Miller, *Volterra integral equations in a Banach space*, Funkcial. Ekvac. 18 (1975), 163-194.

17. R. K. Miller and R. L. Wheeler, *Well-posedness and stability of linear Volterra integrodifferential equations in abstract spaces* (to appear).

18. A. Pazy, *Semi-groups of linear operators and applications to partial differential equations*, Lecture Notes 10, University of Maryland (1974).

19. C. C. Travis and G. F. Webb, *An abstract second order semilinear Volterra integrodifferential equation*, SIAM J. Math. Anal. (to appear).

20. G. F. Webb, *An abstract semilinear Volterra integrodifferential equation*, Proc. Amer. Math. Soc. (to appear).

21. _________, *Exponential representation of solutions to an abstract semi-linear differential equation*, Pac. J. Math. 70 (1977), 269-280.

22. K. Yoshida, *Functional analysis*, Springer-Verlag, New York, 1968.

ASYMPTOTIC BEHAVIOR OF SOLUTIONS OF LINEAR VOLTERRA

INTEGRODIFFERENTIAL EQUATIONS IN HILBERT SPACE

Robert L. Wheeler*
University of Missouri
Columbia, Missouri 65211/USA

Introduction

We consider the stability and asymptotic behavior of solutions of the linear Volterra integrodifferential equation

$$(1)\quad x'(t) = cx(t) + \int_0^t [B(t-\tau)Ax(\tau) + G(t-\tau)x(\tau)]d\tau + F(t),\ x(0) = x_0 ,$$

in a Hilbert space X with inner product (,) and norm $\|\ \|$. Here ' denotes the strong derivative with respect to the variable $t \in R^+ \equiv [0, \infty)$, A is a self-adjoint, negative definite linear operator with dense domain D(A), the kernels B and G are real-valued functions, $F: R^+ \to X$, and c is a real constant. An equation of this form arises in a linearized model for heat flow in materials with memory proposed by Gurtin and Pipkin [6] (see also, Miller [11] and its bibliography).

Here we examine the behavior of solutions of (1) when the kernels B and G have the form

$$B(t) = 1 + \int_0^t b(\tau)\, d\tau, \quad G(t) = \gamma + \int_0^t g(\tau)\, d\tau$$

with b and g both in $C^1(R^+) \cap L^1(R^+)$ and γ a real constant. Then, if F(t) and x(t) are sufficiently smooth, equation (1) may be differentiated and x(t) satisfies

$$(2)\quad x''(t) = cx'(t) + [A + \gamma I]\, x(t) + \int_0^t [b(t-\tau)\, Ax(\tau) + g(t-\tau)\, x(\tau)]d\tau + f(t) ,$$

$x(0) = x_0$, $x'(0) = v_0$, where $f = F'$, $v_0 = cx_0 + F(0)$, and I denotes the identity operator. For the rest of this paper we will work mainly with this second order equation (2).

For related work on the asymptotic behavior of solutions of linear Volterra equations in abstract spaces, we mention, in particular, the papers by Friedman and Shinbrot [4], Miller and Wheeler [12], Hannsgen [7], [8], and Carr and Hannsgen [1]. In Section 1 we briefly compare and contrast our results and techniques with those contained in these papers. We remark that Dafermos [3] and Slemrod [15] have used Lyapunov techniques to study the existence, uniqueness and asymptotic behavior of solutions of second order linear Volterra integrodifferential equations in Hilbert space. Also, the asymptotic behavior of solutions of abstract nonlinear integral equations has been examined by several authors, including work in a Hilbert space setting by MacCamy [10], and in Banach space by Crandall, Londen and Nohel [2] and Gripenberg [5].

This paper is based to a large extent on the asymptotic behavior results contained in the author's forthcoming paper with R. K. Miller [13]. With the

---

*Supported in part by NSF Grant MCS78-01330.

exception of the proof of Theorem 3 and the remarks concerning the proofs of Lemmas 1 and 2 in Section 2, complete proofs of the results may be found in [13].

1. Statement and discussion of results.

By a solution $x(t) = x(t, x_0, v_0, f)$ of equation (2) we mean a function $x\colon R^+ \to D(A)$ so that $x(0) = x_0$, $x'(0) = v_0$, $x(t) \in C^2(R^+; X)$ and $Ax(t) \in C(R^+; X)$ which satisfies (2) for all $t \geq 0$. By a generalized solution we mean the limit uniformly on compact subsets of $0 \leq t < \infty$ of a sequence of solutions.

We assume throughout that the following hypotheses hold:

(H1) The real-valued functions $b(t)$ and $g(t)$ both belong to $C^1(R^+) \cap L^1(R^+)$.

(H2) The linear operator A is self-adjoint, negative definite, and the resolvent operator $R(\mu; A) \equiv (A - \mu I)^{-1}$ is compact when it exists.

We remark that it is well-known that if (H2) holds, then the spectrum of A, $\sigma(A)$, consists entirely of eigenvalues $\{\mu_n\}$ satisfying $0 > \mu_1 \geq \mu_2 \geq \cdots \to -\infty$, and there exists a complete orthonormal set of corresponding eigenvectors $\{\phi_n\}$ associated with A. A motivating example of such an operator is $A = \partial^2/\partial x^2$, the one-dimensional Laplacian, with domain $D(A)$ the subspace of $X = L^2(0, \pi)$ defined by $D(A) = \{u \in X\colon u(0) = u(\pi) = 0$ with u, u', u'' all in X, and with u and u' both absolutely continuous on $[0, \pi]\}$.

We first state the following theorem on existence, uniqueness and dependence upon data of solutions of (2) which is a special case of results in [13].

Theorem 1. *Let (H1) and (H2) hold. Then for each $(x_0, v_0) \in D(A) \times D(A)$ and $f \in C^1(R^+; X)$, there exists a unique solution $x(t)$ of (2) on R^+. Moreover, $x(t)$ depends continuously on (x_0, v_0, f) in the sense that for each $T > 0$ there exists $K = K(T) > 0$ such that*

$$||x(t)|| \leq K\{||x_0|| + ||v_0|| + \int_0^T ||f(t)||dt\} \qquad (0 \leq t \leq T)$$

for all $(x_0, v_0) \in D(A) \times D(A)$ and $f \in C^1(R^+; X)$.

We remark that the well-posedness results in [13] are proved in the more general setting where X is a Banach space and A is the infinitesimal generator of a strongly continuous cosine family.

Let $L^p(R^+; X)$ $(1 \leq p \leq \infty)$ denote the space of measurable functions $f\colon R^+ \to X$ with $||f(t)|| \in L^p(R^+)$. Using Theorem 1 and a standard density argument, we immediately see that equation (2) has a unique generalized solution for each $(x_0, v_0) \in X \times X$ and $f \in L^1(R^+; X)$.

In our principal results (Theorems 2 and 3) we give conditions sufficient to ensure that the generalized solution $x(t)$ of equation (2) belongs to $L^2(R^+; X)$ (or, that $x(t)$ may be expressed as a bounded exponential polynomial plus a remainder term in $L^2(R^+; X)$) for all $(x_0, v_0) \in X \times X$ and $f \times L^1(R^+; X)$.

Let $b^*(\lambda) \equiv \int_0^\infty e^{-\lambda t} b(t)dt$, $\operatorname{Re} \lambda \geq 0$, denote the Laplace transform. Since $b^*(\lambda)$

is analytic in Re $\lambda > 0$, it is easy to see that if $1 + b^*(\lambda_0) = 0$ for some λ_0 with Re $\lambda_0 > 0$, then for all N sufficiently large one can find solutions $\lambda = \lambda_N$ of the equation $\lambda^2 - c\lambda - \gamma - g^*(\lambda) - \mu_N(1 + b^*(\lambda)) = 0$ with Re $\lambda_N > 0$. One can then verify that the function $x_N(t) = \exp(\lambda_N t)\phi_N$ is a solution of (2) with $x_0 = \phi_N$, $v_0 = \lambda_N\phi_N$ and $f(t) = \int_t^\infty \exp(\lambda_N(t - \tau))[\mu_N b(\tau) + g(\tau)]\, d\tau\, \phi_N$. Since $f(t) \in L^1(R^+; X)$ and $||x_N(t)|| \to \infty$ as $t \to \infty$, we have

Proposition 1. *If (H1) and (H2) hold and the generalized solution $x(t)$ of equation (2) belongs to $L^2(R^+; X)$ for all $(x_0, v_0) \in X \times X$ and all $f \in L^1(R^+; X)$, then it follows that $1 + b^*(\lambda) \neq 0$ when Re $\lambda > 0$.*

The following two lemmas, whose proofs are sketched in Section 2, are needed in the proofs of Theorems 2 and 3.

Let B(X) denote the space of bounded linear operators on X. We have

Lemma 1. *Assume that the operator A satisfies hypothesis (H2), let $p(t)$ and $q(t)$ be real-valued functions in $L^1(R^+)$, and let $\alpha < 0$ be a constant. Suppose that $\Gamma(\lambda)$ is a complex-valued function which is bounded and continuous on Re $\lambda \geq 0$ and analytic in Re $\lambda > 0$. Assume that*

(1.1) $\qquad \lambda^2 - \alpha\lambda - \lambda p^*(\lambda) - \Gamma(\lambda) - q^*(\lambda) - \mu_n \neq 0$ for $n = 1,2,\ldots$ and Re $\lambda \geq 0$.

Then the B(X)-valued function $T^*(\lambda) = [\lambda^2 - \alpha\lambda - \lambda p^*(\lambda) - \Gamma(\lambda) - q^*(\lambda) - A]^{-1}$ is defined for Re $\lambda \geq 0$ and satisfies $\sup\left\{\int_{-\infty}^{\infty} ||T^*(\sigma + i\tau)||^2\, d\tau : \sigma \geq 0\right\} < \infty$.

Lemma 2. *Assume that the hypotheses and notation of Lemma 1 hold, and define the B(X)-valued function $S^*(\lambda) = \lambda T^*(\lambda)$ for Re $\lambda \geq 0$. Then there exists a positive K so that* $\sup\left\{\int_{-\infty}^{\infty} ||S^*(\sigma + i\tau)\, x||^2\, d\tau : \sigma \geq 0\right\} \leq K||x||^2$ *for all* $x \in X$.

Conditions which guarantee that generalized solutions of equation (2) lie in $L^2(R^+; X)$ are given by

Theorem 2. *Let (H1) and (H2) hold, and assume that the transform conditions*

(T1) $\qquad 1 + b^*(\lambda) \neq 0$ *for* Re $\lambda \geq 0$,

(T2) $\qquad \lambda^2 - c\lambda - \gamma - g^*(\lambda) - \mu_n(1 + b^*(\lambda)) \neq 0$ *for* $n = 1,2,\ldots$ *and* Re $\lambda \geq 0$

are satisfied. In addition, assume that $b'(t) \in L^1(R^+)$ and that $b(0) + c < 0$. Then the generalized solution $x(t, x_0, v_0, f)$ of equation (2) belongs to $L^2(R^+; X)$ for all $(x_0, v_0) \in X \times X$ and all $f \in L^1(R^+; X)$.

As the discussion preceding Proposition 1 shows, if we have that $x(t)$ belongs to $L^2(R^+; X)$ whenever $(x_0, v_0) \in X \times X$ and $f \in L^1(R^+; X)$, then the transform conditior (T2) must hold for Re $\lambda \geq 0$, and (T1) must hold for Re $\lambda > 0$. The case where $1 + b^*(\lambda)$ has a finite number of simple zeros on Re $\lambda = 0$ is examined in

Theorem 3. *Assume that (H1) and (H2) hold, that $b'(t)$, $t^2b(t)$ and $t^2g(t)$ all belong to $L^1(R^+)$, and that $b(0) + c < 0$. Let ω_m, $1 \leq m \leq M$, be real constants, and assume* *that* (T1) *and* (T2) *hold for all* λ *satisfying* Re $\lambda \geq 0$, $\lambda \neq i\omega_m$. *For* $1 \leq m \leq M$ *assume that the three conditions*

(1.2) $$b^*(i\omega_m) = -1 \text{ and } (b^*)'(i\omega_m) \neq 0 ,$$

(1.3) $$\omega_m^2 + ic\omega_m + \gamma + g^*(i\omega_m) = 0 ,$$

(1.4) $$(2i\omega_m - c - (g^*)'(i\omega_m))/(b^*)'(i\omega_m) \neq \mu_n \quad (n = 1,2, \ldots)$$

hold. Then $x(t, x_0, v_0, f)$ *may be expressed as*

$$x(t) = \sum_{m=1}^{M} x_m e^{i\omega_m t} + x_1(t)$$

with $x_m \in X$ $(1 \leq m \leq M)$ *and* $x_1(t) \in L^2(R^+; X)$ *whenever* $(x_0, v_0) \in X \times X$ *and* $f \in C^1(R^+; X) \cap L^1(R^+; X)$ *with* $f'(t)$ *and* $tf(t)$ *both in* $L^1(R^+; X)$.

We remark that $1 + b^*(\lambda) = \lambda B^*(\lambda)$ must always have a zero at $\lambda = 0$ whenever the kernel $B(t)$ in the integrated form of the equation (i.e., equation (1)) belongs to $L^1(R^+)$. Also, (1.3) and (1.4) are automatically satisfied at $\omega = 0$ in the case where $c = \gamma = 0$ and $g(t) \equiv 0$.

Theorem 3 extends Theorem 5.2 of [13] where we consider the homogeneous case ($f(t) \equiv 0$) when $1 + b^*(\lambda)$ has only one simple zero in Re $\lambda \geq 0$ located at $\lambda = 0$. The proof of Theorem 3 is given in Section 2.

Before proceeding to discuss Theorems 2 and 3 and their relationship to other results on the asymptotic behavior of linear Volterra integrodifferential equations in abstract spaces, we pause to sketch the proof of Theorem 2.

Proof of Theorem 2. Let $r(t)$ be the scalar resolvent function defined by

(1.5) $$r(t) = b(t) - b * r(t), \quad t \geq 0,$$

where $*$ denotes the convolution $b * r(t) \equiv \int_0^t b(t - \tau)\, r(\tau)\, d\tau$. It is well-known that $r(t) \in C(R^+)$; moreover, since (T1) holds, a classical result due to Paley and Wiener [14] gives $r(t) \in L^1(R^+)$. If we now differentiate both sides of equation (1.5), we see that $r'(t) \in C(R^+) \cap L^1(R^+)$ as well.

Assume that $(x_0, v_0) \in D(A) \times D(A)$ and that $f \in C^1(R^+; X) \cap L^1(R^+; X)$ with $||f'(t)||$ of exponential order, and let $x(t)$ be the unique solution of (2) which is guaranteed to exist by Theorem 1. Convolve both sides of equation (2) by $r(t)$. If we use (1.5) on the right-hand side of the resulting equation, and integrate by parts one time on the left-hand side, we see that $x(t)$ satisfies

(1.6) $$x''(t) = \alpha x'(t) + p * x'(t) + [A + \gamma I]\, x(t) + q * x(t) + h(t) , \quad t \geq 0 ,$$

$x(0) = x_0$, $x'(0) = v_0$. Here $\alpha \equiv b(0) + c < 0$, $p(t) \equiv r'(t) - cr(t)$, $q(t) \equiv g(t) - r * g(t) - \gamma r(t)$, and $h(t) \equiv f(t) - r*f(t) - r(t)\, v_0$. Clearly, the real-valued functions $p(t)$ and $q(t)$ lie in $C(R^+) \cap L^1(R^+)$, and $h(t) \in C^1(R^+; X) \cap L^1(R^+; X)$.

Set $\Gamma(\lambda) \equiv \gamma$ for Re $\lambda \geq 0$. An elementary calculation yields

$$\lambda^2 - \alpha\lambda - \lambda p^*(\lambda) - \Gamma(\lambda) - q^*(\lambda) = (\lambda^2 - c\lambda - \gamma - g^*(\lambda))/(1 + b^*(\lambda)) ,$$

and using (T2) it follows that (1.1) holds. Define the X-valued function $z^*(\lambda)$ on Re $\lambda \geq 0$ by

$$z^*(\lambda) = S^*(\lambda)\, x_0 + T^*(\lambda)[-\alpha x_0 - p^*(\lambda)\, x_0 + v_0 + h^*(\lambda)]$$

where $T^*(\lambda)$ and $S^*(\lambda)$ are the B(X)-valued functions defined in Lemmas 1 and 2, respectively. Clearly, $z^*(\lambda)$ is continuous on Re $\lambda \geq 0$ and analytic in Re $\lambda > 0$. Since $|p^*(\lambda)| \leq ||p||_1$ and $||h^*(\lambda)|| \leq ||h||_1$, Lemmas 1 and 2 yield that

$$\sup \left\{ \int_{-\infty}^{\infty} ||z^*(\sigma + i\tau)||^2 \, d\tau : \sigma \geq 0 \right\} < \infty . \tag{1.7}$$

Thus, z^* belongs to the Hardy space $H^2(0; X)$ consisting of all X-valued functions which are analytic in Re $\lambda > 0$ with boundary values defined almost everywhere on Re $\lambda = 0$, and which satisfy inequality (1.7). Since X is a Hilbert space, there is a function $z(t) \in L^2(R^+; X)$ whose Laplace transform is $z^*(\lambda)$ for all λ in Re $\lambda > 0$ (see [4, p. 164]).

Next, we may easily verify that $||x(t)||$, $||x'(t)||$ and $||Ax(t)||$ are all of exponential order; hence, there exists $\sigma_0 \geq 0$ so that we may take Laplace transforms in (1.6) whenever Re $\lambda \geq \sigma_0$. An elementary calculation yields that $x^*(\lambda) = z^*(\lambda)$ for Re $\lambda \geq \sigma_0$, and by uniqueness of Laplace transforms we have that $x(t) = z(t)$ for almost all $t \geq 0$. This completes the proof that $x(t) \in L^2(R^+; X)$ when the initial data and forcing function are sufficiently smooth. Since, by Theorem 1, solutions of (2) depend continuously on data, an elementary density argument may be used to show that the generalized solution $x(t)$ of (2) belongs to $L^2(R^+; X)$ for all $(x_0, v_0) \in X \times X$ and all $f \in L^1(R^+; X)$. This completes the proof of Theorem 2.

We remark that Theorem 2 does not hold without the assumption that $b(0) + c < 0$. For if $b(0) + c > 0$, an elementary argument using Rouche's theorem shows that (1.1) must fail to hold for all sufficiently large n and some λ_n in the open half-plane Re $\lambda > 0$. Thus, (T2) fails, and recalling the discussion preceding Proposition 1, we see that the conclusion of Theorem 2 does not hold in this case. On the other hand, if $b(0) + c = 0$, then an examination of the inequalities used in the proof of Lemma 2 (see Lemma 4.2 of [13]) yields that there must exist $x \in X$ so that $\int_{-\infty}^{\infty} ||S^*(i\tau)x||^2 \, d\tau = \infty$. Thus, Theorem 2 must be false when $b(0) + c = 0$ even though (T1) and (T2) may hold in this case.

We conclude this section by briefly comparing Theorems 2 and 3 with the results in [4], [12] and [1], [7], [8]. We begin by observing that our Theorems 2 and 3 differ in a basic way from the results contained in these papers since Theorems 2 and 3 deal with the asymptotic behavior of individual solutions of (2), whereas the other papers examine the asymptotic behavior of a resolvent function (fundamental solution) associated with the equation. Results concerning a resolvent function are stronger since they imply uniform decay of individual solutions.

More specifically, Friedman and Shinbrot [4] obtain $L^p(1 \le p < \infty)$ estimates for the operator norm of the resolvent function $R(t)$ of the equation

$$(1.8) \qquad x(t) = \int_0^t B(t-\tau)\, Ax(\tau)\, d\tau + F(t)$$

in a Banach space under the assumption that $\sigma(A)$ is contained in a closed subsector of the open left-half plane. Here R is defined by the operator equation

$$R(t) = I + A\int_0^t B(t-\tau)\, R(\tau)d\tau\ ,$$

and R can be used to express the solution of (1.8) as

$$x(t) = R(t)\, F(0) + \int_0^t R(t-\tau)\, F'(\tau)d\tau\ .$$

It is assumed that B satisfies at least $B(0) > 0$ and $B' \in L^1(R^+)$; hence, if $B(0) = 1$ (w.l.o.g.), $b = B'$ and $f = F'$, formal differentiation gives that $x(t)$ satisfies the first order equation

$$(1.9) \qquad x'(t) = Ax(t) + \int_0^t b(t-\tau)\, Ax(\tau)d\tau + f(t), \quad x(0) = F(0)\ .$$

Miller and Wheeler [12] employ techniques similar to those used in [4] to study the resolvent function R of equation (1.9) in a Hilbert space when the operator A satisfies (H2). There we obtain conditions which ensure that R may be written as an exponential polynomial with finite-dimensional projections as coefficients, plus a remainder term whose operator norm lies in $L^p(R^+)$.

The technique used in both [4] and [12] depends on the operational calculus based on contour integrals, and estimates of the form

$$(1.10) \qquad \int_0^\infty |R_\mu(t)|^p dt \le K\, |\mu|^{-\delta} \quad (\delta > 0)$$

which hold uniformly for all μ in an open sector in the complex plane which contains the negative real axis. Here R_μ is the complex-valued resolvent of the related scalar equation which has the same form as (1.9), but with the operator A replaced by the complex parameter μ.

An example showing that the methods of [4] and [12] do not extend to the equations examined here is provided by Carr and Hannsgen in [1]. There they examine the integrated equation (1) when $c = 0$ and $G(t) \equiv 0$, that is,

$$(1.11) \qquad x'(t) = \int_0^t B(t-\tau)\, Ax(\tau)d\tau + F(t)\ , \quad x(0) = x_0\ .$$

In order to use the techniques of [4] and [12] to obtain L^p norm estimates for the resolvent function $U(t)$ of (1.11), defined formally by the operator equation

$$U'(t) = A\int_0^t B(t-\tau)\, U(\tau)d\tau\ , \quad U(0) = I\ ,$$

one would need to obtain estimates having the form (1.10) which hold for the scalar resolvents u_μ defined by

$$(1.12) \qquad u'_\mu(t) = \mu\int_0^t B(t-\tau)\, u_\mu(\tau)d\tau\ , \quad u_\mu(0) = 1\ .$$

Moreover, these estimates must hold uniformly for all μ in a sector of the complex plane which contains the negative real axis in its interior. However, as Carr and Hannsgen observe in [1], equation (1.12) is easily explicitly solved when $B(t) = e^{-t}$, and the resulting u_μ does not belong to $L^p(R^+)$ for any p satisfying $1 \le p < \infty$ whenever $\mu = |\mu|e^{i\phi}$, $e^{i\phi} \neq -1$ with $|\mu|$ large.

By using entirely different methods, Carr and Hannsgen [1] and Hannsgen [7], [8] obtain conditions which ensure that the resolvent function U(t) of equation (1.11) satisfies

$$(1.13) \qquad (i) \quad ||U(t)|| \in L^1(R^+) , \qquad (ii) \quad ||U(t)|| \to 0 \quad \text{as} \quad t \to \infty .$$

In these papers, X is a Hilbert space, and A is a self-adjoint negative definite linear operator. It is not necessary to assume that $\sigma(A)$ consists only of isolated eigenvalues. The scalar kernel B(t) is assumed to be a nonconstant, nonnegative, nonincreasing and convex function. Also, B(t) must satisfy an additional technical assumption. For example, in [7] B(t) is continuous on R^+ and completely monotonic on $(0, \infty)$, while in [1] a weaker frequency condition is assumed to hold. We remark that since the hypotheses of [1], [7] and [8] do not include the requirement that B(t) be differentiable at $t = 0$, one is not allowed to differentiate equation (1) and examine instead the second order equation (2). In fact, in [1] it is not required that B have a finite limit at $t = 0$, but only that B be integrable on (0, 1). The analysis used to obtain (1.13) consists of showing that the scalar solutions u_μ of (1.12) satisfy the following uniform behavior

$$(1.14) \qquad (i) \quad \int_0^\infty u^{(0)}(t)dt < \infty , \qquad (ii) \quad u^{(0)}(t) \to 0 \quad \text{as} \quad t \to 0 .$$

Here $u^{(0)}(t) \equiv \sup \{ |u_\mu(t)| : -\infty < \mu \le \mu_0 \}$ where $\mu_0 < 0$ is chosen so that $\sigma(A) \subseteq (-\infty, \mu_0]$. Once (1.14) is proved, (1.13) follows immediately from the spectral decomposition formula $U(t) = \int_{-\infty}^0 u_\mu(t)d E_\mu$.

The proof that (1.14i) holds depends on obtaining delicate technical estimates and inequalities for certain quantities involving the real and imaginary parts of the Fourier transform of B. These estimates use strongly the monotonicity and convexity of the kernel B. It appears that it is not possible to apply similar techniques to deduce uniform behavior of the scalar resolvents u_μ such as that exhibited in (1.14) in the case where B(t) is neither necessarily monotone nor convex, but has instead the form $B(t) = 1 + \int_0^t b(\tau)d\tau$ with $b \in L^1(R^+)$, and with the appropriate Laplace transform conditions satisfied.

Finally, we remark that if $F \in C(R^+; X)$, then the assumptions of [1], [7] or [8] imply that U(t) may be used to express the solution x(t) of (1.11) as

$$x(t) = U(t)x_0 + \int_0^t U(t - \tau) F(\tau) d\tau .$$

Moreover, in addition to (1.13), the hypotheses in [1], [7] or [8] imply that

$$\int_0^\infty U(t)dt = A^{-1}/\int_0^\infty B(t)dt$$

where the right-hand side is interpreted as zero when $B(t) \notin L^1(R^+)$ (see, e.g., [7]). Therefore, if $F(t)$ is also bounded and tends to a limit F_∞ as $t \to \infty$, the representation formula for the solution $x(t)$ of (1.11) and (1.13) yield that $x(t)$ either tends to zero as $t \to \infty$, or that $x(t)$ can be written as a constant plus a remainder term that tends to zero as $t \to \infty$. Hence, the relationship between Theorems 2 and 3 and the results in [1], [7] and [8] is clear.

2. Remarks on the proofs.

Lemmas 1 and 2 are slight generalizations of the corresponding Lemmas 4.1 and 4.2 in [13] since there we considered only the case where $\Gamma(\lambda) \equiv \gamma$. However, this change does not necessitate essential changes in the proofs; hence, we comment only briefly on their proofs in this Section. Since the proof of Theorem 3 is new, it is given in greater detail.

Proof of Lemma 1. Since (H2) holds, the operator norm of $R(\mu; A)$ satisfies $||R(\mu; A)|| \le [\text{dist}(\mu, \sigma(A))]^{-1}$ (see, e.g., [16, p. 343]). Thus, Lemma 1 holds provided that we show that there exists a $K > 0$ so that

$$\text{dist}(\lambda^2 - \alpha\lambda - \lambda p^*(\lambda) - \Gamma(\lambda) - q^*(\lambda),\ \sigma(A)) \ge K(|\tau| + 1) \tag{2.1}$$

whenever $\lambda = \sigma + i\tau$ with $\sigma \ge 0$ and $-\infty < \tau < \infty$. Now if (2.1) is false for every $K > 0$, there exists $\lambda_n = \sigma_n + i\tau_n$ so that

$$\text{dist}(\lambda_n^2 - \alpha\lambda_n - \lambda_n p^*(\lambda_n) - \Gamma(\lambda_n) - q^*(\lambda_n),\ \sigma(A)) \le (|\tau_n| + 1)/n \ .$$

By using (1.1), $\alpha < 0$, the continuity of p^*, q^* and Γ, the boundedness of q^* and Γ, and the fact that $p^*(\lambda)$ tends to zero as $\lambda \to \infty$ in $\text{Re}\ \lambda \ge 0$, and by considering separately the cases where (i) λ_n has a finite accumulation point, (ii) $\lambda_n \to \infty$ in the subregion $|\tau| \le \sigma/2$, or (iii) $\lambda_n \to \infty$ in the subregion $|\tau| \ge \sigma/2$, it is easy to show that the last inequality must lead to a contradiction.

Proof of Lemma 2. For each positive integer n, define the scalar function

$$s_n^*(\lambda) = \lambda[\lambda^2 - \alpha\lambda - \lambda p^*(\lambda) - \Gamma(\lambda) - q^*(\lambda) - \mu_n]^{-1}\ , \quad \text{Re}\ \lambda \ge 0\ .$$

The hypotheses of Lemma 2 imply that each s_n^* is continuous on $\text{Re}\ \lambda \ge 0$, analytic on $\text{Re}\ \lambda > 0$, and that $|s_n^*(\lambda)| = O(|\lambda|^{-1})$ as $\lambda \to \infty$ in $\text{Re}\ \lambda \ge 0$. Hence, s_n^* belongs to the Hardy space $H^2(0; \mathbb{C})$, and by a theorem of Paley and Wiener [14, p.8], s_n^* is the Laplace transform of a scalar-valued function $s_n(t) \in L^2(R^+)$. Thus

$$\sup\left\{\int_{-\infty}^{\infty} |s_n^*(\sigma + i\tau)|^2\, d\tau : \sigma \ge 0\right\} = \int_{-\infty}^{\infty} |s_n^*(i\tau)|^2\, d\tau \quad (n = 1,2, \ldots)\ ,$$

and Lemma 2 follows at once from the spectral decomposition formula $S^*(\lambda)x = \Sigma\, s_n^*(\lambda)(x, \phi_n)\, \phi_n$ provided we show that there exists $K > 0$ so that

$$\int_{-\infty}^{\infty} |s_n^*(i\tau)|^2\, d\tau \le K \quad (n = 1,2, \ldots)\ . \tag{2.2}$$

Since the integral in (2.2) is finite for each n, it clearly suffices to find a uniform bound for these integrals which holds for all sufficiently large n. Moreover, since p(t) is real, it is enough to show that $\int_0^\infty |s_n^*(i\tau)|^2\,d\tau$ is uniformly bounded for all large n. This is accomplished by writing this integral as

$$\int_0^\infty |s_n^*(i\tau)|^2\,d\tau = \left\{ \int_0^{\eta_n-D-1} + \int_{\eta_n-D-1}^{\eta_n+D+1} + \int_{\eta_n+D+1}^{2\eta_n} + \int_{2\eta_n}^{\infty} \right\} |s_n^*(i\tau)|^2\,d\tau ,$$

where $\eta_n = \sqrt{-\mu_n}$ and $D = \|p\|_1$. One can now find bounds for each of the integrals on the right-hand side of the last line which hold uniformly for all large n. For the details of these estimates, we refer the reader to the proof of Lemma 4.2 of [13] with the remark that the definition of d given there should now be changed to read $d \equiv \max_{0\le\tau<\infty} |\Gamma(i\tau)| + \|q\|_1$.

Proof of Theorem 3. Let r be the scalar resolvent function defined in (1.5). Since the only zeros of $1 + b^*(\lambda)$ in Re $\lambda \ge 0$ are simple zeros at $\lambda = i\omega_m$ $(1 \le m \le M)$, a theorem of Jordan and Wheeler [9] implies that r has the form

$$r(t) = \sum_{m=1}^{M} a_m e^{i\omega_m t} + r_1(t)$$

with the a_m constants and $r_1 \in L^1(R^+)$. By substituting this expression for r into (1.5) and differentiating, we can show that $r_1' \in C(R^+) \cap L^1(R^+)$ as well.

Let p, q and h be as in the proof of Theorem 2, and define $p_1(t) = r_1'(t) - cr_1(t)$, $q_1(t) = g(t) - r_1 * g(t) - \gamma\, r_1(t)$ and $h_1(t) = f(t) - r_1 * f(t) - r_1(t)v_0$. Clearly, p_1 and q_1 both belong to $C(R^+) \cap L^1(R^+)$, and $h_1 \in C^1(R^+; X) \cap L^1(R^+; X)$. Define $\Gamma(\lambda)$ for Re $\lambda \ge 0$, $\lambda \ne i\omega_m$ $(1 \le m \le M)$, by

$$\Gamma(\lambda) = \sum_{m=1}^{M} a_m(\lambda - i\omega_m)^{-1}(\lambda i\omega_m - c\lambda - \gamma - g^*(\lambda)) .$$

Then, $\Gamma(\lambda)$ is analytic in Re $\lambda > 0$, and since (1.3) holds, Γ can be defined at $\lambda = i\omega_m$ $(1 \le m \le M)$ so as to be continuous and bounded in Re $\lambda \ge 0$. An easy calculation shows that

$$\begin{aligned}\lambda^2 - \alpha\lambda - \lambda p_1^*(\lambda) - \Gamma(\lambda) - q_1^*(\lambda) &= \lambda^2 - \alpha\lambda - \lambda p^*(\lambda) - \gamma - q^*(\lambda) \\ &= (\lambda^2 - c\lambda - \gamma - g^*(\lambda)) \,/\, (1 + b^*(\lambda))\end{aligned}$$

for Re $\lambda \ge 0$ $(\lambda \ne i\omega_m)$, and it follows using (T2) and (1.4) that (1.1) holds with p^* and q^* replaced by p_1^* and q_1^*, respectively. Hence, Lemmas 1 and 2 hold for the B(X)-valued functions $T_1^*(\lambda)$ and $S_1^*(\lambda)$ defined on Re $\lambda \ge 0$ as in those lemmas, but with p^* and q^* replaced by p_1^* and q_1^*, respectively.

The density argument used in the proof of Theorem 2 also shows that it suffices to prove Theorem 3 when $(x_0, v_0) \in D(A) \times D(A)$. Making this assumption, define the X-valued function

$$\begin{aligned} z_1^*(\lambda) &= S_1^*(\lambda)x_0 + T_1^*(\lambda)[-\alpha x_0 - p_1^*(\lambda)\, x_0 + v_0 + h_1^*(\lambda)] \\ &\quad + \sum_{m=1}^{M} a_m(\lambda - i\omega_m)^{-1}(T_1^*(\lambda) - T_1^*(i\omega_m))[(c - i\omega_m)x_0 - v_0 - f^*(\lambda)] . \end{aligned} \tag{2.3}$$

From the resolvent formula [16, p. 257] we find that for each m ($1 \le m \le M$)

$$(\lambda - i\omega_m)^{-1}(T_1^*(\lambda) - T_1^*(i\omega_m)) = k_m(\lambda)\ T_1^*(i\omega_m)\ T_1^*(\lambda)\ ,$$

where, for Re $\lambda \ge 0$ ($\lambda \ne i\omega_m$) ,

$$k_m(\lambda) \equiv (\lambda - i\omega_m)^{-1}\{[2i\omega_m - c - (g^*)'(i\omega_m)][(b^*)'(i\omega_m)]^{-1} - [\lambda^2 - c\lambda - \gamma - g^*(\lambda)][1+b^*(\lambda)]^{-1}\}\ .$$

Since $t^2b(t)$ and $t^2g(t)$ both lie in $L^1(R^+)$, and (1.2) and (1.3) hold, it is easy to see that $k_m(\lambda)$ can be extended to be continuous at $\lambda = i\omega_m$. Thus, we may rewrite each term in the sum in (2.3) as

$$a_m T_1^*(i\omega_m) k_m(\lambda)\ \{T_1^*(\lambda) f^*(\lambda) + (\lambda + 1)^{-1}(S_1^*(\lambda) + T_1^*(\lambda))[(c - i\omega_m)x_0 - v_0]\}\ .$$

Since f and f' both belong to $L^1(R^+;\ X)$, $||f^*(\lambda)|| = O(|\lambda|^{-1})$ as $\lambda \to \infty$ in Re $\lambda \ge 0$. Observing that $k_m(\lambda) = O(|\lambda|)$ as $\lambda \to \infty$ in Re $\lambda \ge 0$, it follows using Lemmas 1 and 2, that $z_1^*(\lambda) \in H^2(0;\ X)$; hence $z_1^*(\lambda)$ is the transform of a function $z_1(t) \in L^2(R^+;\ X)$.

Let $x(t) = x(t, x_0, v_0, f)$, take Laplace transforms in (1.6), and use the relations between p, q, h and p_1, q_1, h_1 to find that

$$x^*(\lambda) = z_1^*(\lambda) + \sum_{m=1}^{M} a_m(\lambda - i\omega_m)^{-1}\ T_1^*(i\omega_m)[(c - i\omega_m)x_0 - v_0 - f^*(\lambda)]$$

for all λ with Re λ sufficiently large. Thus, uniqueness of Laplace transforms yields that

$$x(t) = z_1(t) + \sum_{m=1}^{M} a_m\ T_1^*(i\omega_m)[(c - i\omega_m)x_0 - v_0 - \int_0^t \exp(-i\omega_m\tau)\ f(\tau)\ d\tau]\exp(i\omega_m t)$$

$$= x_1(t) + \sum_{m=1}^{M} x_m e^{i\omega_m t}\ ,$$

where

$$x_m = a_m\ T_1^*(i\omega_m)[(c - i\omega_m)x_0 - v_0 - f^*(i\omega_m)] \qquad (1 \le m \le M)\ ,$$

$$x_1(t) = z_1(t) + \sum_{m=1}^{M} a_m T_1^*(i\omega_m) \int_t^\infty \exp(i\omega_m(t - \tau))\ f(\tau)\ d\tau\ .$$

Since $tf(t) \in L^1(R^+;\ X)$, the integral terms in the expression for $x_1(t)$ belong to $L^1(R^+;\ X)$, and the proof of Theorem 3 is complete.

REFERENCES

1. R. W. Carr and K. B. Hannsgen, A nonhomogeneous integrodifferential equation in Hilbert space, Math. Res. Center, Univ of Wisconsin, Tech. Summary Report #1764 1977.

2. M. G. Crandall, S.-O. Londen and J. A. Nohel, An abstract nonlinear Volterra integrodifferential equation, J. Math. Anal. Appl. (to appear) and Math. Res. Center, Univ. of Wisconsin, Tech. Summary Report #1684, 1976.

3. C. M. Dafermos, Asymptotic stability in viscoelasticity, Arch. Rational Mech. Anal. 37(1970), 297-308.

4. A. Friedman and M. Shinbrot, Volterra integral equations in Banach space, Trans. Amer. Math. Soc. 126 (1967), 131-179.

5. G. Gripenberg, On the asymptotic behavior of solutions of nonlinear integral equations in Banach spaces, Helsinki Univ. of Tech. Report - HTKK-MAT-A110 (1977).

6. M. E. Gurtin and A. C. Pipkin, A general theory of heat conduction with finite wave speeds, Arch. Rational Mech. Anal. 31 (1968), 113-126.

7. K. B. Hannsgen, The resolvent kernel of an integrodifferential equation in Hilbert space, SIAM J. Math. Anal. 7 (1976), 481-490.

8. ____________, Uniform L^1 behavior for an integrodifferential equation with parameter, SIAM J. Math. Anal. 8 (1977), 626-639.

9. G. S. Jordan and R. L. Wheeler, Structure of resolvents of Volterra integral and integrodifferential systems (submitted).

10. R. C. MacCamy, Remarks on frequency domain methods for Volterra integral equations, J. Math. Anal. Appl. 55 (1976), 555-575.

11. R. K. Miller, An integrodifferential equation for rigid heat conductors with memory, J. Math. Anal. Appl. (to appear).

12. R. K. Miller and R. L. Wheeler, Asymptotic behavior for a linear Volterra integral equation in Hilbert space, J. Differential Equations 23 (1977), 270-284.

13. ___________, Well-posedness and stability of linear Volterra integrodifferential equations in abstract spaces, Funckcial. Ekvac. (to appear).

14. R. E. A. C. Paley and N. Wiener, "Fourier Transforms in the Complex Domain", Amer. Math. Soc. Colloq. Publ., vol. 19, Amer. Math. Soc. Providence, R. I., 1934.

15. M. Slemrod, A hereditary partial differential equation with applications in the theory of simple fluids, Arch. Rational Mech. Anal. 62(1976b), 303-322.

16. A. E. Taylor, "Introduction to Functional Analysis", Wiley, New York, 1958.